Climate Change and Agriculture—Sustainable Plant Production

Climate Change and Agriculture—Sustainable Plant Production

Guest Editors

Jinlong Dong
Junjie Lin
Yang Yang
Zhongxiu Sun
Nazim S. Gruda

MDPI

Basel • Beijing • Wuhan • Barcelona • Belgrade • Novi Sad • Cluj • Manchester

Guest Editors
Jinlong Dong
Institute of Soil Science
Chinese Academy of Sciences
Nanjing
China

Junjie Lin
Key Laboratory of Water
Environment Evolution and
Pollution Control in Three
Gorges Reservoir
Chongqing Three Gorges
University
Chongqing
China

Yang Yang
Institute of Earth
Environment
Chinese Academy of Sciences
Xi'an
China

Zhongxiu Sun
College of Land and
Environment
Shenyang Agricultural
University
Shenyang
China

Nazim S. Gruda
Division of Horticultural
Sciences
University of Bonn
Bonn
Germany

Editorial Office
MDPI AG
Grosspeteranlage 5
4052 Basel, Switzerland

This is a reprint of the Special Issue, published open access by the journal *Agronomy* (ISSN 2073-4395), freely accessible at: https://www.mdpi.com/journal/agronomy/special_issues/673CK4YJFK.

For citation purposes, cite each article independently as indicated on the article page online and as indicated below:

Lastname, A.A.; Lastname, B.B. Article Title. *Journal Name* **Year**, *Volume Number*, Page Range.

ISBN 978-3-7258-5253-6 (Hbk)
ISBN 978-3-7258-5254-3 (PDF)
https://doi.org/10.3390/books978-3-7258-5254-3

Cover image courtesy of Zhongxiu Sun

Contents

About the Editors

Jinlong Dong

Jinlong Dong is an Assistant Professor at the Institute of Soil Science, Chinese Academy of Sciences. His current research interests include soil science, plant nutrition, sustainable vegetable production, planetary boundaries, human nutrition, and climate change.

Junjie Lin

Junjie Lin is a Doctor, Professor, and Master's Supervisor at Chongqing Three Gorges University. He holds prestigious titles including Chongqing Talent·Leading Entrepreneur and Innovator (Science & Technology Sector)—serving as Head of an Innovation and Entrepreneurship Demonstration Team—and is a Member of the Chongqing Young Scientists and Technological Leaders Association. His honors encompass State-designated S&T Talent for Underdeveloped Regions (Chongqing), Chongqing Municipal Backbone University Teacher, Wanzhou Pinghu Talent, and Intermediate Professional Title Evaluation Expert for Wanzhou District, Chongqing. He serves as an Expert Committee Member of the Chongqing Ecological Civilization Construction Academic Consortium and as a Review Panel Expert for the Environmental and Ecological Science Division of the National Natural Science Foundation of China (NSFC). His academic appointments include being an Innovation Committee Member of the Young Scientists Working Committee (Soil Science Society of China), a Youth Editorial Board Member for the *Chinese Journal of Soil Science*, a Youth Editorial Board Member for *Biochar and Carbon Research*, and an Editorial Board Member for *Ecological Indicators*. His research focuses on climate change and soil ecology, soil carbon–nitrogen cycling, stable isotope ecology, and soil remediation technologies. Representative research projects include NSFC grants (31470527, 41301248), the Ministry of Education "Chunhui Plan" (Z2015133), Chongqing Municipal Science and Technology Commission Key Project (cstc2015jcyjB0332), and Chongqing Municipal Education Commission projects (KJ1601016, KJ121108).

Yang Yang

Yang Yang is a Professor at the State Key Laboratory of Loess and Quaternary Geology. His research focuses on frontier ecological and pedological challenges in the Loess Plateau ecological barrier region, specializing in climate change ecology and soil microbe-mediated organic carbon cycling. His honors include being awarded the National Top Ten Forestry Graduate Students, the National Wang Dong Award for Grassland Science, and a National Scholarship. He was selected as a Member of the Youth Innovation Promotion Association (CAS), serves on the Youth Working Committee of the Soil Science Society of China, and was included in the Institute's "Young Backbone Researcher (BR)" Program. He has led multiple research grants, including the National Natural Science Foundation of China (NSFC) General Program, NSFC Young Scientists Project, provincial-/ministerial-level projects, CAS talent programs, and State Key Laboratory cultivation funds. As first and corresponding author, he has published over 40 papers in international journals such as *Global Change Biology*, *iMeta*, *Soil Biology and Biochemistry*, and *Carbon Neutrality*, including five ESI Highly Cited Papers, one ESI Hot Paper, and one paper featured in "Research Fronts." He received the 2024 Shaanxi Provincial Natural Science and Technology Award (Second Class, ranked second out of four contributors). He has co-authored three academic books and serves as an Editorial Board Member for *Science of The Total Environment and Plant and Soil*, and as a Youth Editorial Board Member for *iMeta*, *Carbon Neutrality*, *Journal of Plant Ecology*, *Acta Pedologica Sinica*, and *Chinese Journal of Ecology*.

Zhongxiu Sun

Zhongxiu Sun is a Doctor, Associate Professor, and Master's Supervisor at Shenyang Agricultural University. He concurrently serves as a Council Member of the ANSO—World Black Soil Federation—China Black Soil Research Association, Deputy Director of the Popularization Working Committee of the Soil Science Society of China (SSSC), Member of the Soil Classification and Soil Geography Committee, Member of Youth Committee of the SSSC and its Young Scientists Working Committee, and Youth Editorial Board Member of both the *Chinese Journal of Soil Science* and the journal *Soil* (published by SSSC). His current research interests include investigating the development and evolution mechanisms of loess deposits using non-traditional stable isotope techniques (such as Fe and Mg). He studies the formation and evolution of soils derived from loess along with the genesis, classification, evolution, and health evaluation of soil restrictive layers within the black soil region of Northeast China. He leads several major research projects, which include National Natural Science Foundation grants and key projects under the National Key R&D Program, alongside four provincial-level projects. His current research projects include the following: (1) the formation mechanisms and evolution of long-term soil chronosequences since 1152 ka BP for accurate paleoclimatic reconstruction; (2) the formation mechanisms and diagnostic criteria of soil restrictive layers; (3) soil carbon sequestration and soil fertility in greenhouse horticultural systems; (4) the development of new methods to gain new insights and improve our understanding of soil evolution; and (5) soil function and health evaluation for sustainable soil resource utilization. His scholarly contributions include twenty publications in journals, featuring fourteen top-tier SCI papers (cumulative IF: 120) with 189 citations; two nationally funded monographs as editor; six patents; and three software copyrights as first inventor.

Nazim S. Gruda

Nazim S. Gruda is a Professor in the Department of Horticultural Sciences, University of Bonn, Germany. His research focuses on agricultural sciences and agronomy. His current research interests include vegetables, climate change, horticulture, soilless culture, growing media, protected cultivation, and greenhouse production. He is an active participant in international scientific conferences: he has been a keynote and invited speaker at more than 30 conferences, is often requested to chair scientific sections, and to participate in scientific committees and dissertation defense boards. He is a member of the standing committee of the doctoral program 'Sustainable Agricultural and Forestry Systems and Food Security at the University of Naples Federico II (Italy). Prof. Nazim Gruda holds memberships in 15 international and national organizations. He is a vice-president and member of the steering committee of the Union of European Academies for Sciences Applied to Agriculture, Food and Nature (UEAA), a vice-chair of the ISHS Vegetable, Root and Tuber Section, a member of the Working Group on the European Research Area of the European Federation of Academies of Sciences and Humanities (ALLEA), a member of the Advisory Council on Plant Plastics Recovery and Recycling System (Germany), the chair and coordinator of the FAO Working Group on Greenhouse Crop Production in South-East Europe, a full member of the Albanian Academy of Sciences, etc. In 2025, he became a foreign member of the Lithuanian Academy of Sciences. He has participated in over 20 international and national research projects. His work includes 250+ publications in international peer-reviewed journals, proceedings, professional magazines, and books.

agronomy

MDPI

Editorial

Climate Change and Agriculture—Sustainable Plant Production

Zhong-Xiu Sun [1,*], Cheng-Cheng Zhang [1], Jin-Long Dong [2] and Ying-Ying Jiang [3,*]

1 College of Land and Environment, Shenyang Agricultural University, Shenyang 110866, China; cczhang624@163.com
2 State Key Laboratory of Soil and Sustainable Agriculture, Institute of Soil Science, Chinese Academy of Sciences, Nanjing 210008, China; jldong@issas.ac.cn
3 Shenyang Institute of Technology, Shenyang 113122, China
* Correspondence: zhongxiusun@syau.edu.cn (Z.-X.S.); yingyingjiang9111@163.com (Y.-Y.J.); Tel.: +86-15734005989 (Z.-X.S.)

Climate change has a great impact on plant growth and agricultural production, especially on the growing season, growth rate, and growth distribution. Climate change can prolong the plant growing season and expand the areas suitable for crop planting [1], as well as promote crop photosynthesis thanks to increased atmospheric carbon dioxide concentrations. However, an excessive carbon dioxide concentration in the atmosphere may lead to unbalanced nutrient absorption in crops and hinder photosynthesis [2], respiration, and transpiration, thus affecting crop yields [3,4]. Irregular precipitation patterns and extreme weather events such as droughts and floods can lead to hypoxia and nutrient loss in the plant roots [5]. An increase in the frequency of extreme weather events directly damages plants and expands the range of diseases and pests [6]. In addition, climate change will also affect soil moisture content, temperature, microbial activity, nutrient cycling, and quality, thus affecting plant growth [7]. Plants are the basis of agricultural production, and the uncertainty of climate change brings challenges and opportunities to the sustainable production of plants. Therefore, it is important to understand the mechanisms through which climate change impacts plants and agriculture and take adaptive measures to promote sustainable agricultural development.

This Special Issue of *Agronomy*, "Climate Change and Agriculture—Sustainable Plant Production", focuses on the interaction between climate change and agriculture. The collection features a total of ten papers, six of which (including two reviews) deal with crops and four of which discuss soils. The research included in this Special Issue covers the impact of temperature/climate change on the yield and quality of rice, wheat, and olive oil; the adaptation strategies of crops in the face of climate change; the impact of climate change on soil nutrient cycling; and the spatial–temporal characteristics and influencing factors of the normalized vegetation index (NDVI), which are analyzed using various models. Moreover, the mechanism of early maturing of short-season cotton, high-yield cultivation methods, and carbon planting strategies of organic vegetables are discussed in detail.

The first two articles included in this Special Issue focus on the impact of environmental factors on soil and vegetation. Based on previous studies on the stoichiometric relationship of C:N:P, Zhang et al. compared the spatial distribution of the stoichiometric relationship of C:N:P:K between farmland and grassland soils at different scales in the agropastoral ecotone of Inner Mongolia. The authors then explored the influence of environmental factors on soil stoichiometry, combining their results with Spearman correlation analysis. They concluded that there are differences in C:N:P:K stoichiometric relationships between farmland and grassland soils. The average annual precipitation has a significant influence on the soil stoichiometric relationship in farmland, while the soil stoichiometric relationship in grassland is more significantly affected by the average annual air temperature. The study reveals the important effects of land use patterns and environmental factors on soil nutrient cycling. In the future, based on this research result and combined with the C:N:P:K stoichiometric relationship between farmland soil and plants, the stoichiometric relationship mechanism of different agricultural management methods and its effects on biomass distribution can be studied to improve agricultural production efficiency [8]. Meanwhile, Hou et al. focused on the climate warming and aridification which have been ongoing in Maowusu Sandland for the past 20 years. The authors explored the impact of climate warming and aridification on vegetation change and predict future vegetation growth trends using a deep learning model. This improved prediction accuracy and provided new perspectives for sustainable agricultural development [9].

With the intensification of global climate change, both high and low temperatures will harm crop yield and quality. It is therefore crucial to study the impact of climate change on crop growth, yield, and quality. Rice, wheat, and olive oil are major agricultural products under threat from climate change. Oh et al. evaluated the growth of and physiological processes in rice using plant height, chlorophyll, the normalized difference vegetation index (NDVI), and maximum photosynthetic rate (Amax). The authors used the heating degree day index to evaluate the effect of heat stress on yield. The growth, ripening, and senescence of rice during the whole growing season were studied under high-temperature conditions. The results show that rice was not affected by heat stress during the tillering stage, but there was a cumulative effect during the booting stage. At 3 °C above AT (AT + 3 °C), the photosynthetic capacity of rice was maintained for a long time in the grain-filling stage [10]. Zhang et al. found that current studies on the growth and development of winter wheat mainly focus on the formation stage of heading and yield after flowering, while relatively few studies discuss the jointing stage. Therefore, Zhang et al. focused on the changes in the physiological characteristics of the typical variety "Ji Mai 22" under different low-temperature conditions and different durations in view of the low-temperature stress endured by winter wheat during the jointing stage. The results show that under low-temperature stress, the photosynthetic parameters, aging characteristics, and endogenous hormone levels of the leaves of "Ji Mai 22" showed varying degrees of change, and it was found that the lower limit of growth temperature during jointing was −3 °C [11]. The influence of climate change on the quality and composition of vegetable oil, a daily necessity, cannot be ignored. Bortoluzzi et al. studied how climatic conditions affect the composition of olive oil produced by centenarian olive trees. In their study, 25 centenarian olive trees located in the Côa Valley region of Northern Portugal were evaluated for two consecutive years. It was found that changes in climate conditions significantly affected the composition of olive oil. Among them, the results of the second-year evaluation showed that the proportion of palmitic acid in the olive oil increased significantly, while the content of stearic acid and arachidonic acid decreased. In addition, the concentrations of β-tocopherol, hydroxytyrosol, and tyrosol also changed significantly. These preliminary results lay a foundation for future studies to explore the response of olive oil components of century-old olive trees to different climate change scenarios [12].

Under conditions exacerbated by climate change, heat stress seriously threatens plant growth and yield in warm environments. It is crucial to study genes related to heat tolerance, especially for plants that typically lack the ability to resist high temperatures. In light of this, Wang et al. chose the bottle gourd (*Lagenaria siceraria* (Mol.) Standl.) as their research object and compared the gene expression characteristics of two varieties—"Mei feng" (heat-resistant) and "Lv long" (heat-sensitive)—under heat stress conditions. Through transcriptome analysis and other methods, the differential expression of a series of key genes was found. Genes related to the MAPK signaling pathway and bHLH transcription factor were up- or down-regulated especially significantly. Subsequently, the expression patterns of these genes were verified by quantitative real-time PCR, further confirming the reliability of the RNA-Seq results. The transcriptome analysis conducted in the study revealed some key genes that may be related to plant heat resistance, providing new research ideas for the future improvement of heat-resistant vegetable varieties and a fresh perspective on the molecular mechanisms of plant heat resistance. The study provides an important foundation for coping with the challenge of heat stress caused by climate change [13].

As an important factor supporting plant growth, soil is crucial to agricultural production and sustainable land use. Jiang et al. took quaternary red soils under different land use patterns as the research object and quantitatively analyzed the evolution characteristics of soil nutrients, thus providing an important reference for selecting and optimizing land use patterns and improving soil productivity and management levels. Nutrient change is related to vegetation type, coverage rate, fertilization method, and species [14]. In addition, based on the Cornell Soil Health Assessment system, the authors established a quaternary red soil health evaluation system and found that land use patterns had significant effects on soil health. In general, the health status of quaternary red soils under different land use patterns was better than that of buried quaternary red soil without human activity influences, showing a trend of evolution towards healthy soil. This indicates that human land use activities promote the healthy development of quaternary red soils to a certain extent [15].

The remaining two articles collated in this Special Issue focus on innovative approaches to agriculture, focusing on short-season cotton and carbon farming, respectively. Qi et al. provided a detailed overview of the early-maturing mechanism of short-season cotton, including its morphology, physiology, and molecular biology, and discussed its applications in the fields of planting pattern optimization, saline–alkali soil planting, machine picking, and cultivation without plastic cover. With the continuous progress of breeding technology, the lint rate and fiber quality of the new generation of short-season cotton varieties have been significantly improved while maintaining early maturity, high yield, and stress resistance [16]. On the other hand, Avasiloaiei et al. systematically reviewed the literature to assess the impact of carbon agriculture strategies for organic vegetable cultivation on carbon sequestration, soil health, and crop productivity, revealing the potential of carbon agriculture practices for improving the sustainability of organic agriculture systems. Carbon sequestration rates can be effectively improved through cover planting, reduced tillage, and composting application. The findings of the systematic review are a key foundation for the promotion of sustainable agricultural development [17].

Collectively, these studies focus on the effects of climate change on soils, vegetation, and crops, as well as exploring adaptation mechanisms in different ecosystems. Through stoichiometric relationship analysis, deep learning modeling, and transcriptome analysis, the challenges imposed by climate change on agricultural production are revealed. The adaptation mechanisms of plant growth and the various physiological processes and gene expression changes in the face of climate change and heat stress are also discussed. In addition, the establishment of a soil health evaluation system and the impact of different land use patterns on soil health are also emphasized, highlighting the importance of evidence-based land management to improve soil productivity and ecosystem stability. Overall, these studies provide important references for the improvement of agricultural production efficiency in the future, emphasize the key role of scientific management in addressing the challenge of climate change, and provide theoretical and practical guidance for achieving sustainable agricultural development.

Acknowledgments: The authors sincerely thank all those who contributed to this Special Issue. Our acknowledgements also extend to the anonymous reviewers for their constructive reviews of the manuscripts.

Conflicts of Interest: The authors declare no conflicts of interest.

References

1. Zhang, H. Impacts of climate change on ecosystem and agricultural production. *Agric. Disaster Res.* **2023**, *13*, 201–203.
2. Körner, C. Paradigm shift in plant growth control. *Curr. Opin. Plant Biol.* **2015**, *25*, 107–114. [CrossRef] [PubMed]
3. Smith, P.; Gregory, P.J. Climate change and sustainable food production. *Proc. Nutr. Soc.* **2013**, *72*, 21–28. [CrossRef] [PubMed]
4. Akter, N.; Islam, M.R. Heat stress effects and management in wheat. A review. *Agron. Sustain. Dev.* **2017**, *37*, 37. [CrossRef]
5. Sánchez-Bermúdez, M.; Del Pozo, J.C.; Pernas, M. Effects of combined abiotic stresses related to climate change on root growth in crops. *Front. Plant Sci.* **2022**, *13*, 918537. [CrossRef] [PubMed]
6. Sutherst, R.W.; Constable, F.; Finlay, K.J.; Harrington, R.; Luck, J.; Zaluckiet, M.P. Adapting to crop pest and pathogen risks under a changing climate. *WIRE Clim. Chang.* **2011**, *2*, 220–237. [CrossRef]

7. Jansson, J.K.; Hofmockel, K.S. Soil microbiomes and climate change. *Nat. Rev. Microbiol.* **2020**, *18*, 35–46. [CrossRef] [PubMed]
8. Zhang, Y.; Liu, M.; Han, L.; Yang, J.; Zhao, X.; Qu, J.; Li, L.; Bai, Y.; Yan, D.; Hou, G. Spatial Distribution Characteristics of Soil C:N:P:K Eco-Stoichiometry of Farmland and Grassland in the Agro-Pastoral Ecotone in Inner Mongolia, China. *Agronomy* **2024**, *14*, 346. [CrossRef]
9. Hou, H.; Li, R.; Zheng, H.; Tong, C.; Wang, J.; Lu, H.; Wang, G.; Qin, Z.; Wang, W. Regional NDVI Attribution Analysis and Trend Prediction Based on the Informer Model: A Case Study of the Maowusu Sandland. *Agronomy* **2023**, *13*, 2882. [CrossRef]
10. Oh, D.; Ryu, J.H.; Jeong, H.; Moon, H.D.; Kim, H.; Jo, E.; Kim, B.K.; Choi, S.; Cho, J. Effect of Elevated Air Temperature on the Growth and Yield of Paddy Rice. *Agronomy* **2023**, *13*, 2887. [CrossRef]
11. Zhang, F.; Jiang, N.; Zhang, H.; Huo, Z.; Yang, Z. Effect of Low Temperature on Photosynthetic Characteristics, Senescence Characteristics, and Endogenous Hormones of Winter Wheat "Ji Mai 22" during the Jointing Stage. *Agronomy* **2023**, *13*, 2650. [CrossRef]
12. Bortoluzzi, L.; Casal, S.; Cruz, R.; Peres, A.M.; Baptista, P.; Rodrigues, N. Influence of Interannual Climate Conditions on the Composition of Olive Oil from Centenarian Olive Trees. *Agronomy* **2023**, *13*, 2884. [CrossRef]
13. Wang, M.; Liu, W.; Peng, Q.; Shi, S.; Wang, Y.; Cao, L.; Jiang, B.; Lin, Y.; Zhao, T.; Cui, X.; et al. Excavation of Genes Response to Heat Resistance by Transcriptome Analysis in Bottle Gourd (*Lagenaria siceraria* (Mol.) Standl.). *Agronomy* **2024**, *14*, 299. [CrossRef]
14. Jiang, Y.Y.; Sun, Z.X.; Wang, R.M.; Wang, H.L.; Wang, J.Q. A Quantitative Reconstruction of Nutrient Changes of Quaternary Red Soils (Luvisols) Affected by Land-Use Patterns. *Agronomy* **2023**, *13*, 2386. [CrossRef]
15. Jiang, Y.Y.; Sun, Z.X.; Zheng, Y.B.; Wang, H.L.; Wang, J.Q. Establishing a Soil Health Assessment System for Quaternary Red Soils (Luvisols) under Different Land Use Patterns. *Agronomy* **2023**, *13*, 2026. [CrossRef]
16. Qi, J.; Feng, K.Y.; Zhang, Y.J.; Dong, H.Z. Early Maturity Mechanism and High-Yielding Cultivation of Short-Season Cotton in China. *Agronomy* **2023**, *13*, 2770. [CrossRef]
17. Avasiloaiei, D.I.; Calara, M.; Brezeanu, P.M.; Gruda, N.S.; Brezeanu, C. The Evaluation of Carbon Farming Strategies in Organic Vegetable Cultivation. *Agronomy* **2023**, *13*, 2406. [CrossRef]

agronomy

MDPI

Review

Early Maturity Mechanism and High-Yielding Cultivation of Short-Season Cotton in China

Jie Qi [1,2], Keyun Feng [1], Yanjun Zhang [2,*] and Hezhong Dong [2,*]

1 Crop Research Institute, Gansu Academy of Agricultural Sciences, Lanzhou 730070, China; qijie8565@163.com (J.Q.); fengkeyun@126.com (K.F.)
2 Institute of Industrial Crops, Shandong Academy of Agricultural Sciences, Jinan 250100, China
* Correspondence: zhangyanjunche@163.com (Y.Z.); donghezhong@163.com (H.D.); Tel.: +86-531-6665-8187 (Y.Z.); +86-531-6665-9255 (H.D.)

Abstract: Short-season cotton is a type of cotton variety characterized by its abbreviated cycle, rapid development, and concentrated flowering and boll setting. Compared with full-season cotton, short-season cotton facilitates an easier attainment of desirable maturation even when sown relatively late. This advantage of late sowing and early maturation eliminates the necessity for plastic film mulching, thereby creating opportunities for diversified double cropping, such as cotton–wheat, cotton–garlic, cotton–rape, and cotton–triticale systems. This paper provides a comprehensive review of the morphological, physiological, and molecular biological mechanisms underlying early maturity in short-season cotton. Furthermore, the significance and application of short-season cotton is discussed in relation to optimizing planting patterns and methods, promoting its cultivation in saline fields, developing machine-harvested cotton, and encouraging plastic mulch-free cotton planting. Based on these analyses and discussions, the paper proposes future strategies aimed at enhancing the breeding and cultivation of short-season cotton. These findings serve as valuable references for global breeding and cultivation research, and application of short-season cotton in the future.

Keywords: short-season cotton; early maturity; physiological mechanism; cultivation modes

1. Introduction

Short-season cotton (*Gossypium hirsutum* L.) is an ecotype that is characterized by a relatively compact growth form, concentrated flowering and boll setting, and a short growth and development period, leading to early maturity upon late sowing [1,2]. Although China began breeding early maturing cotton as early as the mid-19th century, the initial short-season cotton varieties developed in the early stages had shortcomings such as low lint percentage and poor fiber quality [3]. However, recent advances in breeding techniques have resulted in significant improvements in lint percentage and fiber quality of newly developed short-season cotton varieties, while retaining their early maturity, high yield, and stress resistance [4]. Noteworthy examples of such varieties are CCRI 50, CCRI 67, Xinluzao 42, and Lumian 532, which exhibit a lint percentage exceeding 38%, fiber length surpassing 28 mm, breaking specific strength over 28 cN/tex, and micronaire value below 4.8. These fiber quality parameters meet the requirements of cotton production and textile industry.

Moreover, the adoption of short-season cotton has offered new alternatives for regions with potential multi-cropping in China. Particularly, the direct sowing of short-season cotton after garlic (*Allium sativum* L.) in the southwestern cotton region of Shandong [5] and sowing of short season-cotton after rape (*Brassica campestris* L.) or wheat (*Triticum aestivum* L.) in the cotton region of the Yangtze River Basin [6] have emerged as predominant cultivation models. These approaches not only enhance land use efficiency and align with agricultural structural adjustments [7], but also contribute to the reduction in plastic film usage, curbing surface pollution in farmland, and fostering environmentally friendly and sustainable agricultural practices. This review presents insights into the characteristics of short-season cotton varieties, mechanism behind early maturity, current cultivation and utilization status, and future development prospects of this crop. The primary objective is to offer guidance and references for the genetic breeding and cultivation of short-season cotton.

2. Characteristics of Short-Season Cotton

Short-season cotton, a distinctive variety, exhibits a growth period lasting 100–115 days, which is notably shorter 20–30 days than full-season cotton types [1,8–10].

The cultivation characteristics of short-season cotton mainly manifest in various planting modes. For instance, in the Huang-Huai-Hai cotton region, cotton-wheat intercropping is utilized, whereas the double-season planting mode of wheat (rape) transplanting or direct seeding is adopted in the cotton region of the Yangtze River Basin. In the northwest inland cotton region, the one-season cotton planting mode is favored. Early-maturing cotton offers flexibility in sowing time, with the Huang-huai-Hai cotton area typically concentrated between 15 May and 5 June. Emphasis is placed on dense planting and dwarfing cultivation, involving high density (75,000–120,000 plants ha^{-1}), early topping (9–11 fruit branches left per plant before and after 20 July), and appropriate chemical regulation to control plant height at 75–95 cm. This approach shapes a rational plant structure and regulates per plant production, maximizing population advantages and achieving high yield and efficiency [11].

According to Yu et al. [12] early maturity in cotton harnesses local light and heat resources optimally, thereby yielding enhanced productivity and economic benefits within specific climatic cultivation conditions. Consequently, the morphological attributes of early-maturing cotton or short-season cotton, can be summarized as follows: Firstly, the initial fruit branch on the main stem is positioned lower and 4–6 or lower, thereby promoting early development. Secondly, the internode between the main stem and fruit branch is short, rendering the plant compact with a low height, and a minimal angle between the main stem and fruiting branch, thus efficiently utilizing light energy, reducing nutrients consumption, and facilitating the transportation of photosynthetic products to reproductive organs. Thirdly, an abundance of moderate-sized bolls is present in the middle and lower parts of a plant, featuring a short maturation period, early and concentrated boll opening, and easy dehydration post-maturity [4,12].

The northwest inland cotton-growing region offers favorable conditions for high-density cultivation of short-season cotton due to its limited frost-free period, ample light resources, and well-established cotton mechanization, aimed by water-saving irrigation and continuous promotion of new varieties. In the northern Xinjiang region of this cotton-growing area, short-season cotton displays a growth period of 100–112 days, with a plant height ranging from 75 to 85 cm and the first fruiting branch positioned more than 15 cm above the node [13,14]. These environment factors and cultivation practices create an advantageous environment for production short-season cotton in northern Xinjing.

3. The Early-Maturity Mechanism of Short-Season Cotton

Early maturity is the most significant feature of short-season cotton, which has contributed to its widespread adoption in multiple cropping systems and regions with super early-maturity conditions. The research and utilization of short-season cotton commenced with an exploration of its early-maturity mechanism. In recent times, advancements in molecular biology technology have enabled partial disclosure of the molecular mechanism responsible for early maturity in short-season cotton. Building upon current research, the early-maturity mechanism in short-season cotton primarily manifests in the following aspects.

3.1. Characteristics of Photosynthesis and Carbon and Nitrogen Metabolism in Short-Season Cotton

The intensity of photosynthesis and carbon and nitrogen metabolism plays a pivotal role in both the accumulation of cotton plant biomass and the transition from vegetative growth to reproductive growth [15]. Research by Deng et al. [16] indicates that the short-season cotton variety CCRI 10 exhibits higher chlorophyll content than the medium-mature cotton variety, thereby enhancing its photosynthesis intensity. This observation might be attributed to the thicker leaves and longer palisade tissue cells in CCRI 10. Furthermore, for summer-sown short-season cotton, the photosynthetic peak occurs at the initial flowering stage, resulting in an advancement of both the production and accumulation of photosynthetic products. Consequently, this advancement serves as a foundational element in facilitating early square, flowering and boll setting. In addition, Shen et al. [17] argued that the early-maturing CCRI 10 exhibits a shorter transport period of photosynthetic products to reproductive organs than the later-maturity CCRI 16 and CCRI 36. In their investigation using ^{14}C isotope tracer technology, Guo et al. [18] analyzed the distribution of carbon assimilation products among cotton varieties with different maturities. The findings revealed that the proportion of photosynthate transport in short-season cotton varieties during the seedling stage, with a predominant allocation to the terminal bud and a smaller allocation to the root, whereas the opposite pattern was observed in medium-maturity varieties. During peak squaring, the proportion of carbon assimilates transported from short-season cotton varieties to terminal buds was significantly higher than that of medium-maturity varieties, which contributes to early maturity. Following squaring, the carbohydrate (soluble sugar and starch) allocated to the reproductive organs in short-season cotton was significantly higher than that in medium-maturity cotton, benefiting the fiber development of short-season cotton [16]. Overall, the earlier appearance of the photosynthesis peak and the increased photoassimilate partitioning to the reproductive organs are essential mechanisms driving the early maturity of short-season cotton.

Nitrogen metabolism plays a crucial role in both the growth of cotton plants and the development of fibers. Prior to flowering, early-maturing cotton primarily directs its resources towards trophosome expansion, involving root development, stem elongation, and leaf growth, subsequently transitioning to squares. During this period, nitrogen absorption and utilization are relatively low but accelerate after square formation. The peak of nitrogen absorption and utilization occurs during the flowering and boll-setting stage. Notably, early-maturing varieties exhibits an earlier peak in nitrogen absorption compared with mid-maturing varieties [19]. When short-season cotton is sown in spring, its leaves consistently maintain higher levels of non-protein nitrogen throughout various growth stages, particularly during the flowering stage, compared to medium-maturity cotton. For instance, the non-protein nitrogen content in short-season cotton CCRI 10 is 47.0% higher than that of medium-maturity cotton CCRI 31, indicating the significance of non-protein nitrogen as a determinant of material accumulation [16]. Conversely, when short-season cotton is sown in summer, the total nitrogen content in the leaves remains consistent with that in spring, but the protein nitrogen content increases significantly. The content of 12, 10, and 15 amino acids in CCRI 10 is higher than those in CCRI 31 during the squaring, flowering and boll-setting stage, respectively. Notably, aspartic acid experiences the most significant increase, and the higher amino acid content plays a crucial role in ammonia transport and protein synthesis, thereby ensuring vigorous nitrogen metabolism, rapid protein transformation, and accelerated organic development. Yu et al. [12] highlight that early-maturing varieties exhibits higher catalase and sulfhydryl compounds activities than late-maturing varieties. Additionally, the protein and amino acids levels in early-maturing cotton varieties are significantly higher than those in medium-maturing cotton throughout the growth period, providing a material foundation for its high yield within a limited time.

The early peak of nitrogen absorption rate and the vigorous nitrogen metabolism constitute another important mechanism for the early maturity of short-season cotton. During peak squaring, the buds of short-season cotton display significantly higher carbohydrate allocation (soluble sugar and starch) to reproductive organs compared to medium-maturity varieties, facilitating early maturity. Furthermore, after squaring, short-season cotton allocates a significantly higher number of photosynthetic products to reproductive organs than medium-maturity cotton, which contributes significantly to fiber development in short-season cotton [16]. These observations indicate that the early appearance of the photosynthesis peak and the preferential distribution of photosynthetic products to reproductive organs are essential mechanisms driving the early maturity of short-season cotton.

3.2. Flower Bud Differentiation and Hormone Regulation Characteristics of Short-Season Cotton

Cotton exhibits an indeterminate growth habit encompassing both vegetative growth, such as rooting, leafing, and branching, and reproductive growth, including budding, flowering, and boll setting. It is commonly acknowledged that cotton's reproductive growth commences with the process known as "squaring"; however, it should be noted that cotton flower buds have already undergone differentiation before this stage, making it the true initiation of cotton's reproductive growth. This differentiation occurs after complex physiological, biochemical, and morphological changes in the cotton seedling under suitable environmental conditions, such as light and temperature. The development of cotton flower bud serves as the foundation for the subsequent formation of cotton bolls and fiber yield. The timing and number of flower bud differentiations significantly influence boll suitability, as well as the overall yield, making flower bud differentiation vital to cotton yield and quality [20]. Ren et al. [20] conducted research on early and mid-maturing cotton cultivar, "CCRI 16" and "CCRI 12", respectively. They found that morphological differentiation of flower buds, transitioning from vegetative to reproductive growth, occurred when the second true leaf unfolded for the early-maturing cultivar and when the fourth true leaf expanded for the mid-maturing cultivar. Chen [21] also observed that early-maturing cotton varieties exhibited earlier flower bud differentiation compared to medium- and late-maturing varieties. The flower bud differentiation is influenced not only by external factors but also by intricate internal factors. Nutrients serve as the basis for flower bud formation, while plant hormones play a crucial role in regulating this process [22].

The flower bud differentiation in upland cotton is under the comprehensive regulation of multiple hormones [22,23]. The endogenous hormone content and dynamic balance significantly impact the bud differentiation in upland cotton. Higher concentrations of gibberellic acid (GA3), cytokinin (CTKs), and abscisic acid (ABA), as well as lower concentrations of indoleacetic acid (IAA), are conducive to the flower bud differentiation in upland cotton. Additionally, higher levels of CTK/IAA, ABA/IAA, and GA3/IAA, along with lower levels of ZR/GA3 and ABA/GA3, promote flower bud differentiation in upland cotton [20,23–25]. Moreover, the regulation mechanism of hormones on flower bud differentiation varies across different parts of the plant. For instance, low levels of IAA promote flower bud differentiation in cotton stem tips, whereas high levels promote differentiation in main stem leaves [24]. Shen et al. [17] observed that the contents of IAA and Zeatin riboside (ZR) in the functional leaves of three short-season cotton varieties peak during the transition vegetative to reproductive growth, from squaring to the flowering stage.

Ren et al. [20] demonstrated that high levels of IAA in stem tips have an inhibitory effect on flower bud differentiation in cotton. In both early-maturing and medium-maturing cotton varieties, the IAA content decreases significantly at the onset of flower bud differentiation, supporting the notion that the reduced IAA content in stem tips benefits cotton flower bud differentiation. Cytokinin's high levels enhance nutrient transport to the reproductive organs while reducing distribution to the vegetative organs, thereby promoting flower bud differentiation [26]. In conclusion, the regulation of endogenous hormones in different parts of the cotton plant plays a crucial role in promoting early differentiation and flower bud formation, contributing to the early maturity of short-season cotton.

3.3. Molecular Mechanism of Early Maturity in Short-Season Cotton

The early maturity of cotton is a complex trait controlled by multiple genes, and it is characterized by metabolic changes and material transformation rate [4,27]. Flowering time, which is closely associated with early maturity, is regulated by various pathways such as the autonomous pathway, photoperiodic pathway, vernalization pathway, GA pathway, temperature pathway, and age pathway. Therefore, studying the function of key genes involved in the flowering pathway in cotton is highly significant [28–30]. In *Arabidopsis thaliana* L., the *FLOWERING LOCUS T (FT)* gene plays a crucial role in flowering. It acts as a transcription factor, integrating signals from internal genetic information and external environmental cues. The *FT* gene triggers a flowering transition by transmitting signals downstream from the apical meristem to promote the expression of flowering genes, ultimately leading to flower bud differentiation and flowering [31,32]. The gene *GhFLP1*, associated with cotton flowering time, was isolated using genomic DNA and cDNA from the leaves of CCRI 36. Its expression is primarily detected in flower buds, suggesting a specific role in the regulation of cotton flowering time [33]. Moreover, the expression level of *GhFLP1* in early-maturing varieties (CCRI 36, CCRI 74) was significantly higher than medium-maturing varieties (CCRI 60, SCRC 28), implying a potential link between the gene and the early maturity of cotton [34].

Previous studies have indicated the influence of gibberellin (GA) and salicylic acid (SA) on plant flowering [34–38]. Zhang et al. [33] observed that the expression of *GhFLP1* was induced by exogenous GA and SA in cotton seedlings. In *Arabidopsis thaliana*, functional verification of the gene through *Agrobacterium tumefaciens* mediated transformation resulted in reduced rosette leaf number and earlier flowering. An analysis of endogenous flowering gene expression in transgenic *Arabidopsis* revealed up-regulation of certain flowering-promoting genes (e.g., *AtFT, AtAP1, AtLFY, AtSOC1*) and down-regulation of the flowering-inhibiting gene *AtFLC* [39,40]. These findings suggest that *GhFLP1* potentially regulates flowering time in upland cotton, offering a basis for the development of new transgenic early-maturing cotton varieties.

Wang et al. [41] examined the expression patterns of the *GhFLP5* gene in different cotton varieties and observed that it was primarily expressed in leaves. The peak expression of *GhFLP5* in the early-maturing variety CCRI50 occurred earlier than in the mid-maturing variety SCRC 28, aligning with previous studies on cotton flower bud differentiation. This suggests a potential role of *GhFLP5* in flower primordia development. The expression of *GhFLP5* increased after stimulation with SA and ABA, while the application of Jasmonic acid (JA) suppressed its expression. Heterologous expression of *GhFLP5* in *Arabidopsis thaliana* resulted in an early flowering phenotype, indicating its involvement in regulating flowering through the IAA and GA pathways [42].

In a genome-wide association study (GWAS), Ma et al. [42] found that the expression of the *GhCIP1* gene was significantly higher in early flowering varieties (6–8 true leaves) during the flowering decision stage, than that in late flowering varieties. Further verification using virus-induced gene silencing (VIGS) demonstrated that the silencing of *GhCIP1* in cotton plants led to the absence of fruit branches and flower buds, indicating its major role in regulating cotton flowering time. Additionally, the *GhUCE* gene in early flowering cotton was found to be involved in the regulation of cotton fiber initiation and development. Jia et al. [43] observed significantly higher expression levels of the candidate gene *EMF2*, related to early maturity, in the early-maturing variety CCRI 74 compared to the late-maturing variety Bomian1.

Utilizing molecular markers closely associated with quantitative trait locus (QTL) for assisted selection enables the identification of single-nucleotide polymorphism (SNP) or QTL alleles linked to prematurity, thus reducing breeding time and enhancing breeding efficiency [44]. Fan et al. [45,46] generated an F2 mapping population comprising 207 individual plants derived from the parental lines CCRI 36 and TM-1. They detected 12 QTLs associated with early-maturity traits in short-season cotton. Two of these QTLs were specifically related to flowering period and pre-frost flowering rate, factors that contribute to early maturity. These QTLs accounted for 38.45% and 39.73% of the phenotypic variance, respectively. Ai [47] conducted crosses between upland cotton varieties Xinluzao 8 and Xinluzao 10 with upland cotton TM-1 to establish F2 and F2:3 family populations.

Two genetic maps of early-maturing upland cotton varieties were constructed using simple sequence repeats (SSR) markers and composite interval mapping (CIM), and a total of 61 significant QTLs were detected. Yang [48] constructed an F2:9 recombinant inbred population using the short-season cotton variety CCRI 36 and the island cotton introgressive line material G2005 with a background of upland cotton as parents, and detected a total of 43 QTLs related to early maturity. QTL mapping of cotton early maturity based on high-density genetic maps revealed a total of 247 QTLs related to early-maturity traits (squaring, flowering, whole growth stage, plant height, fruit branch beginning node, fruit branch beginning node height), and 55 QTL overlap regions were found on 22 chromosomes, accounting for more than 60% of the total QTLs [49]. Li et al. [49] detected 47 early-maturity-related QTLs on 26 chromosomes, with a phenotypic variation explained (PVE) of 2.61% to 32.57% for each QTL. They predicted and annotated 112 genes within the PVE interval and found that the expression levels of *Gh-D03G0885* and *Gh-D03G0922* candidate genes in early-maturity variety CCRI 213 were significantly higher than those in mid-maturity variety SCRC 28, suggesting that the *Gh-D03G0885* and *Gh-D03G0922* genes may play a role in controlling cotton flowering.

Su et al. [50] developed a multitude of SNP markers through genome sequencing and identified 13 associations between 8 SNP loci and 5 early-maturing traits. They observed a significant increase in the expression level of the CotAD-01947 gene during flowering in early-maturing varieties (CCRI 50 and CCRI 74) compared to mid-maturing varieties (SCRC 28 and CCRI 41). In another study by Wang et al. [51], the regulatory mechanism of miRNA in flower bud differentiation of upland cotton was investigated. It was discovered that miRNA plays a role in inducing flower bud differentiation by regulating the expression of target genes. Specifically, miR164 and miR166 facilitate shoot tip meristem development and shoot differentiation in upland cotton by targeting NAC and HD-Zip III transcription factors, respectively. The regulation of gibberellin is governed by the SPL transcription factor of the miR156 target gene. Ghrmirn16 targets the F-box protein and contributes to the ubiquitin-mediated degradation pathway, leading to a decrease in gibberellin content and an increase in IAA content. Another miRNA, GhrmiRn2, indirectly regulates sugar signals by targeting β-Hexosaminidase, thereby providing energy for flower bud differentiation and promoting the expression of genes associated with ABA synthesis and signal transduction.

4. Cultivation of Short-Season Cotton in China

Short-season cotton in China primarily grows in three traditional ecological areas: the northern super early-maturing ecological area, the Yellow River basin ecological area, and the Yangtze River basin ecological area. The northern super early-maturing ecological area is mainly located in Northern Xinjiang and the Hexi Corridor, where short-season cotton is cultivated as full-season cotton. In the Yellow River valley, which spans the Huang-Huai-Hai cotton region, double cropping is the primary planting method, including relay intercropping of cotton–wheat or cotton–garlic and direct seeding after wheat or garlic harvest. The ecological area of the Yangtze River basin consists mainly of two types of planting: direct seeding after garlic harvest and transplanting after wheat harvest (Figure 1). In recent years, there has been a significant expansion in the cultivation and utilization of short-season cotton, driven by the need to restructure the planting industry and promote environmentally friendly production practices. This expansion includes the adoption of techniques such as non-plastic mulching cotton and machine-picked cotton varieties [52].

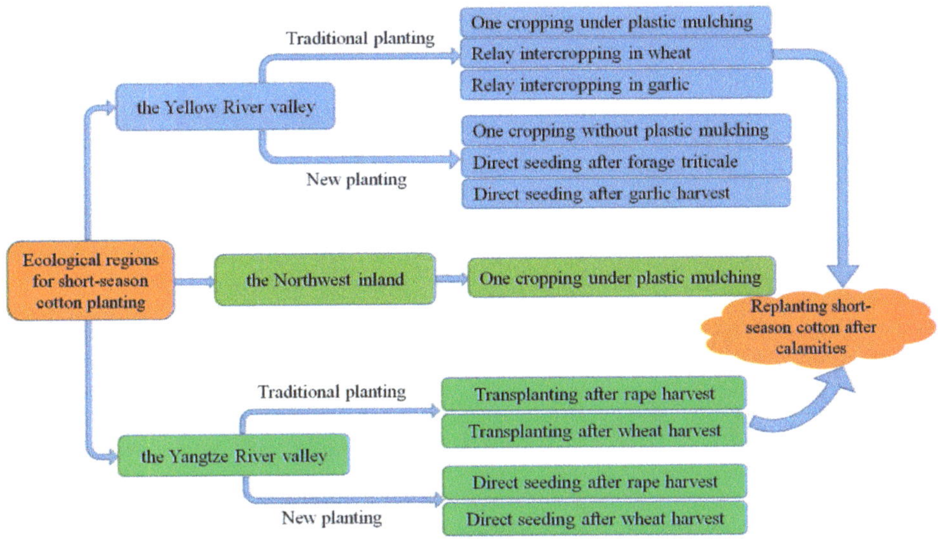

Figure 1. Schematic diagram of cultivation models for short-season cotton in different ecological regions.

4.1. Direct Seeding of Short-Season Cotton following Preceding Crop Harvest

Implementing direct seeding of short-season cotton following garlic (wheat or rape) harvest is a prevalent cultivation method in cotton-growing regions along the Yellow River and the Yangtze River Basin. However, this approach requires significant material and labor inputs, leading to increased costs and limited mechanization. With rising labor costs in China's progressing economy and society, the traditional intercropping method of garlic (wheat, rape) and cotton has become incompatible with modern agricultural production. Therefore, reforming conventional cotton planting techniques and adopting a lighter and more streamlined cotton production process are crucial technical solutions to address current challenges in the cotton industry. Transitioning from interplanting full-season cotton in garlic (wheat) fields to direct seeding of short-season cotton after garlic (wheat) harvest provides a viable perspective for achieving this goal. This transition involves shifting from sparse planting to dense dwarfing, from fine pruning to no pruning, from multiple fertilization rounds to one-time fertilization, and from multiple harvesting episodes to centralized boll formation and single harvest [53,54].

A light and simplified cultivation technique for direct seeding of short-season cotton after garlic has been developed in recent years. This technique involves high-density planting to achieve concentrated boll formation. Its adoption has significantly improved the mechanization level of double cropping in Southwest Shandong Province [54]. It entails selecting short-season cotton varieties with growth periods of no more than 110 days, compact plant structures, and strong boll-setting capabilities. The sowing process is performed mechanically before the end of May, without thinning the seedlings, thus maintaining a plant density of 90,000–120,000 plants per hectare. Consequently, in late July, topping of the plants is needed, eliminating the need to remove vegetative branches. Plant height is controlled between 80 and 90 cm using a combination of chemical control and fertilization. Defoliation is conducted in late September to facilitate centralized boll opening and enable a one-time picking process [55].

In certain regions within the Yangtze River Basin and Yellow River Basin, the adoption of light simplification and mechanized cotton planting has been increasing. This trend has led to research and demonstrations of mechanical direct seeding techniques after the harvest of rape and wheat. Short-season cotton's brief growth period, concentrated boll formation, and ease of management facilitate efficient soil utilization and environmentally friendly production [56].

4.2. Cultivating Short-Season Cotton without Plastic Mulching

The primary cotton-producing region in northwest China faces significant challenges during the sowing period, including low soil temperature, a short frost-free period, and water scarcity. Plastic film mulching has effectively addressed these issues by increasing soil temperature, conserving water, and maintaining moisture. However, its use has resulted in numerous problems in agriculture and the ecological environment. These problems include soil structure damage, reduced cultivated land quality, hindered cotton seedling emergence, and compromised cotton fiber quality [52,57–59]. To address these challenges, Yu [12,52] introduced the concept of "short-season cotton with non-mulching cultivation" and conducted a six-year demonstration using the early-maturing cotton variety "Zhongmian 619" in Shaya County, Xinjiang. Through late sowing, increased planting density, shallow burial, and drip irrigation, successful non-mulching cultivation was achieved. Preliminary results demonstrate a seed cotton production of 4800–5250 kg ha^{-1}, eliminating film residue pollution and showcasing promising prospects for further development [52].

China has a considerable amount of saline soil, both inland and coastal. Exploiting cotton's robust salt tolerance by cultivating it in saline soil proves to be a promising approach to enhance cotton production stability [56]. Previous studies have shown the effectiveness of furrow border plastic film mulching for full-season cotton in severe coastal saline alkali soil and flat plastic film mulching for short-season cotton in mild saline alkali soil, optimizing land utilization and achieving positive economic outcomes [60,61]. However, the pollution caused by residual film and the variations in heat and rainfall conditions between the Yellow River Basin and Northwest inland cotton regions make non-mulched short-season cotton more conducive in the former. Therefore, late sowing of non-mulched short-season cotton in mildly saline soil or low-yield cotton fields presents a viable solution to residual film pollution.

In a five-year study by Qi et al. [62,63], it was demonstrated that non-mulched short-season cotton yielded comparable results to mulched short-season cotton in terms of seed cotton yields, boll density, average boll weight, lint percentage, and fiber quality parameters such as fiber length, strength, micronaire value, elongation rate, and uniformity. Furthermore, the earliness of non-mulched short-season cotton, which was achieved through early flower bud differentiation, rapid flowering and boll setting, and concentrated boll opening, was desirable and comparable to that of mulched short-season cotton. An important advantage of cultivating short-season cotton without mulching is the avoidance of plastic residual pollution, resulting in cost savings and environmental benefits. Therefore, this approach proves to be more economical and efficient. Considering the implications for sustainable cotton production in medium- and low-yield fields of the Yellow River Basin cotton region, the cultivation of short-season cotton without mulching emerges as a crucial option.

4.3. Double Cropping of Short-Season Cotton and Triticale

In recent years, there has been a shift in cotton cultivation towards the northwest inland region, where it now accounts for over 83% of the country's cotton planting area. Conversely, the cotton planting area in the Yangtze River and Yellow River valleys has been in decline. The traditional cotton cultivation practices in these valleys are labor-intensive, relying on intensive cultivation to achieve high yields, resulting in high management costs and significant labor input. However, due to rapid economic development and urbanization, the rural labor force in China has experienced a sharp decline. Consequently, the traditional cotton planting methods are no longer suitable for the current environment in the Yangtze and Yellow River valleys. Therefore, it is crucial to reform and simplify the cotton cultivation system in these regions [64].

In the Yangtze and Yellow River valleys, there have been a variety of cotton planting systems for a long time, including both one cropping and two cropping systems. Currently, it is highly significant to develop flexible and diverse cotton planting models in order to stabilize the overall cotton planting area, increase yield and efficiency per unit area, and revitalize the cotton industry in these regions. However, the limited thermal resources in the cotton region of the Yellow River Basin pose a major obstacle to the development of double-cropping crops. In order to address this limitation, relay interplanting or intercropping methods are often employed to compensate for the lack of heat, thereby enhancing the intercropping index and economic benefits. Nevertheless, it should be noted that relay interplanting or intercropping requires high labor inputs and lacks mechanization convenience [65–67].

Utilizing the significant advantages of short-season cotton such as its short growth period, concentrated flowering and boll setting, as well as the favorable characteristics of forage crops like triticale and oats including high harvest elasticity and nutritional value, the implementation of a double-cropping system for short-season cotton and forage crops in the Yellow River Basin cotton region proves to be an effective approach to achieve enhanced productivity efficiency. Cui et al. [68] have demonstrated this double-cropping mode and produced promising results. Other researchers, including Mao [69], Wang et al. [70], Yao et al. [71], and Zheng et al. [72], have also corroborated the productivity and benefits associated with the double cropping of short-season cotton and forage crops in the Yellow River Basin. This model has been instrumental in the development of high-quality cotton production in the region.

4.4. Replanting Short-Season Cotton after Calamities

The prolonged growth period of cotton makes it vulnerable to hail disasters in regions with unpredictable cold and hot airflows. Effective measures must be implemented to minimize or eliminate losses resulting from these calamities [73]. Cotton plants exhibit strong compensatory effects on both biological and economic yield due to their indeterminate growth habit following physical damage, such as hailstorms during the squaring stage [74]. Mild to moderate damage in cotton fields requires immediate and enhanced field management practices to mitigate yield reduction. Conversely, severely affected fields necessitate prompt replanting. The choice of replanting strategy should be context-specific. In Shandong Province, China, the appropriate action depends on the timing of the disaster. If the disaster occurs before 5 May, it is advisable to consider planting medium- and early-maturity cotton varieties. If the disaster occurs before 5 June, replanting using short-season cotton varieties becomes a viable option. However, after 5 June, cultivating crops with shorter growth cycles, such as mung beans and edible beans, becomes the only feasible approach.

5. Prospects of Short-Season Cotton in China

China's main cotton-producing areas exhibit complexity and diversity in terms of planting systems, modes, and varieties due to national conditions such as a large population, limited land, and significant ecological and production variations among these regions. As the development of light and simplified cotton and mechanized production advances, the importance of short-season cotton is expected to grow. In cotton regions with abundant resources like water, fertilizer, light, and heat, the adoption of mechanical direct seeding of short-season cotton after stubble can promote light and simplified cotton cultivation in multi-cropping and efficient cotton fields. In regions with limited light and heat resources, high salinity, or poor soil fertility, short-season cotton can be cultivated once a year without plastic film planting. By increasing density and fully utilizing the advantages of smaller individual plants and larger groups, the cultivation can bring about diversification through double cropping with forage crops like triticale and oats annually. This approach reduces pesticide and labor input, minimizes plastic film usage, and ensures high and stable yields, making it a cost-effective, environmentally friendly, and efficient method for cotton production. In future research and application of short-season cotton, it is important to focus on the following aspects:

Investigating the early-maturing mechanism of short-season cotton: Collaborating across multiple disciplines and employing techniques such as gene editing in physiology and molecular biology, researchers can unveil the mechanism of early maturity in short-season cotton from the perspective of flower bud differentiation. This will provide valuable theoretical support for molecular breeding of short-season cotton.

Developing new short-season cotton varieties: Short-season cotton also has its own drawbacks, such as lower yield potential than full-season cotton, and some short-season cotton varieties have lower fiber quality than full-season cotton. By utilizing a combination of molecular breeding technology and conventional breeding methods, researchers can create short-season cotton varieties with characteristics such as extra early maturity, salt alkali tolerance, low temperature tolerance, high yield potential, and excellent fiber quality. These new varieties will serve as an essential guarantee for achieving high yield and superior cotton quality.

In-depth study of mechanized production technology for short-season cotton without mulching: Expanding the research on agronomic technology for diversified planting of short-season cotton without mulching in suitable cotton regions is crucial. The focus should be on enhancing trial production and developing supporting machinery for sowing, plant protection, and harvesting. This, in turn, will facilitate the integration of agricultural machinery and agronomy to establish a diversified planting technology system for short- season cotton. The promotion of this technology can be facilitated through demonstrations, technical training, media publicity, and other means, enabling the widespread adoption of light, cost-effective, green, and efficient cotton production.

In conclusion, short-season cotton offers numerous advantages, including its rapid growth cycle, early maturation, and potential for diversified double-cropping systems. Moreover, mechanical sowing is achieved via adoption of short-season cotton. In addition, due to high plant density and good earliness, short-season cotton is also conducive to mechanic cotton picking. Therefore, the use of short-season cotton has promoted the mechanization of cotton production in China. This paper highlights the morphological, physiological, and molecular mechanisms underlying its early maturity, making it a promising option for optimizing planting patterns and cultivation in different environments. As cotton production becomes more mechanized and simplified, short-season cotton is expected to play a crucial role. Further research is recommended to explore its early-maturation mechanisms, develop new varieties, and advance comprehensive mechanized production technology without mulching. These efforts will ultimately promote cost-effective, environmentally friendly, and efficient cotton production. These findings serve as valuable references for global breeding and cultivation research, and the application of short-season cotton in the future.

Author Contributions: Conceptualization: H.D. and Y.Z.; Writing—Original Draft: J.Q.; Writing—Review and Editing: H.D., K.F. and Y.Z.; Visualization: J.Q., K.F., H.D. and Y.Z.; Funding Acquisition: J.Q. and K.F. All authors have read and agreed to the published version of the manuscript.

Funding: This work was financially supported by the Doctoral Foundation of Gansu Academy of Agricultural Sciences (2023GAAS45), Modern Biological Breeding of Gansu Academy of Agricultural Sciences (2022GAAS04), China Agricultural Research System (CARS-15–15), Dong Hezhong Studio for Popularization of Science and Technology in Salt Tolerant Industrial Crops (202228297), and Tianchi Talent Program.

Conflicts of Interest: The authors declare no conflict of interest.

References

1. Cao, T.V.; Oumarou, P.; Gawrysiak, G.; Klassou, C.; Hau, B. Short-season cotton (*Gossypium hirsutum* L.) may be a suitable response to late planting in sub-Saharan regions. *Field Crops Res.* **2011**, *120*, 9–20. [CrossRef]
2. Yu, S.X.; Song, M.Z.; Fan, S.L.; Wang, W.; Yuan, R.H. Biochemical genetics of short-Season cotton cultivars that express early maturity without senescence. *J. Integr. Plant Biol.* **2005**, *47*, 334–342. [CrossRef]
3. Cui, R.M.; Liu, C.J.; Liu, S.E.; Guo, B.S.; Geng, J.Y.; Wang, Z.X.; Yin, X.L. Analysis on the main characters of the new short-season cotton varieties in Yellow River Cotton Region. *J. Hebei Agric. Sci.* **2007**, *11*, 1–4. (In Chinese)

4. Yu, S.X.; Wang, H.T.; Wei, H.L.; Su, J.J. Research progress and application of early maturity in upland cotton. *Cotton Sci.* **2017**, *29* (Suppl. S1), 1–10. (In Chinese)
5. Zhao, Z.T.; Zhang, G.H.; Li, S.; Wang, H.M. Direct seeding and light-simplified cultivation techniques for short-season cotton and garlic in southwestern Shandong cotton planting area. *Cotton Sci.* **2019**, *41*, 29–31. (In Chinese)
6. Yang, C.Q.; Zhou, Z.G.; Chen, D.H.; Zheng, S.F.; Zhang, Z.G. Light-simplified and highly efficient cultivation technologies of cotton with increased planting density and reduced fertilizer application for wheat/rape-cotton cropping system in the Yangtze valley. *China Cotton* **2018**, *45*, 1–4. (In Chinese)
7. Yu, S.S. Research and appraisal of short-season upland and cotton breeding achievements in China. *Cotton Sci.* **2005**, *17*, 232–339. (In Chinese)
8. Li, R.Z.; Wang, J.H.; Wang, Z.W.; Shen, G.F.; Zhao, F.T.; Han, Z.F. Breeding and trait analysis of short-season Bt cotton variety SCRC19. *Shandong Agric. Sci.* **2011**, *2*, 18–20. (In Chinese) [CrossRef]
9. Zhao, H.W.; Li, Y.Y.; Zhai, L.F.; Li, J.H.; Wang, Y.L.; Zhai, X.J. Breeding of super early mature cotton varieties. *J. Hebei Agric. Sci.* **2012**, *16*, 49–51, 55. (In Chinese)
10. Dong, C.G.; Wang, J.; Zhou, X.F.; Ma, X.M.; Li, S.X.; Wang, X.W.; Xiao, G.S.; Li, B.C. Inheritance of earliness traits in Xinjiang early-maturity upland Cotton (*G. hirsutum* L.). *Acta Agric. Boreali-Occident. Sin.* **2014**, *23*, 96–101. (In Chinese)
11. Yue, J.F.; Gao, W.; Zhai, J. Growth characteristics and the key high quality and yield technologies of short-season cotton under wheat-cotton double cropping system. *Mod. Seed Ind.* **2005**, *17*, 12. (In Chinese)
12. Yu, S.X.; Huang, Z.M. Present situation and development prospect of short season cotton production in our country. *China Cotton* **1989**, *16*, 6–8. (In Chinese)
13. Dong, C.G.; Wang, J.; Zhou, X.F.; Ma, X.M.; Li, S.X.; Li, B.C. Correlation study on major breeding target traits in Beijiang early-maturity upland cotton. *Southwest China J. Agric. Sci.* **2014**, *27*, 2255–2257. (In Chinese)
14. Zhao, H.W. Status of machine-harvested cotton breeding. *China Seed Ind.* **2013**, *9*, 18–19. (In Chinese)
15. Yu, S.X.; Zhang, C.X. *Overview of Short Season Cotton in China*; China Agriculture Press: Beijing, China, 2003. (In Chinese)
16. Deng, S.H.; Jiang, G.Z. A study on the developmental regularity and the physiological specificity of the short-season cottons (*Gossypium hirsutum* L.). *Sci. Agric. Sin.* **1987**, *20*, 15–22. (In Chinese)
17. Shen, F.F.; Yu, S.X.; Fan, S.L.; Li, J.; Huang, Z.M. Changes of endogenous hormone in stem leaves of different short season cotton varieties in development processes. *Sci. Agric. Sin.* **2003**, *36*, 1014–1019. (In Chinese)
18. Guo, Q.Z.; Dong, H.Z.; Liu, Q.H.; Wang, Z.F. Comparative studies on 14C-assimilates partition in cotton plants between short-season varieties and Mid-maturation varieties. *Acta Agric. Nucleatae Sin.* **1998**, *12*, 57–59. (In Chinese)
19. Li, J.Y.; Liu, R.R.; Dong, H.L.; Wang, R.Z.; Liu, A.Z.; Li, R.Y.; Cheng, M.T. Researches on the characteristics of nutrient Absorption and application time of nitrogenous fertilizer of cotton variety with early-maturity. *Cotton Sci.* **1992**, *2*, 47–51. (In Chinese)
20. Ren, G.J.; Chen, Y.Z.; Dong, H.Z.; Chen, S.Y. Studies on flower bud differentiation and changes of endogenous hormones of *Gossypium hirsutism*. *Acta Bot. Boreali-Occident. Sin.* **2000**, *20*, 847–851, 903. (In Chinese)
21. Chen, B.S. Morphogenesis and physiology of cotton organs. *Hubei Agric. Sci.* **1982**, *7*, 37–40. (In Chinese)
22. Cui, W.; Lv, Z.S. The 12th International Plant Growth Substances Association Conference. *Plant Physiol. Commun.* **1986**, *2*, 66–72. (In Chinese)
23. Gao, Y.S.; Wang, Y.; Zhu, Y.M.; Lin, S.Q. Molecular mechanism of TFL1 on the regulation of flowering time in Ro-saceae. *J. Fruit Sci.* **2016**, *33*, 1007–1013. (In Chinese)
24. Ren, G.J.; Dong, H.Z.; Chen, Y.Z.; Zhuang, Y.L.; Shao, F.Z.; Liu, Z.F. Studies on endogenous hormone changes in the stem terminal of *Gossypium hirsutum* during flower bud differentiation. *Acta Bot. Boreali-Occident. Sin.* **2002**, *22*, 113–118. (In Chinese)
25. Qi, J. *Leaf Senescence and Yield Formation of Field-Grown Short-Season Cotton without Plastic Mulching*; Shandong Agricultural University: Taian, China, 2022. (In Chinese)
26. He, Z.P. *Hormonal Physiology and Chemical Control of Field Crop*; China Agricultural University Press: Beijing, China, 1997. (In Chinese)
27. Cheng, H.L.; Yu, S.X. Studies on the earliness inheritance of upland cottons (*Gossypium hirsutum* L.). *Cotton Sci.* **1994**, *6*, 9–15. (In Chinese)
28. Wickland, D.P.; Hanzawa, Y. The FLOWERING LOCUS T/TERMINAL FLOWER 1 gene family: Functional evolution and molecular mechanisms. *Mol. Plant.* **2015**, *8*, 983–997. [CrossRef]
29. Wang, J.W. Regulation of flowering time by the miR156-mediated age pathway. *J. Exp. Bot.* **2014**, *65*, 4723–4730. [CrossRef]
30. Duan, Y.H.; Yan, Y.Y.; Chen, L.T.; Li, Q.; Zhang, D.M.; Sun, Z.W.; Zhang, Y.; Ma, Z.Y.; Wang, S.F. Cloning and functional validation of the flowering time related gene *GhMYB44* in upland cotton. *China J. Agric. Sci. Technol.* **2020**, *22*, 10. (In Chinese)
31. Kobayashi, Y.; Weigel, D. Move on up, it's time for change mobile-signals controlling photoperiod-dependent flowering. *Genes Dev.* **2007**, *21*, 2371–2384. [CrossRef]
32. Li, W. *Transcriptome Profiling of Upland Cotton Shoot Apex during Floral Induction*; Huazhong Agricultural University: Wuhan, China, 2011. (In Chinese)
33. Zhang, P.; Fan, S.L.; Song, M.Z.; Pang, C.Y.; Wei, H.L.; Yu, S.X. Cloning and functional analysis of the flowering-related gene *GhFLP1* from upland cotton (*Gossypium hirsutum* L.). *Cotton Sci.* **2016**, *28*, 199–207. (In Chinese)
34. Nowak, G.A.; Czapla, J. Response of soybean to gibberellin A3 application under conditions of high boron availability. *J. Plant Nutr.* **1995**, *18*, 2179–2190. [CrossRef]

35. Chen, L.G.; Zhang, L.P.; Yu, D.Q. Wounding-induced *WRKY8* is involved in basal defense in *Arabidopsis*. *Mol. Plant-Microbe Interact.* **2010**, *23*, 558–565. [CrossRef] [PubMed]
36. Li, Q.F.; Wang, C.; Jiang, L.; Li, S.; Sun, S.S.; He, J.X. An interaction between BZR1 and DELLAs mediates direct signaling crosstalk between brassinosteroids and gibberellins in *Arabidopsis*. *Sci. Signal.* **2012**, *5*, ra72. [CrossRef] [PubMed]
37. Li, Y.; Chang, Y.; Zhao, C.C.; Yang, H.L.; Ren, T.D. Expression of the inactive *ZmMEK1* induces salicylic acid accumulation and salicylic acid-dependent leaf senescence. *J. Integr. Plant Biol.* **2016**, *58*, 724–736. [CrossRef] [PubMed]
38. Liang, C.Z.; Wang, Y.Q.; Zhu, Y.N.; Tang, J.Y.; Hu, B.; Liu, L.H.; Chu, C.C. *OsNAP* connects abscisic acid and leaf senescence by fine-tuning abscisic acid biosynthesis and directly targeting senescence-associated genes in rice. *Proc. Natl. Acad. Sci. USA* **2014**, *111*, 10013–10018. [CrossRef] [PubMed]
39. Zhang, X.H.; Wei, J.H.; Fan, S.L.; Song, M.Z.; Pang, C.Y.; Wei, H.L.; Wang, C.S.; Yu, S.X. Functional characterization of *GhSOC1* and *GhMADS42* homologs from upland cotton (*Gossypium hirsutum* L.). *Plant Sci.* **2016**, *242*, 178–186. [CrossRef]
40. Zhang, X.H.; Dou, L.L.; Pang, C.Y.; Song, M.Z.; Wei, H.L.; Fan, S.L.; Wang, C.S.; Yu, S.X. Genomic organization, differential expression, and functional analysis of the SPL gene family in *Gossypium hirsutum*. *Mol. Genet. Genom.* **2015**, *290*, 115–126. [CrossRef]
41. Wang, C.C.; Zhang, X.H.; Wang, X.Y.; Zhang, P.; Fan, S.L.; Pang, C.Y.; Ma, Q.F.; Wei, H.L.; Wang, H.T.; Su, J.J. The expression patterns and function analysis of *GhFLP5*, a gene related to flowering in upland cotton (*Gossypium hirsutum* L.). *Sci. Agric. Sin.* **2017**, *50*, 2220–2233. (In Chinese)
42. Ma, Z.Y.; He, S.P.; Wang, X.F.; Sun, J.L.; Zhang, Y.; Zhang, G.Y.; Du, X.M. Resequencing a core collection of upland cotton identifies genomic variation and loci influencing fiber quality and yield. *Nat. Genet.* **2018**, *50*, 803–813. [CrossRef]
43. Jia, X.Y.; Pang, C.Y.; Wei, H.L.; Wang, H.L.; Ma, Q.F.; Yang, J.L.; Yu, S.X. High-density linkage map construction and QTL analysis for earliness-related traits in *Gossypium hirsutum* L. *BMC Genom.* **2016**, *17*, 909. [CrossRef]
44. Liang, B. *Association Analysis of Early Maturity, Yield and Fiber Quality Traits in Upland Cotton Using SSR Markers*; Northwest A&F University: Xi'an, China, 2014. (In Chinese)
45. Fan, S.L.; Yu, S.X.; Song, M.Z.; Yuan, R.H. Construction of molecular linkage map and QTL mapping for earliness in short-season cotton. *Cotton Sci.* **2006**, *18*, 135–139. (In Chinese)
46. Fan, S.L.; Yu, S.X.; Song, M.Z. Research progress and direction of genetic improvement of short-season cotton in China. *Chin. Agric. Sci. Bull.* **2008**, *24*, 164–167. (In Chinese)
47. Ai, N.J. *Inheritance and QTL Mapping for Earliness in Upland Cotton*; Nanjing Agricultural University: Nanjing, China, 2010. (In Chinese)
48. Yang, J.L. *QTL Location of Earliness and Fiber Quality in Cotton Using RILs*; Huazhong Agricultural University: Wuhan, China, 2013. (In Chinese)
49. Li, L.B.; Zhao, S.Q.; Su, J.J.; Fan, S.L.; Pang, C.Y.; Wei, H.L.; Yu, S.X. High-density genetic linkage map construction by F2 populations and QTL analysis of early-maturity traits in upland cotton (*Gossypium hirsutum* L.). *PLoS ONE* **2017**, *12*, e0182918. [CrossRef] [PubMed]
50. Su, J.J.; Pang, C.Y.; Wei, H.L.; Li, L.B.; Liang, B.; Wang, C.X.; Yu, S.X. Identification of favorable SNP alleles and candidate genes for traits related to early maturity via GWAS in upland cotton. *BMC Genom.* **2016**, *17*, 687. [CrossRef] [PubMed]
51. Wang, Q.L.; Li, F.; Sun, R.R.; Zhang, J.B.; Wang, Y.Y.; Chao, M.N. Study on the regulation machanism flower bud differentiation by endognous hormones and microRNAs in *Gossypium hirsutum* L. In *The 2016 Annual Conference of China Agricultural Association Cotton Section*; China Cotton Journal Office: Anyang, China, 2016.
52. Lian, W.M.; Yu, S.X.; Tai, H.Z.; Mao, S.C.; Su, J.J.; Li, K.F.; Li, W.P.; Wu, B. Production demonstration results of no-mulching cultivation techniques of early maturing cotton cultivars in southern Xinjiang. *China Cotton* **2017**, *44*, 20–23. (In Chinese)
53. Xie, Z.H.; Li, W.J.; Su, M.; Miu, L.; Cheng, L.J.; Dong, H.Z. Effects of plant pruning and plant population density on yield and fiber quality of cotton in a garlic-cotton intercropping system. *Cotton Sci.* **2014**, *26*, 459–465. (In Chinese)
54. Dong, H.Z. A new alternative of extensive farming under garlic-cotton double cropping, direct seeding of short-season cotton after garlic. *China Cotton* **2016**, *43*, 8–9. (In Chinese)
55. Dong, H.Z.; Yang, G.Z.; Tian, L.W.; Zheng, S.F. *Light and Simplified Cultivation of Cotton*; Science Press: Beijing, China, 2016. (In Chinese)
56. Dai, J.L.; Dong, H.Z.; Duan, L.S. Technology and mechanism in control of salt injury in cotton. *Cotton Sci.* **2010**, *22*, 486–494. (In Chinese)
57. Steinmetz, Z.; Wollmann, C.; Schaefer, M.; Buchmann, C.; David, J.; Troger, J.; Munoz, K.; Fror, O.; Schaumann, G.E. Plastic mulching in agriculture. Trading short-term agronomic benefits for long-term soil degradation? *Sci. Total Environ.* **2016**, *550*, 690–705. [CrossRef]
58. Wang, X.R.; Hou, Y.R.; Du, M.W.; Xu, D.Y.; Lu, H.Y.; Tian, X.L.; Li, Z.H. Effect of planting date and plant density on cotton traits as relating to mechanical harvesting in the Yellow River valley region of China. *Field Crops Res.* **2016**, *198*, 112–121. [CrossRef]
59. Wang, L.; Li, X.G.; Lv, J.T.; Fu, T.T.; Ma, Q.J.; Song, W.Y.; Wang, Y.P.; Li, F.M. Continuous plastic-film mulching increases soil aggregation but decreases soil pH in semiarid areas of China. *Soil Tillage Res.* **2017**, *167*, 46–53. [CrossRef]
60. Dong, H.Z.; Li, W.J.; Xin, C.S.; Tang, W.; Zhang, D.M. Late planting of short-season cotton in saline field of the Yellow River Delta. *Crop Sci.* **2010**, *50*, 292–300. [CrossRef]

61. Dong, H.Z. Underlying mechanisms and related techniques of stand establishment of cotton on coastal saline-alkali soil. *Chin. J. Appl. Ecol.* **2012**, *23*, 566–572. (In Chinese)
62. Qi, J.; Zhang, Y.J.; Dai, J.L.; Xu, S.Z.; Zhang, D.M.; Nie, J.J.; Sun, X.Z.; Dong, H.Z. Late-planted short-season cotton without plastic mulching is an alternative to early-planted mulched full-season cotton. *Ind. Crops Prod.* **2021**, *163*, 113325. [CrossRef]
63. Qi, J.; Nie, J.J.; Zhang, Y.J.; Xu, S.Z.; Li, Z.H.; Zhang, D.M.; Cui, Z.P.; Li, W.J.; Dai, J.L.; Tian, L.W.; et al. Plastic film mulching does not increase the seedcotton yield due to the accelerated late-season leaf senescence of short-season cotton compared with non-mulching. *Field Crops Res.* **2022**, *287*, 108660. [CrossRef]
64. Dong, H.Z.; Yang, G.Z.; Li, Y.B.; Tian, L.W.; Dai, J.L.; Kong, X.Q. Key technologies for light and simplified cultivation of cotton and their eco-physiological mechanisms. *J. Crops* **2017**, *43*, 631–639. (In Chinese) [CrossRef]
65. Liu, H.; Gao, Y.; Sun, J.S.; Wu, X.L.; Li, Y. Responses of yield, water use efficiency and quality of short-season cotton to irrigation management: Interactive effects of irrigation methods and deficit irrigation. *Irrig. Sci.* **2017**, *35*, 125–139. (In Chinese) [CrossRef]
66. Zhang, H.; Liu, H.; Wang, S.S.; Guo, X.; Ge, L.; Sun, J.S. Variations in growth, water consumption and economic benefit of transplanted cotton after winter wheat harvest subjected to different irrigation methods. *Sci. Rep.* **2019**, *9*, 14972. (In Chinese) [CrossRef]
67. Wang, S.L.; Qi, H.; Wang, Y.; Zhang, Q.; Feng, G.Y.; Lei, X.P.; Liu, Y.Z.; Liang, Q.L.; Wang, G.P. Benefit analysis of wheat-cotton intercropping system. *Tianjin Agric. Sci.* **2017**, *23*, 86–89. (In Chinese)
68. Cui, Z.P.; Dai, J.L.; Li, W.J.; Dong, H.Z. Optimizing the cotton system and innovating the planting mode, to plug in the wings of science and technology for the development of Shandong cotton industry. *China Cotton* **2021**, *48*, 13–16. (In Chinese)
69. Mao, S.C. The Application and Development of Cotton Planting Technology in China. *China Cotton* **2009**, *39*, 17–22. (In Chinese)
70. Miao, X.W.; Wang, Z.H. Matching Techniques for Cotton and Forage Triticale Rotation in the Yellow River Delta. *China Cotton* **2019**, *46*, 2. (In Chinese)
71. Yao, Y.; Yao, Y.G.; Liu, J.; Wang, M.; Cui, J.D.; Yang, X.L. Integrated Technology research on the mechanized planting model of grass-cotton double cropping in the Yellow River Delta. *China Cotton* **2020**, *47*, 34–37. (In Chinese)
72. Zheng, S.S.; Liu, X.L.; Wang, W.; Xu, D.Q.; Kan, H.H.; Chen, M.; Li, S.Y. On the green and light-simplified and mechanized cultivation of cotton in a cotton-based double cropping system. *Acta Agron. Sin.* **2022**, *48*, 541–552. (In Chinese) [CrossRef]
73. Feng, B.K. How to remedy after hail damage of cotton. *China Cotton* **2005**, *32*, 29–30. (In Chinese)
74. Lu, H.Q.; Qi, J.; Dai, J.L.; Zhang, Y.J.; Kong, X.Q.; Li, Z.H.; Li, W.J.; Xu, S.Z.; Tang, W.; Zhang, D.M.; et al. Adjustment and compensation of cotton to physical damage at early squaring stage. *Acta Agron. Sin.* **2019**, *45*, 904–911.

agronomy

MDPI

Review

The Evaluation of Carbon Farming Strategies in Organic Vegetable Cultivation

Dan Ioan Avasiloaiei [1,*], Mariana Calara [1], Petre Marian Brezeanu [1], Nazim S. Gruda [2] and Creola Brezeanu [1,*]

[1] VRDS Bacău—Vegetable Research and Development Station, Calea Bârladului, 220, 600388 Bacău, Romania; calara.mariana@legumebac.ro (M.C.); brezeanumarian@legumebac.ro (P.M.B.)

[2] INRES—Institute of Crop Science and Resource Conservation, Department of Horticultural Science, University of Bonn, Auf dem Hügel 6, 53121 Bonn, Germany; ngruda@uni-bonn.de

* Correspondence: avasiloaiei_dan_ioan@yahoo.com (D.I.A.); creola.brezeanu@yahoo.com (C.B.)

Abstract: The urgent need to mitigate greenhouse gas (GHG) emissions has prompted the exploration of various strategies, including the adaptation of carbon farming practices, to achieve sustainability in agricultural systems. In this research, we assess the viability of carbon farming practices for organic vegetable growing in Europe. The study explores the potential benefits of these practices, including GHG emissions' mitigation and improved soil health, biodiversity, and ecosystem services, while also acknowledging the need for further research to optimize implementation strategies and foster widespread adoption. However, the suitability and effectiveness of carbon farming practices in organic vegetable production systems remain uncertain. The analysis considers the measurement and estimation methods employed to assess changes in soil carbon stocks and the potential environmental and economic implications for farmers. Despite a substantial body of data demonstrating the sustainable attributes of carbon farming and its multifaceted advantages, a degree of hesitancy persists. Considering this, we propose undertaking a concise strengths, weaknesses, opportunities, and threats (SWOT) analysis to evaluate multiple aspects of carbon farming. The findings reveal that carbon farming practices can be viable and advantageous in organic vegetable production. Carbon farming practices, such as cover cropping, reduced tillage, compost application, and agroforestry, can significantly enhance the sustainability of organic farming systems. Implementing these practices can mitigate greenhouse gas emissions, improve soil health and fertility, and promote biodiversity conservation. Farmer education and support, policy measures, and continued research are crucial for maximizing the potential of these practices for a sustainable future. These practices also contribute to developing climate-friendly agricultural systems, promoting environmental resilience, and reducing the ecological footprint of organic vegetable production. However, further research is needed to optimize implementation strategies, address site-specific challenges, and foster widespread adoption of carbon farming practices in organic vegetable production.

Keywords: soil carbon storage; soil fertility; ecological vegetable system; GHG mitigation; regenerative practices

1. Introduction

Currently, humanity can only effectively address environmental and climate change issues if it fully embraces ecological principles for genuine sustainability. However, history has repeatedly shown that we tend to take substantial action only during crises and that we often fall short of complete commitment. In this framework, the widely endorsed DNSH (Do No Significant Harm) principles, which have recently gained popularity among political factors, signify nothing more than a return to fundamental ecological values by striving to "not significantly harm" the environment.

Carbon, as an essential element, plays a fundamental role in life on Earth. It forms the building blocks of human DNA and is pervasive in our food [1]. However, using carbon-based fossil fuels over the past century for energy generation and industrial processes has led to the accumulation of greenhouse gas (GHG) emissions, primarily carbon dioxide (CO_2), in the atmosphere. This accumulation has caused various detrimental effects, including global climate warming, biodiversity loss, increased ocean acidity, and severe meteorological phenomena (droughts, heatwaves, floods, wildfires, severe thunderstorms, mudslides, and landslides) [2].

Therefore, the situation has become critical, and it has led to the imposition of extreme measures. At the European level, this means the achievement of zero greenhouse gas emissions by 2050, with an intermediate stage aiming at a 55% decrease by 2030 compared to the currently recorded value, according to the EU Climate Law [3]. It is said that farming represents the "single biggest cause" of the worst air pollution not only in Europe but also in China, the US, and Russia. Hence, restrictive measures regarding a series of agricultural practices need to be applied, and a new approach in farming techniques and technologies has thus become imperative. Paustian et al. and Mattila et al. [4,5] stressed the crucial role of carbon sequestration in agricultural soils to mitigate climate change.

Carbon farming encompasses a range of agricultural techniques (cover cropping, no-till farming, agroforestry, crop rotation, organic farming, use of perennial crops, managed grazing, wetland restoration, use of biochar, compost and mulching, reforestation and afforestation) designed to capture atmospheric carbon and store it within the soil and in crop roots, wood, and leaves. Additionally, Tariq et al. [6] define carbon farming as a comprehensive approach to maximize the uptake of carbon dioxide from the atmosphere and enhance its storage in plant matter and soil organic matter across working landscapes. This holistic approach involves implementing practices that are proven to accelerate the removal of CO_2 from the atmosphere and facilitate its sequestration in vegetation and soil.

A study conducted by Fageria [7] illustrates that the intake of soil organic matter and composition, which is inextricably linked with the amount of carbon stored at the soil level and water retention capacity, is impacted by management practices. A direct correlation between soil organic matter and its ability to retain water while enhancing structure and productivity has been outlined by several studies [8–10], with an inverse relationship between drought and disease occurrence [9,11,12]. Certain studies [13,14] have emphasized the significance of promoting agricultural practices that facilitate organic matter sequestration in the soil, thus reducing environmental CO_2 levels. Carbon farming is pivotal in aligning with new ecological standards concerning climate change resilience and impact [4,5,15,16].

Considering these factors, our study aimed to investigate and establish a symbiotic connection between ecological vegetable cultivation methods and carbon farming practices. This objective stems from the inherent similarities in the operational principles of both approaches, which emerged naturally and highlighted the potential for synergies between these two critical fields.

2. Materials and Methods

Systematic Literature Review

In this review article, we performed a comprehensive systematic literature review (SLR) following the guidelines outlined in the Preferred Reporting Items for Systematic Reviews and Meta-Analysis (PRISMA) methodology (Table 1). Our main objective was to systematically review and synthesize the existing literature on carbon farming strategies in organic vegetable cultivation, assess their impact on carbon sequestration, soil health, and crop productivity, and identify gaps in the research. Mainly, the present study was developed by means of no fewer than nine questions of paramount importance for the subject, as follows: (1) How have carbon farming strategies been defined and conceptualized in the context of organic vegetable cultivation? (2) What are the primary objectives and goals of implementing carbon farming strategies in organic vegetable cultivation? (3) What are the key carbon sequestration practices employed in organic vegetable cultivation? (4) How do carbon farming strategies impact soil health parameters? (5) How do carbon farming practices influence crop productivity (yield) and quality (nutrient content, pest resistance) in organic vegetable cultivation? (6) What economic benefits or costs are associated with the adoption of carbon farming strategies in organic vegetable cultivation? (7) What are the key barriers and challenges faced by farmers when implementing these strategies? (8) What areas require further investigation to enhance our understanding of these strategies? (9) How do government policies and regulations impact the adoption and effectiveness of carbon farming strategies in organic vegetable cultivation?

Table 1. Data source and selection activities according to PRISMA methodology.

Phase Number	Activity Description
Phase 1	Research database identification using Web of Science, Scopus, Google Scholars, ScienceDirect, MDPI, and Springer.com
Phase 2	Assessment of the research papers with relevance for the subject published in prominent journals
Phase 3	Removal of papers that were found irrelevant to the subject or could cause scientific incoherence
Phase 4	Draft the actual paper, including the relevant literature

Some of the keywords used to find the relevant scientific papers were: carbon farming, carbon sequestration, carbon in vegetable organic farming, and soil management in organic farming. In a subsequent stage, a qualitative data analysis was performed using Covidence systematic review software (Veritas Health Innovation, Melbourne, Australia. Available online www.covidence.orgmanagement accessed on 20 August 2023)–studies were identified, analyzed, and ranked, focusing on their content and thematics to ensure their significance and relevance to the selected topic. With this in mind, we conducted a systematic literature review, examining 273 scientific papers related to the subject (Table 2).

Table 2. Keywords used and the number of scientific papers generated.

1st Keyword Used	2nd Keyword Used	Number of Research Papers	
		Initial Search	After Filtering
Carbon farming	Vegetable cultivation	855	46
Carbon sequestration		1054	73
Organic biochar		723	21
Organic conservation tillage		162	7
Agroforestry		136	6
Organic cover crops		398	19
Organic nutrient management		103	11
Organic soil management		908	34
Permaculture		50	11
Organic carbon footprint		206	6
Urban farming system		218	12
Conservation farming		309	14
Regenerative practices		188	11
Zero budget farming		7	2
Total number of research papers		5317	273

The studies selected for this synthesis met the following criteria: to be relevant so that the data provided has practical applicability—given the novelty degree of the subject studied, this condition is implied; all the data presented have a solid scientific background; and the opinions of the authors are well argued and find approval within the scientific community.

Furthermore, to highlight the potential feasibility of carbon farming measures practical to organic vegetable cultivation, we conducted a SWOT analysis, highlighting the four essential aspects: strengths, weaknesses, opportunities, and potential threats.

Similarly, to highlight the trend of specific aspects such as emissions generated by human activities, particularly in agriculture, or the areas occupied by organic farming at the European level, relevant databases for these fields were consulted, such as Statista, European Environment Agency, or IFOAM Organics Europe.

3. Results and Discussion

3.1. Overview

Our goal is to evaluate the potential of carbon farming to reduce agricultural greenhouse gas emissions, including carbon dioxide, methane, and nitrous oxide, which are significant contributors to climate change. By implementing sustainable land management practices, maximizing carbon sequestration, and minimizing carbon emissions, we consider carbon farming as an opportunity to mainly transform the organic vegetable cultivation system into effective carbon sinks and contribute to a more sustainable future.

Efforts to combat GHG emissions should prioritize a combination of mitigation and adaptation strategies [17,18]. This involves shifting towards low-carbon energy sources, enhancing energy efficiency, advocating for sustainable transportation and production methods, improving waste management approaches, and implementing nature-based solutions into action in order to manage pollution in key industrial sectors, as shown in Figure 1. International cooperation and collaboration are vital to accelerate the deployment of clean technologies and knowledge sharing to achieve significant emission reductions [19].

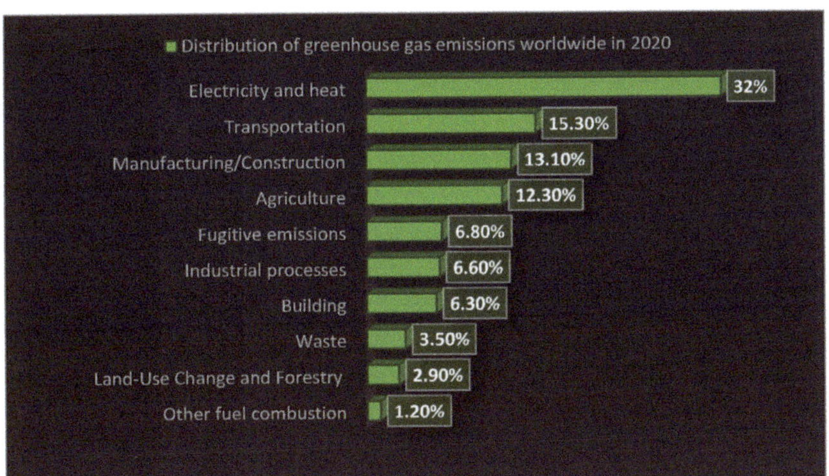

Figure 1. Distribution of greenhouse gas emissions worldwide in 2020 by sector (https://www.statista.com/statistics/241756/proportion-of-energy-in-global-greenhouse-gas-emissions/, accessed on 23 June 2023).

Equity and fairness are crucial factors to be considered in the scientific examination of global greenhouse gas (GHG) emissions [20]. Developed countries bear a historical responsibility due to their substantial contributions to cumulative emissions, necessitating supporting developing nations as they transition towards low-carbon economies [21].

The sources of agricultural emissions are diverse and significant, contributing to the global challenge of climate change (Figure 2). Livestock production, particularly enteric fermentation and manure left on pasture, is a significant source of agricultural greenhouse gas emissions. Addressing this issue requires promoting efficient feed conversion, adopting improved manure management systems, and exploring alternative protein sources [22].

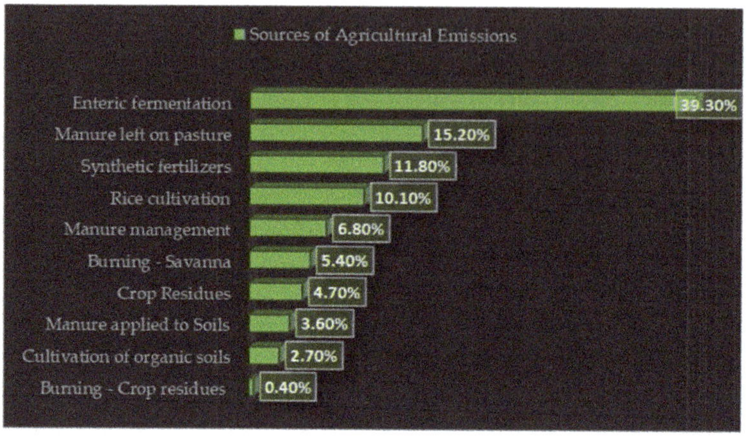

Figure 2. Sources of Agricultural Emissions (https://www.wri.org/insights/5-questions-about-agricultural-emissions-answered, accessed on 12 July 2023).

Furthermore, at the European level, the emissions from agriculture represent a significant concern in terms of environmental impact and climate change (Figure 3). Policy frameworks at the regional and national levels and financial incentives could play a crucial role in driving the adoption of sustainable agricultural practices [23]. Riccaboni et al. [24] emphasize the preeminence of supporting research and innovation in agriculture, facilitating knowledge exchange among farmers, and raising awareness among consumers about the environmental impact of their food choices.

Thus, Verschuuren [25] highlights the significant progress that can be made in reducing greenhouse gas emissions and contributing to building a more resilient and sustainable agricultural sector by addressing emissions from agriculture at the European level. Fytili and Zabaniotou [26] discuss the need for a holistic and collaborative approach involving farmers, policymakers, researchers, and consumers to successfully transition towards a low-carbon and environmentally responsible agriculture industry.

GHG net emissions/removals by land use, land use change, and forestry (LULUCF) encompass alterations in atmospheric concentrations of all greenhouse gases linked to changes in forests and land use practices, comprising, yet not restricted to, (1) the discharge and sequestration of CO_2 resulting from variations in biomass stocks due to forest administration, logging, fuelwood collection, etc.; (2) the transformation of existing forests and natural grasslands into alternative land uses; (3) the sequestration of CO_2 resulting from the abandonment of previously managed lands (e.g., croplands and pastures); and (4) CO_2 emissions and removals [27].

Organic agriculture has experienced significant expansion at the European level, driven by consumer demand for healthier and more sustainably produced food (Table 3). The sector's core principles, including prohibiting synthetic pesticides and fertilizers, promote biodiversity conservation, soil health, and reduced environmental impact. These practices can potentially mitigate greenhouse gas emissions, protect water resources, and preserve natural habitats [28].

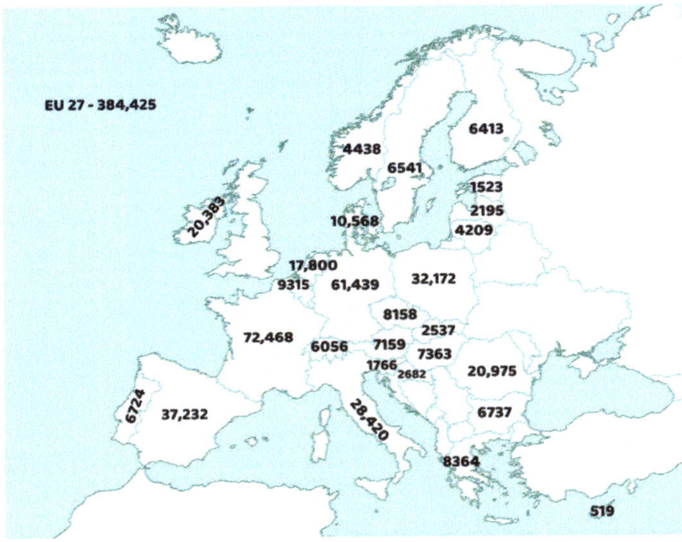

Figure 3. GHG projections with existing measures (WEM scenario) from the agricultural sector (kilotons of CO_2 equivalents) at the EU-27 level in 2023 (excluding LULUCF) (https://www.eea.europa.eu/data-and-maps/data/data-viewers/eea-greenhouse-gas-projections-data-viewer, accessed on 2 July 2023).

Table 3. Organic agriculture in 2021 at the European level (adaptation based on IFOAM Organics Europe).

COUNTRY	Percentage of Organic Agricultural Land (%)	Organic Land Area (1000 Hectares)
Liechtenstein	40.20	1
Austria	26.50	679
Estonia	23.00	227
Sweden	20.20	607
Switzerland	17.40	181
Italy	16.70	2186
Czech Republic	15.80	558
Latvia	14.80	291
Finland	14.40	328
Slovakia	11.70	223
Denmark	11.40	300
Germany	10.80	1802
Spain	10.80	2635
Slovenia	10.80	52
Greece	10.10	535
EU-27	9.60	15,600
France	9.60	2777
Lithuania	8.90	262
Croatia	8.10	122
Portugal	7.80	308
Belgium	7.40	102
Hungary	5.90	294
Cyprus	5.70	8
Luxembourg	5.20	7
Norway	4.60	45
Romania	4.30	579
Netherlands	4.20	76

Table 3. *Cont.*

COUNTRY	Percentage of Organic Agricultural Land (%)	Organic Land Area (1000 Hectares)
Poland	3.50	509
United Kingdom	2.80	489
Ireland	1.90	87
Montenegro	1.70	4
Bulgaria	1.70	86
Turkey	0.90	328
Serbia	0.70	24
Republic of North Macedonia	0.60	8
Malta	0.60	0.1
Kosovo	0.50	2
Iceland	0.40	6
Bosnia and Herzegovina	0.10	2
Albania	0.10	1

The vegetable cultivation sector has a two-fold impact on the environment. Firstly, it contributes to climate change (CC) through various activities (soil tillage, fertilization, irrigation, fossil fuel consumption). Secondly, the resulting environmental changes, in turn, affect vegetable production. Thus, a reciprocal relationship exists between horticultural activities and their ecological consequences, impacting the climate and fresh food production sector [17,18].

Organic farming (OF) embraces a comprehensive strategy for agricultural operations and food cultivation, emphasizing eco-friendly and climate-conscious methodologies, preserving natural resources, the fostering of biodiversity, and adhering to stringent standards of animal welfare and production [29,30]. Regarding the eco-friendly practices, OF prioritizes environmentally friendly methods by avoiding the use of synthetic pesticides, herbicides, and chemical fertilizers. Instead, it relies on natural alternatives and sustainable practices to manage pests, weeds, and soil fertility. In terms of the climate-conscious methods, OF aims to mitigate climate change by reducing greenhouse gas emissions and promotes practices such as crop rotation, cover cropping, and reduced tillage, which help sequester carbon in the soil and enhance soil health. For the preservation of natural resources, OF places a strong emphasis on protecting natural resources such as soil and water, promoting practices that prevent soil erosion, improve soil structure, and conserve water resources. Furthermore, OF encourages biodiversity in agricultural landscapes. This includes planting diverse crop varieties, creating habitats for beneficial insects and wildlife, and avoiding monoculture farming, which can be detrimental to biodiversity. In organic livestock farming, animals are typically raised under higher welfare standards compared to conventional methods. This includes providing access to the outdoors, pasture grazing, and adhering to strict regulations regarding animal health and well-being. Not least, OF adheres to strict production standards and certification processes. These standards encompass everything from soil management and pest control to animal husbandry and food processing. Third-party organizations certify farms and products as organic to ensure compliance [29,30].

Lal [31] defines carbon sequestration and carbon farming extensively. Carbon sequestration involves transferring carbon dioxide (CO_2) from the atmosphere and long-lasting, secure storage at the soil level by improving the accumulation of organic and inorganic carbon stocks. This is achieved through the adoption of appropriate land use practices and a set of recommended management techniques, such as mulch farming [32,33], conservation tillage [34], agroforestry [35], diverse cropping systems [36–39], cover crops [40,41], and integrated nutrient management, including the use of manure, biosolids, compost, sustainable forest management, and improved grazing [42]. The sequestration of soil organic carbon (SOC) is influenced by agricultural systems that promote the incorporation of significant amounts of biomass into the soil, minimize soil disturbance, protect soil and water resources, enhance soil structure, increase the activity and diversity of soil organisms, and strengthen elemental cycling processes.

Thus, carbon farming is presently a prominent aspect of sustainable agricultural practices due to its climate-related advantages, primarily through carbon sequestration in agricultural soils [43,44]. Soil carbon levels are closely linked to soil organic matter [45], which significantly influences the soil's structure, health, and nutrient content. Alterations in soil organic matter resulting from climate change and modifications in management approaches can impact water retention capabilities.

A series of very well-documented studies highlighted the preeminence of the organic farming system in terms of carbon sequestration at the soil level. Thus, Gattinger et al. [46] revealed that over 14 years, the soil organic carbon stocks in the upper 20 cm were 3.50 ± 1.08 Mg C ha^{-1} higher in organic than in nonorganic systems. In the same way, Leifeld and Fuhrer [47] determined in their review an average annual increase in the SOC concentration in organic systems of 2.2% compared to conventional methods, where it did not vary significantly. Furthermore, using calculations based on the combination of single practices, such as extensification, improved rotations, residue incorporation, and manure use, Freibauer et al. [48] assessed the C sequestration potential of organic farming in Europe to be 0–500 kg ha^{-1} y^{-1}.

Recently, Palayukan et al. [49] emphasized that the implementation of ecological farming methods resulted in notable enhancements in the physical and chemical characteristics of the soil. Additionally, these practices increased the accessibility of the organic carbon fraction. Likewise, Sardiana [50] demonstrated the preeminence of an ecological farming system in terms of carbon sequestration, highlighting an annual increase of 1.13 tons per hectare compared to conventional farming practices. Thus, most findings emphasize the organic farming potential for increasing C stocks in agricultural soils [31,51,52].

Moreover, Tuomisto et al. [53] reported a distinct correlation between organic vegetable growing and external carbon inputs, which were higher than the conventional system. Blair et al. [54] and Tejada et al. [55] certified an improved soil structure among the apparent benefits.

However, adopting a holistic approach when considering organic materials in agriculture is essential. Although organic materials can influence soil's physical and chemical properties, it becomes crucial to address the specific type and origin of these materials. For instance, peat is commonly used as an organic horticultural substrate, but its production involves the destruction of peatlands, which has profound environmental and climate change implications [56,57]. Therefore, it is imperative to carefully evaluate the sustainability and ecological consequences of using organic materials in horticultural practices.

Furthermore, considering a comprehensive approach that addresses the interplay between soil, plants, and the environment is of utmost importance. Recently biostimulants and biofertilizers, such as plant-growth-promoting rhizobacteria (PGPR) and arbuscular mycorrhizal fungi (AMF), have gained popularity in horticulture due to their positive effects on plant nutrition [58–60]. Utilizing the natural activity of soil microbes, biostimulants hold immense potential for breaking down toxins in the soil and harnessing the in situ soil microbial activity [61]. Similarly, biofertilizers can promote plant growth by aiding nitrogen fixation and phosphorus solubilization. Additionally, they synthesize plant hormones such as indole acetic acid and cytokinins and promote gibberellin synthesis within plants [62,63]. Moreover, they contribute organic acids and enzymes, enhancing water and nutrient uptake and bolstering systemic resistance against diseases [64,65]. Optimizing the synergistic interactions between soil and environmental factors, along with implementing good agricultural practices, including the strategic utilization of effective plant-beneficial microorganisms in the root zone, epitomizes the concept of smart agriculture. This integrated framework effectively harnesses the potential of these combined approaches [66–68].

3.2. Environmental Impact of Organic Vegetable Cultivation on Carbon Emissions

Regarding the carbon footprint (CF) in organic vegetable production, Adewale et al. [69] conducted a study on a small organic vegetable farm. They highlighted soil emissions, irrigation, fuel use, and organic fertilizer as the main hotspots. A viable option in this matter is replacing gasoline and diesel with biodiesel and the alternative of solar-powered irrigation systems instead of grid-powered irrigation systems that could determine a reduction in the farm's CF of 34%. Regarding vegetable crops, cauliflower, potato, and pepper displayed the highest CF ha^{-1}.

The organic farming system emits significantly less CO_2 per unit of land than the conventional method. On the other hand, when reported on per unit of production, both the organic and the conventional systems demonstrate similar emissions, primarily due to the lower yield levels associated with organic farming [70].

According to Küstermann and Hülsbergen [71], the energy consumption per hectare in the organic farming system is significantly lower, approximately 2.75 times less, compared to the conventional method. Furthermore, the organic system exhibited substantially reduced nitrous oxide (852 CO_2 eq kg^{-1} compared with 1307 CO_2 eq kg^{-1} recorded in conventional farming) and carbon dioxide emissions (349 CO_2 eq kg^{-1} as opposed to 707 CO_2 eq kg^{-1}), along with significantly higher carbon sequestration rates (+110 kg C ha^{-1} a^{-1} = reduction in the greenhouse potential by 415 kg CO_2 eq ha^{-1} a^{-1}) [71].

Over the past two decades, numerous studies have examined the environmental impact of agricultural practices on the production of organic vegetables and their carbon footprint. Thus, Smith et al. [72] revealed that organic vegetable production had an energy input of 50% compared to conventional carrots, 65% for onions, and 27% for broccoli. Fritsche and Eberle [73] found that GHG emissions (CO_2 eq kg^{-1}) were between 15% and 31% lower for organic tomatoes than conventional ones. De Backer et al. [74] stated that an organic leek crop releases only 33% of conventional system emissions. The same trend was recorded for organic potato crops [75–77]. Antithetically, Ziesemer [78] highlighted that organic carrot and potato production had 43% higher energy inputs per unit of output due to mechanical weeding. Overall, Wood et al. [79] showed that organic vegetable production has about 50% of the energy intensity of conventional systems. It was observed that various organic crop rotations generated lower N_2O emissions per hectare compared to conventional rotations. Regarding energy usage for fertilization, a critical aspect of organic vegetable cultivation, the scientific community agrees that applying compost and manure entails relatively low fuel and energy costs [80,81].

Organic farming has a higher potential for carbon sequestration than conventional farming systems [82]. Additionally, organic fertilization, the incorporation of crop residues into the soil, and the cultivation of cover crops have the potential to increase N_2O emissions. However, cultivating deep-rooted crops can help mitigate NO_3 leaching [83].

3.3. Exploring the Connection between Organic Vegetable Farming and Carbon Sequestration

The exclusion of agrochemicals' utilization with a substantial carbon footprint in organic agriculture (OA) production can yield notable advantages for the climate and natural resources. Synthetic nitrogen (N), phosphorus (P), and potassium (K) fertilizers, pesticides, and other agrochemicals contribute to the deterioration in soil organic matter, leading to its depletion, as well as the pollution of water sources [29], and having a direct contribution to climate change [17,18].

Numerous strategies for increasing carbon sequestration in organic vegetable cultivation have been evidenced by the scientific literature. For example, adopting a perennial vegetable system can have a beneficial impact on mitigating climate change, promoting biodiversity, and improving nutritional value. Regarding carbon sequestration, it is estimated that the large-scale cultivation of perennial vegetables would induce an intake between 22.7 and 280.5 MMT CO_2 eq/year by 2050 [84].

Soil management strategies in organic farming have a significant role in global environmental conservation, particularly in increasing soil carbon levels, e.g., by recycling organic matter. According to Sánchez-Marañón et al. [85], assessing soil organic carbon in organic farming systems is crucial for understanding their impact on carbon sequestration. Other studies have highlighted that organically managed soils tend to have higher soil carbon than conventional farming systems [46,86]. Organic farms have the potential to mitigate greenhouse gas (GHG) emissions due to practices such as legume cultivation for nitrogen supply and reduced reliance on external inputs such as fertilizers and agrochemicals. Hence, organic vegetable cultivation could potentially reduce global GHG emissions by approximately 20% by avoiding industrial nitrogen fertilizers [87]. Furthermore, the potential carbon sequestration capacity of organic farming could offset 40–72% of the world's annual agricultural GHG emissions. These findings have been confirmed by other studies, indicating that organic fertilizers can increase soil carbon content more than chemical fertilizers [88–90].

The utilization of cover crops represents another crucial technological link for enhancing carbon sequestration at the soil level. For instance, Adewale et al. [69] found that about 498 kg C ha^{-1} yr^{-1} were lost from the soil's top 30 cm without cover crops as opposed to 57 kg with cover crops, which means 1826 and 209 kg CO_2 eq ha^{-1} yr^{-1}, respectively. Additionally, incorporating cover crops significantly reduced estimated soil greenhouse gas (GHG) emissions by 41%.

Using legumes in crop rotation and multiple cropping systems offers the following advantages: it preserves and potentially enhances soil quality, boosts the yields of vegetable crops, and exerts a positive influence on the functioning of the ecosystem [91,92]. Furthermore, Jensen et al. [93] indicated other benefits of cutting N fertilizers, such as mitigating the farmer's costs and environmental risks related to greenhouse gas emission release.

Biochar research is experiencing rapid growth, primarily due to its potential for carbon sequestration [94]. Additionally, biochar holds promise as a technology for immobilizing pollutants [95], waste management, enhancing soil fertility [96], and use as a sustainable growing media and peat replacement [56,97]. Previous studies have linked the impact of biochar on crop yield to various factors, including increased cation exchange capacity and nutrient retention [56,98], raised pH and base saturation [97,99], elevated availability of phosphorus [98], and improved plant-accessible water content [100].

Multiple strategies are available for carbon sequestration in organic vegetable cultivation soils, referred to as "adaptive restorative practices", as outlined in Table 4. These practices include no-tillage or reduced tillage intensity [101], the increased input of crop residues [102], the application of organic manure [103], the utilization of cover crops [104,105], the implementation of organic mulching [32,33], nutrient management [106,107], a reduction in or the elimination of fallow periods [108,109], and the restoration of permanent forests and grasslands [31,110–114]. These strategies are primarily characterized by minimizing soil disturbance and increasing the application of organic matter. Furthermore, Nishimura et al. [115] underlined the effectiveness of the legume intercropping system and crop rotations [116] for soil fertility enhancement, with nitrogen and phosphorous having a pivotal role in the process.

Regarding soil tillage practices in organic vegetable cultivation, some studies suggest the potential of maintaining soil organic carbon by implementing no-tillage techniques [101,117].

An alternative and feasible approach is the adoption of natural farming (NF) [118], a low-input, no-tillage system incorporating weed residue mulching. This system, initially developed by Fukuoka [119], involves cutting weeds using a brush cutter and utilizing the resulting residues as a natural cover for the area.

Table 4. Carbon farming measures suitable for ecological vegetable growing.

Measures	Expected Results	References
Rethink tillage management	Shifting conventional tillage to no-till farming cuts emissions by 30 to 35 kg C/ha per season	[31]
	Higher SOC after 5 years of organic vegetable production by adopting no tillage	[101]
	Formation of macroaggregates under long-term conservation tillage	[120]
Use of cover crops	Provides nutrients to the soil	[40]
	Decreases soil NO_3 by 30%	[41,104,121]
	C sequestration outweighs N_2O emissions	[105]
Crop rotation and no tillage	Higher SOC compared to conventional tillage system	[117,122]
	Higher mineralized soil N compared with tilled systems	[123,124]
	An overall decline of up to 7.6% in GWP (net global warming potential)	[125]
Use of grain crop residues as fertilizers	1000 kg of cereal residue generates 12 to 20 kg N, 1 to 4 kg P, 7 to 30 kg K, 4 to 8 kg Ca, and 2 to 4 kg Mg	[102]
Adapting a natural farming system	low-input NT with weed residue mulching increases soil carbon sequestration by 0–7.5 cm for 8 years	[126]
Use of biochar—the carbonaceous product obtained from the organic material pyrolysis	Adjusts soil N cycle and reduces N losses	[127]
	An average 63% increase in symbiotic biological dinitrogen (N_2) fixation in crops and an 11% enhancement of plant N uptake	[127]
	Supplies nutrients to soils that stimulate biological N_2 fixation	[128,129]
	Enhances crop yields	[130]
	Reduces N_2O emissions	[131,132]
Soil erosion control	Measurement of emissions compared with burial of C under erosional processes	[72,133]
Management of farming practices: - Long-term organic matter application	Mitigate organic matter degradation that impacts the atmosphere similar to fossil fuel combustion	[42]
	Enhances soil organic carbon level and fertility	[106,107]
	Reduces N_2O emissions	[105,134]
	Magnifies free-living N fixation rates in the soil and/or nodulation in N-fixing crops	[135]
- Use of silicate rock amendments	Speeds up the rate of chemical weathering and consumes atmospheric CO_2	[136]
	Decreases both methane and N_2O emissions	[137,138]
Water resources conservation	Water-efficient farming systems = water conservation = water harvesting	[31,139]
Carbon trading market scope	Short-term (3–5 years) measure of proficiency of SOC level changes	[140,141]

3.4. Advantages and Disadvantages of Carbon Farming: Exploring the Upsides and Downsides

Many research studies [142–146] have highlighted several co-benefits of carbon farming, including but not limited to the promotion of food and nutritional security, the purification and renewal of water resources, the enhancement of biodiversity, and the restoration of degraded soils and ecosystems.

Several meta-analyses comparing organic and conventional farming systems have consistently concluded that organic farming systems exhibit superiority in soil organic matter content [46,53,147].

Organic vegetable systems are foremost in terms of carbon sequestration due to the utilization of organic amendments and cover crops [148,149], as well as the implementation of more diverse crop rotations [150,151]. It is essential to underline that these technological aspects are not limited solely to certified organic systems.

From a financial standpoint, carbon markets offer farmers the opportunity to generate additional income by implementing recommended management practices (RMP) that focus on sequestering soil organic carbon (SOC) and reducing emissions [152,153].

All of the recommended practices for carbon farming offer clear environmental advantages. For instance, adopting no-till practices promotes the formation of soil aggregates, facilitating long-term storage of carbon while reducing CO_2 emissions associated with disturbance [154]. Cover cropping enhances both above- and below-ground plant biomass, thereby increasing carbon inputs to the soil [154]. Although it can contribute to carbon storage in both above- and below-ground biomass, agroforestry may compete with crops for land unless intercropped [155]. Various soil amendments, such as biochar, silicate rock, and organic amendments, are gaining momentum as carbon sequestration practices. While they may have similar practical uses, their pathways for carbon sequestration differ significantly. Biochar, for example, adds and stabilizes carbon in the soil, silicate rock amendments accelerate weathering processes, and organic amendments enhance crop productivity and subsequent carbon formation in the soil [127,136].

Organic vegetable cultivation practices are distinguished by reduced reliance on synthetic inputs, resulting in lower usage of chemically synthesized products and a decreased demand for primary energy compared to conventional systems [66].

The vegetable farmers were willing to adapt practices such as retaining crop residue, using no-till techniques, and applying organic mulch [142]. Gruda [33] reported that organic mulch can improve water retention, balance soil temperature, increase vegetable growth, and reduce weed growth by withdrawing light [32]. Field and greenhouse trials confirmed these results, showing up to 63% fewer weeds and significant growth improvements in crops such as head lettuce and sugar melons, even when it was impossible to suppress all kinds of weeds [32]. In terms of potential drawbacks, it is crucial to highlight the findings from the studies of Leifeld and Fuhrer [47] and Powlson et al. [156] that suggest the beneficial impact of organic vegetable cultivation on soil organic carbon content may be attributed to the higher application of organic fertilizers compared to conventional systems. Consequently, increasing soil organic carbon content may not necessarily represent actual carbon sequestration.

An additional concern is represented by lower average yields recorded in organic vegetable cultivation systems. This implies that a larger land area would be required to ensure a comparable availability of energy and protein for human consumption in the scenario of a global shift to 100% organic agriculture by 2050 [157]. A similar viewpoint was discussed by Smith et al. [158], who highlighted the necessity for increased land use to compensate for the predicted 40% lower yields in organic production. However, this is only sometimes realizable. Converting more natural habitats for agriculture could lead to deforestation and the loss of biodiversity, thereby compromising delicate ecosystems [159]. Moreover, such an approach may exacerbate land degradation and soil erosion issues [160]. In regions with limited arable land, expanding cultivation areas might prove impractical and could trigger conflicts over land ownership [161]. In addition, expanding organic farming through increased land use might have socioeconomic implications [162]. Small-scale farmers, who predominantly practice organic agriculture, could face challenges acquiring more land due to escalating costs and limited availability [160]. Large-scale land conversion might lead to the displacement of local communities, altering their traditional ways of life and threatening their livelihoods [163].

A distinct correlation exists between the sequestration of organic soil carbon and the corresponding nitrogen intake. According to Hungate et al. [164], approximately 10 g of organic carbon requires around 1 g of additional nitrogen. Thus, the potential consequences of increased N_2O emissions and NO_3^- leaching from agricultural soils in organic systems due to the heightened demand for additional nitrogen fertilizers should be considered and further evaluated [108].

Another constraint is that, on a production unit basis, the energy use and carbon footprint do not frequently favor organic systems [165,166]. A summary of the main potential advantages and disadvantages carried out as a SWOT analysis is presented in Table 5, as evidenced by numerous reviewed studies.

Table 5. SWOT analysis of carbon farming practices for ecological vegetable cultivation.

STRENGTHS	References	WEAKNESS	References
The organic vegetable system uses less energy and stores higher values of C per hectare	[53,166,167]	On a production unit bases, both energy use and carbon footprint are higher in organic vegetable growing	[165,166]
Reduced soil erosion improves soil structure and water quality, and reduces sedimentation	[168,169]	Carbon lasting/persistence and stabilization	[170]
Reducing tillage minimizes the use of irrigation water by increasing soil water-holding capacity	[168,169]	Political, economic, and social factors	[171]
Improves water quality and ecology	[168,169,172,173]	C sequestration may lead to a cut in food and fiber production causing higher food prices and reducing exports.	[168,169]
Soil health upgrade	[174,175]	Heightened N_2O emission	[105,176]
Food safety	[177]		
Public health welfare	[178]		
Carbon sequestration at the soil level is a process characterized by three features: naturalness, environment friendly, and cost-effective	[31,141,179]		
Enhances soil CH_4 oxidization capacity	[180–182]		
OPPORTUNITIES	**References**	**THREATS**	**References**
Integration of SOC monitoring and Carbon footprint into the organic farming certification process	[183]	Policymakers and farmers' divergent targets and interests	[184]
Use of stacking environmental credits so that payments overcome costs	[185]	Unattractive carbon contracts	[186,187]
Facilitation of carbon farming contracts for leased farmland	[188–190]	Land ownership and the challenge of farm leasing	[185]
Farmers could receive payments from the government grounded in the area of land enrolled in a GHG mitigation program	[191–193]		
Biodiversity preservation	[194]		
Poverty mitigation	[195]		
Infrastructure upgrading	[196]		

As a future perspective, Adewale et al. [183] suggest that including carbon footprint and soil organic carbon monitoring in the organic farming certification process would increase compliance with the final rule and upgrade the available information about carbon sequestration and the carbon footprint of organic agricultural systems and practices.

3.5. The Synergy between Organic Vegetable Cultivation and Some Emerging Systems within the Context of Carbon Farming

Recently, there has been an increasing interest in developing innovative agricultural systems prioritizing sustainability, productivity, and environmental stewardship. Several low-input agricultural systems, including permaculture, urban gardening, agroforestry, conservation agriculture, and Zero Budget Natural Farming (ZBNF), exhibit genuine potential for transforming our approach to food production, emphasizing efficiency, ecological balance, and resilience. These innovative approaches offer promising solutions for reducing greenhouse gas emissions, sequestering carbon, and enhancing overall environmental sustainability.

Urban gardening can significantly reduce the carbon footprint of food production and distribution by enabling local cultivation. This approach minimizes transportation emissions and lowers energy consumption, which is commonly associated with long-distance food supply chains [197].

Urban gardening, combined with organic farming, represents a sustainable and environmentally friendly approach that encompasses various benefits. This practice promotes local food production, improves food security, and contributes to climate change mitigation. Moreover, it addresses public health issues and provides a means of adapting to geopolitical challenges in urban areas. Overall, urban gardening and organic farming are integrated solutions supporting multiple sustainability aspects in urban environments [29].

The agroforestry system increases organic matter accumulation in soil surface residues. It promotes more effective conservation of biodiversity. Sequestering carbon in trees and soil contributes to mitigating CO_2 emissions and addresses the challenges posed by climate change [35].

Recent studies on perennialization highlight that the intentional integration of perennial species can have positive effects on various ecosystem services, including provisioning (agricultural yields), regulating (pest control, hydrological cycles, water quality, carbon sequestration, and storage), and supporting (soil quality, pollination) services [198]. Furthermore, permaculture's focus on enhancing yield through beneficial interactions has anticipated the emergence of the functional diversity field, where ecologists now refer to this phenomenon as overyielding driven by complementarity or facilitation [199].

A brief description of the main characteristics of low-input systems is presented in Table 6.

Table 6. New types of low-input vegetable growing systems and the carbon farming approach.

A New Type of Vegetable Cultivation System	Synergy with Organic Vegetable Cultivation	Expected Outcomes	References
Urban farming	Make use of recycled materials sourced from the local area, such as compost produced from bio-waste	Alleviates environmental footprint and actively contributes to the development of a sustainable bioeconomy	[39]
	Supply green areas within urban environments	Adjusts temperatures, mitigates air pollution, and enhances air quality, thereby fostering healthier and sustainable urban environments in response to climate change	[37]
Conservation farming	Precision farming and minimum tillage, crop rotation, and residue retention	Enhances crop yield in sandy, acidic soils	[200]
	Preservation and restoration of crucial soil characteristics, including organic carbon content, structure, and biological diversity and activity	Initiates a soil quality restoration process that addresses microbiological activity and soil fauna diversity	[34]

Table 6. *Cont.*

A New Type of Vegetable Cultivation System	Synergy with Organic Vegetable Cultivation	Expected Outcomes	References
Agroforestry system	Integration of trees into annual food crop systems (both perennial and annual species): trees and vegetable crops	Tree leaves offer sufficient nutrients for building and sustaining soil fertility and supplying nutrients to plants	[35,201]
		Enhance carbon storage in both above-ground and below-ground components	
Zero Budget Natural Farming (ZBNF)	Natural resources and implementation of diverse cropping systems, and utilizing products derived from cow dung and urine to enhance soil biology	Significant positive contribution to preserving the ecosystem and mitigating the detrimental effects caused by agrochemicals	[38]
Permaculture	The pursuit of enhancing beneficial connections between elements to attain optimal design and harness their synergistic potential	Minimizes waste, human labor, energy, and resource inputs while establishing systems that maximize benefits and achieve high holistic integrity and resilience	[36,202]

The emergence of new low-input agricultural systems reflects society's growing recognition of the urgent need for sustainable and resilient food production. Permaculture, urban gardening, agroforestry, conservation agriculture, and Zero Budget Natural Farming are just a few innovative approaches that promise to transform the agricultural landscape. Integrating ecological principles, minimizing external inputs, and prioritizing long-term sustainability contribute to a more food-secure, environmentally friendly, and socially inclusive future. Embracing and further developing these approaches will pave the way for a more resilient, ecologically conscious, and carbon neutral agricultural sector.

The presented study should be perceived as consistent with conventional vegetable production. The era of pitting these approaches against each other has passed. Smith et al. [158] conducted a life-cycle assessment to evaluate the impact of a complete shift to organic food production in England and Wales on net greenhouse gas (GHG) emissions. Findings suggest significant deficiencies in the production of most agricultural products compared to a conventional baseline. While organic farming reduces direct GHG emissions, compensating for domestic supply shortfalls through increased overseas land use results in higher net emissions. To effectively address climate change, combining both approaches with a prioritization of sustainable practices would be beneficial. These two approaches have the potential to complement each other greatly. While this has been true in the past, we believe it will continue to be the case.

To summarize, Figure 4 presents an antithetical exposition of the main recommended management practices (RMPs) for carbon farming in contrast to the principal polluting elements from agriculture.

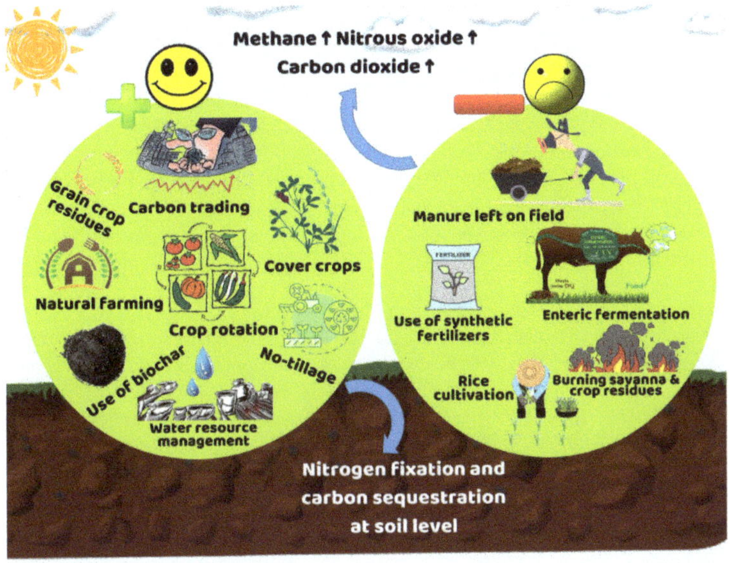

Figure 4. The influence of carbon farming recommended management practices compared to the main agricultural polluters.

4. Conclusions

Carbon farming practices demonstrate significant potential for enhancing the sustainability of organic farming systems. Implementing these practices can contribute to carbon sequestration, mitigate greenhouse gas emissions, and improve overall soil health and fertility. Incorporating cover cropping, reduced tillage, compost application, and agroforestry in organic farming can effectively increase carbon sequestration rates. These practices enhance soil organic matter content, promote nutrient cycling, and support organic agricultural systems' long-term productivity and resilience.

Carbon farming practices in organic farming systems mitigate climate change impacts and offer co-benefits such as improved water infiltration, reduced soil erosion, and increased biodiversity. These practices enhance ecosystem services through biodiversity conservation and contribute to the sustainability of organic farming, promoting habitat diversity, supporting beneficial organisms, and strengthening the overall ecological balance of agricultural landscapes.

Further improving soil structure and moisture retention, these practices help mitigate the effects of extreme weather events, such as droughts and floods, on crop productivity.

The sustainability benefits of carbon farming practices extend beyond the farm level. By sequestering carbon in soils, organic farming systems contribute to reducing atmospheric carbon dioxide levels and addressing global climate change challenges.

Adopting carbon farming practices requires careful consideration of site-specific factors such as soil type, climate, and available resources. Tailoring these practices to local conditions and organic farming systems is crucial for maximizing their effectiveness and sustainability. The successful implementation of carbon farming practices in organic farming relies on farmer education and support, as well as access to technical expertise and financial incentives. Policy measures and collaborative efforts among stakeholders are essential for promoting the widespread adoption of these practices.

Long-term monitoring and research are necessary to assess the persistence and stability of carbon sequestration achieved through carbon farming practices in organic farming. Continued scientific investigation will provide insights into these practices' scalability, economic viability, and environmental benefits. Integrating carbon farming practices into organic farming represents a promising pathway towards sustainable agriculture. These practices align with organic farming principles, enhance soil carbon stocks, and contribute to climate change mitigation and adaptation goals.

In conclusion, scientific evidence underscores the potential of carbon farming practices to enhance the sustainability of organic farming systems. These practices offer a valuable approach to achieving a more sustainable and resilient agricultural future by sequestering carbon, improving soil health, and providing additional ecological benefits. Efforts should focus on knowledge dissemination, policy support, and further research to optimize the implementation and maximize the potential of carbon farming practices in organic farming for a sustainable future.

Author Contributions: Conceptualization, D.I.A., P.M.B. and M.C.; methodology, D.I.A. and M.C.; software, D.I.A.; validation, P.M.B., M.C. and C.B.; formal analysis, D.I.A.; investigation, D.I.A., M.C. and P.M.B.; resources, D.I.A., M.C., P.M.B. and C.B.; data curation, D.I.A.; writing—original draft preparation, D.I.A.; writing—review and editing, D.I.A., N.S.G. and C.B.; visualization, M.C.; supervision, P.M.B.; project administration, P.M.B. and D.I.A.; funding acquisition, P.M.B. All authors have read and agreed to the published version of the manuscript.

Funding: This research received no external funding; the authors acknowledge the in-kind and cash support of the Vegetable Research and Development Station (VRDS), Bacau, Romania.

Data Availability Statement: No new data were created in this study. Data sharing is not applicable to this article.

Acknowledgments: The authors acknowledge the Sectorial Plan of the Romanian Ministry of Agriculture and Rural Development, implemented by VRDS Bacau through ADER projects during 2023–2026 for its objectives aiming to improve and implement climate-friendly agricultural systems, to promote environmental resilience, and to reduce the ecological footprint of vegetable production discussed in this review.

Conflicts of Interest: The authors declare no conflict of interest.

References

1. Noble, R.; Noble, D. *Understanding Living Systems*; Cambridge University Press: Cambridge, UK, 2023; ISBN 1-00-927736-7.
2. Bhavsar, A.; Hingar, D.; Ostwal, S.; Thakkar, I.; Jadeja, S.; Shah, M. The Current Scope and Stand of Carbon Capture Storage and Utilization ~A Comprehensive Review. *Case Stud. Chem. Environ. Eng.* **2023**, *8*, 100368. [CrossRef]
3. Filipović, S.; Lior, N.; Radovanović, M. The Green Deal—Just Transition and Sustainable Development Goals Nexus. *Renew. Sustain. Energy Rev.* **2022**, *168*, 112759. [CrossRef]
4. Mattila, T.J.; Hagelberg, E.; Söderlund, S.; Joona, J. How Farmers Approach Soil Carbon Sequestration? Lessons Learned from 105 Carbon-Farming Plans. *Soil Tillage Res.* **2022**, *215*, 105204. [CrossRef]
5. Paustian, K.; Larson, E.; Kent, J.; Marx, E.; Swan, A. Soil C Sequestration as a Biological Negative Emission Strategy. *Front. Clim.* **2019**, *1*, 8. [CrossRef]
6. Tariq, S.; Mubeen, M.; Hammad, H.M.; Jatoi, W.N.; Hussain, S.; Farid, H.U.; Ali, M.; Javeed, H.M.R.; Sabagh, A.E.; Fahad, S. Mitigation of Climate Change Through Carbon Farming. In *Climate Change Impacts on Agriculture: Concepts, Issues and Policies for Developing Countries*; Springer: New York, NY, USA, 2023; pp. 381–391. [CrossRef]
7. Fageria, N. Role of Soil Organic Matter in Maintaining Sustainability of Cropping Systems. *Commun. Soil Sci. Plant Anal.* **2012**, *43*, 2063–2113. [CrossRef]
8. Karami, A.; Homaee, M.; Afzalinia, S.; Ruhipour, H.; Basirat, S. Organic Resource Management: Impacts on Soil Aggregate Stability and Other Soil Physico-Chemical Properties. *Agric. Ecosyst. Environ.* **2012**, *148*, 22–28. [CrossRef]
9. Sharma, M.; Kaushal, R.; Kaushik, P.; Ramakrishna, S. Carbon Farming: Prospects and Challenges. *Sustainability* **2021**, *13*, 11122. [CrossRef]
10. Tisdall, J. Formation of Soil Aggregates and Accumulation of Soil Organic Matter. In *Structure and Organic Matter Storage in Agricultural Soils*; CRC Press: Boca Raton, FL, USA, 2020; pp. 57–96. [CrossRef]
11. Altieri, M.A.; Nicholls, C.I.; Henao, A.; Lana, M.A. Agroecology and the Design of Climate Change-Resilient Farming Systems. *Agron. Sustain. Dev.* **2015**, *35*, 869–890. [CrossRef]
12. Garrett, K.A.; Forbes, G.A.; Savary, S.; Skelsey, P.; Sparks, A.H.; Valdivia, C.; van Bruggen, A.H.; Willocquet, L.; Djurle, A.; Duveiller, E. Complexity in Climate-change Impacts: An Analytical Framework for Effects Mediated by Plant Disease. *Plant Pathol.* **2011**, *60*, 15–30. [CrossRef]
13. Hartmann, J.; West, A.J.; Renforth, P.; Köhler, P.; De La Rocha, C.L.; Wolf-Gladrow, D.A.; Dürr, H.H.; Scheffran, J. Enhanced Chemical Weathering as a Geoengineering Strategy to Reduce Atmospheric Carbon Dioxide, Supply Nutrients, and Mitigate Ocean Acidification. *Rev. Geophys.* **2013**, *51*, 113–149. [CrossRef]

14. Mcleod, E.; Chmura, G.L.; Bouillon, S.; Salm, R.; Björk, M.; Duarte, C.M.; Lovelock, C.E.; Schlesinger, W.H.; Silliman, B.R. A Blueprint for Blue Carbon: Toward an Improved Understanding of the Role of Vegetated Coastal Habitats in Sequestering CO_2. *Front. Ecol. Environ.* **2011**, *9*, 552–560. [CrossRef] [PubMed]
15. Bradford, M.A.; Carey, C.J.; Atwood, L.; Bossio, D.; Fenichel, E.P.; Gennet, S.; Fargione, J.; Fisher, J.R.; Fuller, E.; Kane, D.A. Soil Carbon Science for Policy and Practice. *Nat. Sustain.* **2019**, *2*, 1070–1072. [CrossRef]
16. Chenu, C.; Angers, D.A.; Barré, P.; Derrien, D.; Arrouays, D.; Balesdent, J. Increasing Organic Stocks in Agricultural Soils: Knowledge Gaps and Potential Innovations. *Soil Tillage Res.* **2019**, *188*, 41–52. [CrossRef]
17. Gruda, N.; Bisbis, M.; Tanny, J. Influence of Climate Change on Protected Cultivation: Impacts and Sustainable Adaptation Strategies-A Review. *J. Clean. Prod.* **2019**, *225*, 481–495. [CrossRef]
18. Gruda, N.; Bisbis, M.; Tanny, J. Impacts of Protected Vegetable Cultivation on Climate Change and Adaptation Strategies for Cleaner Production—A Review. *J. Clean. Prod.* **2019**, *225*, 324–339. [CrossRef]
19. Yin, C.; Gu, H.; Zhang, S. Measuring Technological Collaborations on Carbon Capture and Storage Based on Patents: A Social Network Analysis Approach. *J. Clean. Prod.* **2020**, *274*, 122867. [CrossRef]
20. Van den Berg, N.J.; van Soest, H.L.; Hof, A.F.; den Elzen, M.G.; van Vuuren, D.P.; Chen, W.; Drouet, L.; Emmerling, J.; Fujimori, S.; Höhne, N. Implications of Various Effort-Sharing Approaches for National Carbon Budgets and Emission Pathways. *Clim. Chang.* **2020**, *162*, 1805–1822. [CrossRef]
21. Lenzi, D.; Jakob, M.; Honegger, M.; Droege, S.; Heyward, J.C.; Kruger, T. Equity Implications of Net Zero Visions. *Clim. Chang.* **2021**, *169*, 20. [CrossRef]
22. Kauffman, J.B.; Beschta, R.L.; Lacy, P.M.; Liverman, M. Livestock Use on Public Lands in the Western USA Exacerbates Climate Change: Implications for Climate Change Mitigation and Adaptation. *Environ. Manag.* **2022**, *69*, 1137–1152. [CrossRef] [PubMed]
23. Greiner, R.; Gregg, D. Farmers' Intrinsic Motivations, Barriers to the Adoption of Conservation Practices and Effectiveness of Policy Instruments: Empirical Evidence from Northern Australia. *Land Use Policy* **2011**, *28*, 257–265. [CrossRef]
24. Riccaboni, A.; Neri, E.; Trovarelli, F.; Pulselli, R.M. Sustainability-Oriented Research and Innovation in 'Farm to Fork'Value Chains. *Curr. Opin. Food Sci.* **2021**, *42*, 102–112. [CrossRef]
25. Verschuuren, J. Achieving Agricultural Greenhouse Gas Emission Reductions in the EU Post-2030: What Options Do We Have? *Rev. Eur. Comp. Int. Environ. Law* **2022**, *31*, 246–257. [CrossRef]
26. Fytili, D.; Zabaniotou, A. Organizational, Societal, Knowledge and Skills Capacity for a Low Carbon Energy Transition in a Circular Waste Bioeconomy (CWBE): Observational Evidence of the Thessaly Region in Greece. *Sci. Total Environ.* **2022**, *813*, 151870. [CrossRef]
27. Lorenz, K.; Lal, R. Soil Organic Carbon Sequestration. In *Soil Organic Carbon Sequestration in Terrestrial Biomes of the United States*; Springer: New York, NY, USA, 2022; pp. 55–145. [CrossRef]
28. Gomiero, T.; Pimentel, D.; Paoletti, M.G. Environmental Impact of Different Agricultural Management Practices: Conventional vs. Organic Agriculture. *Crit. Rev. Plant Sci.* **2011**, *30*, 95–124. [CrossRef]
29. Ondrasek, G.; Horvatinec, J.; Kovačić, M.B.; Reljić, M.; Vinceković, M.; Rathod, S.; Bandumula, N.; Dharavath, R.; Rashid, M.I.; Panfilova, O. Land Resources in Organic Agriculture: Trends and Challenges in the Twenty-First Century from Global to Croatian Contexts. *Agronomy* **2023**, *13*, 1544. [CrossRef]
30. European Parliament, Council of the European Union. Regulation (EU) 2018/848 of the European Parliament and of the Council of 30 May 2018 on Organic Pro-Duction and Labelling of Organic Products and Repealing Council Regulation No 834/2007. *Off. J. L* **2018**, *150*, 1–92.
31. Lal, R. Soil Carbon Sequestration Impacts on Global Climate Change and Food Security. *Science* **2004**, *304*, 1623–1627. [CrossRef]
32. Gruda, N. Weed Suppression in Vegetable Crops Using Wood Fibre Mulch. *Ber. Uber Landwirtsch.* **2007**, *85*, 329–334.
33. Gruda, N. The Effect of Wood Fiber Mulch on Water Retention, Soil Temperature and Growth of Vegetable Plants. *J. Sustain. Agric.* **2008**, *32*, 629–643. [CrossRef]
34. Pittarello, M.; Chiarini, F.; Menta, C.; Furlan, L.; Carletti, P. Changes in Soil Quality through Conservation Agriculture in North-Eastern Italy. *Agriculture* **2022**, *12*, 1007. [CrossRef]
35. Timsina, J. Can Organic Sources of Nutrients Increase Crop Yields to Meet Global Food Demand? *Agronomy* **2018**, *8*, 214. [CrossRef]
36. Fiebrig, I.; Zikeli, S.; Bach, S.; Gruber, S. Perspectives on Permaculture for Commercial Farming: Aspirations and Realities. *Org. Agric.* **2020**, *10*, 379–394. [CrossRef]
37. Janowska, B.; Łój, J.; Andrzejak, R. Role of Community Gardens in Development of Housing Estates in Polish Cities. *Agronomy* **2022**, *12*, 1447. [CrossRef]
38. Saharan, B.S.; Tyagi, S.; Kumar, R.; Vijay; Om, H.; Mandal, B.S.; Duhan, J.S. Application of Jeevamrit Improves Soil Properties in Zero Budget Natural Farming Fields. *Agriculture* **2023**, *13*, 196. [CrossRef]
39. Zawadzińska, A.; Salachna, P.; Nowak, J.S.; Kowalczyk, W.; Piechocki, R.; Łopusiewicz, Ł.; Pietrak, A. Compost Based on Pulp and Paper Mill Sludge, Fruit-Vegetable Waste, Mushroom Spent Substrate and Rye Straw Improves Yield and Nutritional Value of Tomato. *Agronomy* **2021**, *12*, 13. [CrossRef]
40. Marcillo, G.; Miguez, F. Corn Yield Response to Winter Cover Crops: An Updated Meta-Analysis. *J. Soil Water Conserv.* **2017**, *72*, 226–239. [CrossRef]

41. Thapa, R.; Mirsky, S.B.; Tully, K.L. Cover Crops Reduce Nitrate Leaching in Agroecosystems: A Global Meta-analysis. *J. Environ. Qual.* **2018**, *47*, 1400–1411. [CrossRef]
42. Lehmann, J.; Hansel, C.M.; Kaiser, C.; Kleber, M.; Maher, K.; Manzoni, S.; Nunan, N.; Reichstein, M.; Schimel, J.P.; Torn, M.S. Persistence of Soil Organic Carbon Caused by Functional Complexity. *Nat. Geosci.* **2020**, *13*, 529–534. [CrossRef]
43. Kincl, D.; Formánek, P.; Vopravil, J.; Nerušil, P.; Menšík, L.; Janků, J. Soil-Conservation Effect of Intercrops in Silage Maize. *Soil Water Res.* **2022**, *17*, 180–190. [CrossRef]
44. Poláková, J.; Janků, J.; Holec, J.; Soukup, J. Soil-Water Effects of Good Agricultural and Environmental Conditions Should Be Weighed in Conjunction with Carbon Farming. *Agronomy* **2023**, *13*, 1002. [CrossRef]
45. Barão, L.; Alaoui, A.; Hessel, R. Identifying and Comparing Easily Accessible Frameworks for Assessing Soil Organic Matter Functioning. *Agronomy* **2022**, *13*, 109. [CrossRef]
46. Gattinger, A.; Muller, A.; Haeni, M.; Skinner, C.; Fliessbach, A.; Buchmann, N.; Mäder, P.; Stolze, M.; Smith, P.; Scialabba, N.E.-H. Enhanced Top Soil Carbon Stocks under Organic Farming. *Proc. Natl. Acad. Sci. USA* **2012**, *109*, 18226–18231. [CrossRef] [PubMed]
47. Leifeld, J.; Fuhrer, J. Organic Farming and Soil Carbon Sequestration: What Do We Really Know about the Benefits? *Ambio* **2010**, *39*, 585–599. [CrossRef]
48. Freibauer, A.; Rounsevell, M.D.; Smith, P.; Verhagen, J. Carbon Sequestration in the Agricultural Soils of Europe. *Geoderma* **2004**, *122*, 1–23. [CrossRef]
49. Palayukan, G.; Hanudin, E.; Purwanto, B. Stable and Unstable Carbon Fraction under Different Vegetable Farming System on Mt. Merbabu's Andisols, Central Java. In *IOP Conference Series: Earth and Environmental Science, Proceedings of the 2nd International Conference on Organic Agriculture in the Tropics (ORGATROP), Online, 28–29 October 2021*; IOP Publishing: Bristol, UK, 2022; Volume 1005, p. 012014.
50. Sardiana, I. Organic Vegetable Farming System Enhancing Soil Carbon Sequestration in Bali, Indonesia. In *IOP Conference Series: Earth and Environmental Science, Proceedings of the 5th International Conference on Climate Change 2020, Bali, Indonesia, 24–25 September 2020*; IOP Publishing: Bristol, UK, 2021; Volume 724, p. 012025.
51. Liao, Y.; Wu, W.; Meng, F.; Smith, P.; Lal, R. Increase in Soil Organic Carbon by Agricultural Intensification in Northern China. *Biogeosciences* **2015**, *12*, 1403–1413. [CrossRef]
52. Matsuura, E.; Komatsuzaki, M.; Hashimi, R. Assessment of Soil Organic Carbon Storage in Vegetable Farms Using Different Farming Practices in the Kanto Region of Japan. *Sustainability* **2018**, *10*, 152. [CrossRef]
53. Tuomisto, H.L.; Hodge, I.; Riordan, P.; Macdonald, D.W. Does Organic Farming Reduce Environmental Impacts?—A Meta-Analysis of European Research. *J. Environ. Manag.* **2012**, *112*, 309–320. [CrossRef]
54. Blair, N.; Faulkner, R.D.; Till, A.R.; Korschens, M.; Schulz, E. Long-Term Management Impacts on Soil C, N and Physical Fertility: Part II: Bad Lauchstadt Static and Extreme FYM Experiments. *Soil Tillage Res.* **2006**, *91*, 39–47. [CrossRef]
55. Tejada, M.; Gonzalez, J.; García-Martínez, A.; Parrado, J. Application of a Green Manure and Green Manure Composted with Beet Vinasse on Soil Restoration: Effects on Soil Properties. *Bioresour. Technol.* **2008**, *99*, 4949–4957. [CrossRef]
56. Gruda, N.S. Advances in Soilless Culture and Growing Media in Today's Horticulture—An Editorial. *Agronomy* **2022**, *12*, 2773. [CrossRef]
57. Gruda, N.S. Increasing Sustainability of Growing Media Constituents and Stand-Alone Substrates in Soilless Culture Systems. *Agronomy* **2019**, *9*, 298. [CrossRef]
58. Dasgan, H.Y.; Aldiyab, A.; Elgudayem, F.; Ikiz, B.; Gruda, N.S. Effect of Biofertilizers on Leaf Yield, Nitrate Amount, Mineral Content and Antioxidants of Basil (*Ocimum basilicum* L.) in a Floating Culture. *Sci. Rep.* **2022**, *12*, 20917. [CrossRef] [PubMed]
59. Dasgan, H.Y.; Yilmaz, M.; Dere, S.; Ikiz, B.; Gruda, N.S. Bio-Fertilizers Reduced the Need for Mineral Fertilizers in Soilless-Grown Capia Pepper. *Horticulturae* **2023**, *9*, 188. [CrossRef]
60. Dasgan, H.Y.; Kacmaz, S.; Arpaci, B.B.; İkiz, B.; Gruda, N.S. Biofertilizers Improve the Leaf Quality of Hydroponically Grown Baby Spinach (*Spinacia oleracea* L.). *Agronomy* **2023**, *13*, 575. [CrossRef]
61. Matthews, S.; Ali, A.; Siddiqui, Y.; Supramaniam, C.V. Plant Bio-Stimulant: Prospective, Safe and Natural Resources. *J. Soil Sci. Plant Nutr.* **2022**, *22*, 2570–2586. [CrossRef]
62. Baum, C.; El-Tohamy, W.; Gruda, N. Increasing the Productivity and Product Quality of Vegetable Crops Using Arbuscular Mycorrhizal Fungi: A Review. *Sci. Hortic.* **2015**, *187*, 131–141. [CrossRef]
63. El-Tohamy, W.; El-Abagy, H.; El-Greadly, N.; Gruda, N. Hormonal Changes, Growth and Yield of Tomato Plants in Response to Chemical and Bio-Fertilization Application in Sandy Soils. *J. Appl. Bot. Food Qual.* **2012**, *82*, 179–182.
64. Emmanuel, O.C.; Babalola, O.O. Productivity and Quality of Horticultural Crops through Co-Inoculation of Arbuscular Mycorrhizal Fungi and Plant Growth Promoting Bacteria. *Microbiol. Res.* **2020**, *239*, 126569. [CrossRef] [PubMed]
65. Kilic, N.; Dasgan, H.Y.; Gruda, N.S. A Novel Approach for Organic Strawberry Cultivation: Vermicompost-Based Fertilization and Microbial Complementary Nutrition. *Horticulturae* **2023**, *9*, 642. [CrossRef]
66. Agrimonti, C.; Lauro, M.; Visioli, G. Smart Agriculture for Food Quality: Facing Climate Change in the 21st Century. *Crit. Rev. Food Sci. Nutr.* **2021**, *61*, 971–981. [CrossRef]
67. Balliu, A.; Zheng, Y.; Sallaku, G.; Fernández, J.A.; Gruda, N.S.; Tuzel, Y. Environmental and Cultivation Factors Affect the Morphology, Architecture and Performance of Root Systems in Soilless Grown Plants. *Horticulturae* **2021**, *7*, 243. [CrossRef]

68. Gruda, N.; Bisbis, M.; Katsoulas, N.; Kittas, C. Smart Greenhouse Production Practices to Manage and Mitigate the Impact of Climate Change in Protected Cultivation. *Int. Soc. Hortic. Sci.* **2021**, *1320*, 189–196. [CrossRef]
69. Adewale, C.; Higgins, S.; Granatstein, D.; Stöckle, C.O.; Carlson, B.R.; Zaher, U.E.; Carpenter-Boggs, L. Identifying Hotspots in the Carbon Footprint of a Small Scale Organic Vegetable Farm. *Agric. Syst.* **2016**, *149*, 112–121. [CrossRef]
70. Gomiero, T.; Paoletti, M.G.; Pimentel, D. Energy and Environmental Issues in Organic and Conventional Agriculture. *Crit. Rev. Plant Sci.* **2008**, *27*, 239–254. [CrossRef]
71. Küstermann, B.; Hülsbergen, K.-J. Emission of Climate-Relevant Gases in Organic and Conventional Cropping Systems. In Proceedings of the 16th IFOAM Organic World Congress, Modena, Italy, 16–20 June 2008.
72. Smith, P.; Powlson, D.S.; Smith, J.U.; Falloon, P.; Coleman, K. Meeting Europe's Climate Change Commitments: Quantitative Estimates of the Potential for Carbon Mitigation by Agriculture. *Glob. Chang. Biol.* **2000**, *6*, 525–539. [CrossRef]
73. Fritsche, U.R.; Eberle, U.; Wiegmann, K.; Schmidt, K. *Treibhausgasemissionen Durch Erzeugung Und Verarbeitung von Lebensmitteln*; Arbeitspapier; Öko-Institut eV Darmstadt: Darmstadt, Germany, 2007; Volume 13.
74. De Backer, E.; Aertsens, J.; Vergucht, S.; Steurbaut, W. Assessing the Ecological Soundness of Organic and Conventional Agriculture by Means of Life Cycle Assessment (LCA): A Case Study of Leek Production. *Br. Food J.* **2009**, *111*, 1028–1061. [CrossRef]
75. Alonso, A.M.; Guzmán, G.J. Comparison of the Efficiency and Use of Energy in Organic and Conventional Farming in Spanish Agricultural Systems. *J. Sustain. Agric.* **2010**, *34*, 312–338. [CrossRef]
76. Lynch, D.H.; Zheng, Z.; Zebarth, B.J.; Martin, R.C. Organic Amendment Effects on Tuber Yield, Plant N Uptake and Soil Mineral N under Organic Potato Production. *Renew. Agric. Food Syst.* **2008**, *23*, 250–259. [CrossRef]
77. Williams, A.; Audsley, E.; Sandars, D. Determining the Environmental Burdens and Resource Use in the Production of Agricultural and Horticultural Commodities: Defra Project Report IS0205. 2006. Available online: http://randd.defra.gov.uk/Default.aspx (accessed on 20 August 2023).
78. Ziesemer, J. *Energy Use in Organic Food Systems*; Natural Resources Management and Environment Department Food and Agriculture Organization of the United Nations: Rome, Italy, 2007; pp. 1–28.
79. Wood, R.; Lenzen, M.; Dey, C.; Lundie, S. A Comparative Study of Some Environmental Impacts of Conventional and Organic Farming in Australia. *Agric. Syst.* **2006**, *89*, 324–348. [CrossRef]
80. Dyer, J.; Desjardins, R. A Simple Meta-Model for Assessing the Contribution of Liquid Fossil Fuel for on-Farm Fieldwork to Agricultural Greenhouse Gases in Canada. *J. Sustain. Agric.* **2005**, *27*, 71–90. [CrossRef]
81. Bos, J.F.; de Haan, J.; Sukkel, W.; Schils, R.L. Energy Use and Greenhouse Gas Emissions in Organic and Conventional Farming Systems in the Netherlands. *NJAS—Wageningen J. Life Sci.* **2014**, *68*, 61–70. [CrossRef]
82. Autret, B.; Beaudoin, N.; Rakotovololona, L.; Bertrand, M.; Grandeau, G.; Gréhan, E.; Ferchaud, F.; Mary, B. Can Alternative Cropping Systems Mitigate Nitrogen Losses and Improve GHG Balance? Results from a 19-Yr Experiment in Northern France. *Geoderma* **2019**, *342*, 20–33. [CrossRef]
83. Hansen, S.; Berland Frøseth, R.; Stenberg, M.; Stalenga, J.; Olesen, J.E.; Krauss, M.; Radzikowski, P.; Doltra, J.; Nadeem, S.; Torp, T. Reviews and Syntheses: Review of Causes and Sources of N_2O Emissions and NO_3 Leaching from Organic Arable Crop Rotations. *Biogeosciences* **2019**, *16*, 2795–2819. [CrossRef]
84. Toensmeier, E.; Ferguson, R.; Mehra, M. Perennial Vegetables: A Neglected Resource for Biodiversity, Carbon Sequestration, and Nutrition. *PLoS ONE* **2020**, *15*, e0234611. [CrossRef]
85. Sanchez-Maranon, M.; Soriano, M.; Delgado, G.; Delgado, R. Soil Quality in Mediterranean Mountain Environments: Effects of Land Use Change. *Soil Sci. Soc. Am. J.* **2002**, *66*, 948–958. [CrossRef]
86. Leithold, G.; Hülsbergen, K.; Brock, C. Organic Matter Returns to Soils Must Be Higher under Organic Compared to Conventional Farming. *J. Plant Nutr. Soil Sci.* **2015**, *178*, 4–12. [CrossRef]
87. Scialabba, N.E.-H.; Müller-Lindenlauf, M. Organic Agriculture and Climate Change. *Renew. Agric. Food Syst.* **2010**, *25*, 158–169. [CrossRef]
88. Brar, B.; Singh, K.; Dheri, G. Carbon Sequestration and Soil Carbon Pools in a Rice–Wheat Cropping System: Effect of Long-Term Use of Inorganic Fertilizers and Organic Manure. *Soil Tillage Res.* **2013**, *128*, 30–36. [CrossRef]
89. Leifeld, J.; Reiser, R.; Oberholzer, H. Consequences of Conventional versus Organic Farming on Soil Carbon: Results from a 27-year Field Experiment. *Agron. J.* **2009**, *101*, 1204–1218. [CrossRef]
90. Manna, M.C.; Swarup, A.; Wanjari, R.; Mishra, B.; Shahi, D. Long-Term Fertilization, Manure and Liming Effects on Soil Organic Matter and Crop Yields. *Soil Tillage Res.* **2007**, *94*, 397–409. [CrossRef]
91. Luce, M.S.; Grant, C.A.; Zebarth, B.J.; Ziadi, N.; O'Donovan, J.T.; Blackshaw, R.E.; Harker, K.N.; Johnson, E.N.; Gan, Y.; Lafond, G.P. Legumes Can Reduce Economic Optimum Nitrogen Rates and Increase Yields in a Wheat–Canola Cropping Sequence in Western Canada. *Field Crop. Res.* **2015**, *179*, 12–25. [CrossRef]
92. Unkovich, M.; Herridge, D.; Peoples, M.; Cadisch, G.; Boddey, B.; Giller, K.; Alves, B.; Chalk, P. *Measuring Plant-Associated Nitrogen Fixation in Agricultural Systems*; Australian Centre for International Agricultural Research (ACIAR): Bruce, Australia, 2008; ISBN 978-1-921531-26-2.
93. Jensen, E.S.; Peoples, M.B.; Boddey, R.M.; Gresshoff, P.M.; Hauggaard-Nielsen, H.; JR Alves, B.; Morrison, M.J. Legumes for Mitigation of Climate Change and the Provision of Feedstock for Biofuels and Biorefineries. A Review. *Agron. Sustain. Dev.* **2012**, *32*, 329–364. [CrossRef]

94. Lehmann, J. A Handful of Carbon. *Nature* **2007**, *447*, 143–144. [CrossRef] [PubMed]
95. Beesley, L.; Moreno-Jiménez, E.; Gomez-Eyles, J.L.; Harris, E.; Robinson, B.; Sizmur, T. A Review of Biochars' Potential Role in the Remediation, Revegetation and Restoration of Contaminated Soils. *Environ. Pollut.* **2011**, *159*, 3269–3282. [CrossRef] [PubMed]
96. Jeffery, S.; Verheijen, F.G.; van der Velde, M.; Bastos, A.C. A Quantitative Review of the Effects of Biochar Application to Soils on Crop Productivity Using Meta-Analysis. *Agric. Ecosyst. Environ.* **2011**, *144*, 175–187. [CrossRef]
97. Prasad, M.; Chrysargyris, A.; McDaniel, N.; Kavanagh, A.; Gruda, N.S.; Tzortzakis, N. Plant Nutrient Availability and PH of Biochars and Their Fractions, with the Possible Use as a Component in a Growing Media. *Agronomy* **2019**, *10*, 10. [CrossRef]
98. Hossain, M.K.; Strezov, V.; Chan, K.Y.; Nelson, P.F. Agronomic Properties of Wastewater Sludge Biochar and Bioavailability of Metals in Production of Cherry Tomato (*Lycopersicon esculentum*). *Chemosphere* **2010**, *78*, 1167–1171. [CrossRef]
99. Major, J.; Rondon, M.; Molina, D.; Riha, S.J.; Lehmann, J. Maize Yield and Nutrition during 4 Years after Biochar Application to a Colombian Savanna Oxisol. *Plant Soil* **2010**, *333*, 117–128. [CrossRef]
100. Novak, J.M.; Busscher, W.J.; Watts, D.W.; Amonette, J.E.; Ippolito, J.A.; Lima, I.M.; Gaskin, J.; Das, K.; Steiner, C.; Ahmedna, M. Biochars Impact on Soil-Moisture Storage in an Ultisol and Two Aridisols. *Soil Sci.* **2012**, *177*, 310–320. [CrossRef]
101. Hashimi, R.; Kaneda, S.; Kaneko, N.; Ohta, H.; Komatsuzaki, M. Aggregate Distribution and Substrate-Induced Respiration under Different Tillage and Mulching Management Systems in Organic Farming. *Soil Sci. Plant Nutr.* **2020**, *66*, 878–888. [CrossRef]
102. Sharma, S.; Thind, H.; Sidhu, H.; Jat, M.; Parihar, C. Effects of Crop Residue Retention on Soil Carbon Pools after 6 Years of Rice–Wheat Cropping System. *Environ. Earth Sci.* **2019**, *78*, 296. [CrossRef]
103. Sosulski, T.; Srivastava, A.K.; Ahrends, H.E.; Smreczak, B.; Szymańska, M. Carbon Storage Potential and Carbon Dioxide Emissions from Mineral-Fertilized and Manured Soil. *Appl. Sci.* **2023**, *13*, 4620. [CrossRef]
104. Alvarez, R.; Steinbach, H.S.; De Paepe, J.L. Cover Crop Effects on Soils and Subsequent Crops in the Pampas: A Meta-Analysis. *Soil Tillage Res.* **2017**, *170*, 53–65. [CrossRef]
105. Guenet, B.; Gabrielle, B.; Chenu, C.; Arrouays, D.; Balesdent, J.; Bernoux, M.; Bruni, E.; Caliman, J.; Cardinael, R.; Chen, S. Can N_2O Emissions Offset the Benefits from Soil Organic Carbon Storage? *Glob. Chang. Biol.* **2021**, *27*, 237–256. [CrossRef] [PubMed]
106. Iwasaki, S.; Endo, Y.; Hatano, R. The Effect of Organic Matter Application on Carbon Sequestration and Soil Fertility in Upland Fields of Different Types of Andosols. *Soil Sci. Plant Nutr.* **2017**, *63*, 200–220. [CrossRef]
107. Li, S.; Li, Y.; Li, X.; Tian, X.; Zhao, A.; Wang, S.; Wang, S.; Shi, J. Effect of Straw Management on Carbon Sequestration and Grain Production in a Maize–Wheat Cropping System in Anthrosol of the Guanzhong Plain. *Soil Tillage Res.* **2016**, *157*, 43–51. [CrossRef]
108. Almaraz, M.; Wong, M.Y.; Geoghegan, E.K.; Houlton, B.Z. A Review of Carbon Farming Impacts on Nitrogen Cycling, Retention, and Loss. *Ann. N. Y. Acad. Sci.* **2021**, *1505*, 102–117. [CrossRef] [PubMed]
109. Williams, A.; Kay, P.; Stirling, G.; Weng, X.; Bell, L. Impacts of Reducing Fallow Periods on Indicators of Soil Function in Subtropical Dryland Farming Systems. *Agric. Ecosyst. Environ.* **2022**, *324*, 107727. [CrossRef]
110. Abbas, F.; Hammad, H.M.; Fahad, S.; Cerdà, A.; Rizwan, M.; Farhad, W.; Ehsan, S.; Bakhat, H.F. Agroforestry: A Sustainable Environmental Practice for Carbon Sequestration under the Climate Change Scenarios—A Review. *Environ. Sci. Pollut. Res.* **2017**, *24*, 11177–11191. [CrossRef] [PubMed]
111. Hillel, D.; Rosenzweig, C. Introduction–Climate Change and Agroecosystems: Key Issues. In *Handbook of Climate Change and Agroecosystems: Impacts, Adaptation, and Mitigation*; World Scientific: Singapore, 2011; pp. 1–5.
112. Hutchinson, J.; Campbell, C.; Desjardins, R. Some Perspectives on Carbon Sequestration in Agriculture. *Agric. For. Meteorol.* **2007**, *142*, 288–302. [CrossRef]
113. Lu, F.; Wang, X.; Han, B.; Ouyang, Z.; Duan, X.; Zheng, H.; Miao, H. Soil Carbon Sequestrations by Nitrogen Fertilizer Application, Straw Return and No-tillage in China's Cropland. *Glob. Chang. Biol.* **2009**, *15*, 281–305. [CrossRef]
114. Zahra, S.; Abbas, F.; Ishaq, W.; Ibrahim, M.; Hammad, H.; Akram, B.; Salik, M. Carbon Sequestration Potential of Soils under Maize Production in Irrigated Agriculture of the Punjab Province of Pakistan. *J. Anim. Plant Sci.* **2016**, *26*, 706–715.
115. Nishimura, S.; Yonemura, S.; Sawamoto, T.; Shirato, Y.; Akiyama, H.; Sudo, S.; Yagi, K. Effect of Land Use Change from Paddy Rice Cultivation to Upland Crop Cultivation on Soil Carbon Budget of a Cropland in Japan. *Agric. Ecosyst. Environ.* **2008**, *125*, 9–20. [CrossRef]
116. Yfantopoulos, D.; Ntatsi, G.; Gruda, N.; Bilalis, D.; Savvas, D. Effects of the Preceding Crop on Soil N Availability, Biological Nitrogen Fixation, and Fresh Pod Yield of Organically Grown Faba Bean (*Vicia faba* L.). *Horticulturae* **2022**, *8*, 496. [CrossRef]
117. Sokolowski, A.C.; McCormick, B.P.; De Grazia, J.; Wolski, J.E.; Rodríguez, H.A.; Rodríguez-Frers, E.P.; Gagey, M.C.; Debelis, S.P.; Paladino, I.R.; Barrios, M.B. Tillage and No-Tillage Effects on Physical and Chemical Properties of an Argiaquoll Soil under Long-Term Crop Rotation in Buenos Aires, Argentina. *Int. Soil Water Conserv. Res.* **2020**, *8*, 185–194. [CrossRef]
118. Kautsar, V.; Cheng, W.; Tawaraya, K.; Yamada, S.; Toriyama, K.; Kobayashi, K. Carbon and Nitrogen Stocks and Their Mineralization Potentials Are Higher under Organic than Conventional Farming Practices in Japanese Andosols. *Soil Sci. Plant Nutr.* **2020**, *66*, 144–151. [CrossRef]
119. Fukuoka, M. *The One-Straw Revolution: An Introduction to Natural Farming*; New York Review of Books: New York, NY, USA, 2009; ISBN 1-59017-313-9.
120. Kahlon, M.S.; Lal, R.; Ann-Varughese, M. Twenty Two Years of Tillage and Mulching Impacts on Soil Physical Characteristics and Carbon Sequestration in Central Ohio. *Soil Tillage Res.* **2013**, *126*, 151–158. [CrossRef]
121. Tonitto, C.; David, M.; Drinkwater, L. Replacing Bare Fallows with Cover Crops in Fertilizer-Intensive Cropping Systems: A Meta-Analysis of Crop Yield and N Dynamics. *Agric. Ecosyst. Environ.* **2006**, *112*, 58–72. [CrossRef]

122. Gong, Y.; Li, P.; Sakagami, N.; Komatsuzaki, M. No-Tillage with Rye Cover Crop Can Reduce Net Global Warming Potential and Yield-Scaled Global Warming Potential in the Long-Term Organic Soybean Field. *Soil Tillage Res.* **2021**, *205*, 104747. [CrossRef]
123. Mahal, N.K.; Castellano, M.J.; Miguez, F.E. Conservation Agriculture Practices Increase Potentially Mineralizable Nitrogen: A Meta-analysis. *Soil Sci. Soc. Am. J.* **2018**, *82*, 1270–1278. [CrossRef]
124. Zhao, X.; Liu, S.-L.; Pu, C.; Zhang, X.-Q.; Xue, J.-F.; Ren, Y.-X.; Zhao, X.-L.; Chen, F.; Lal, R.; Zhang, H.-L. Crop Yields under No-till Farming in China: A Meta-Analysis. *Eur. J. Agron.* **2017**, *84*, 67–75. [CrossRef]
125. Shakoor, A.; Shahbaz, M.; Farooq, T.H.; Sahar, N.E.; Shahzad, S.M.; Altaf, M.M.; Ashraf, M. A Global Meta-Analysis of Greenhouse Gases Emission and Crop Yield under No-Tillage as Compared to Conventional Tillage. *Sci. Total Environ.* **2021**, *750*, 142299. [CrossRef] [PubMed]
126. Dewi, R.K.; Fukuda, M.; Takashima, N.; Yagioka, A.; Komatsuzaki, M. Soil Carbon Sequestration and Soil Quality Change between No-Tillage and Conventional Tillage Soil Management after 3 and 11 Years of Organic Farming. *Soil Sci. Plant Nutr.* **2022**, *68*, 133–148. [CrossRef]
127. Liu, Q.; Zhang, Y.; Liu, B.; Amonette, J.E.; Lin, Z.; Liu, G.; Ambus, P.; Xie, Z. How Does Biochar Influence Soil N Cycle? A Meta-Analysis. *Plant Soil* **2018**, *426*, 211–225. [CrossRef]
128. Chan, K.; Van Zwieten, L.; Meszaros, I.; Downie, A.; Joseph, S. Using Poultry Litter Biochars as Soil Amendments. *Soil Res.* **2008**, *46*, 437–444. [CrossRef]
129. Rondon, M.A.; Lehmann, J.; Ramírez, J.; Hurtado, M. Biological Nitrogen Fixation by Common Beans (*Phaseolus vulgaris* L.) Increases with Bio-Char Additions. *Biol. Fertil. Soils* **2007**, *43*, 699–708. [CrossRef]
130. Jeffery, S.; Abalos, D.; Prodana, M.; Bastos, A.C.; Van Groenigen, J.W.; Hungate, B.A.; Verheijen, F. Biochar Boosts Tropical but Not Temperate Crop Yields. *Environ. Res. Lett.* **2017**, *12*, 053001. [CrossRef]
131. Cayuela, M.L.; Van Zwieten, L.; Singh, B.; Jeffery, S.; Roig, A.; Sánchez-Monedero, M. Biochar's Role in Mitigating Soil Nitrous Oxide Emissions: A Review and Meta-Analysis. *Agric. Ecosyst. Environ.* **2014**, *191*, 5–16. [CrossRef]
132. Borchard, N.; Schirrmann, M.; Cayuela, M.L.; Kammann, C.; Wrage-Mönnig, N.; Estavillo, J.M.; Fuertes-Mendizábal, T.; Sigua, G.; Spokas, K.; Ippolito, J.A. Biochar, Soil and Land-Use Interactions That Reduce Nitrate Leaching and N_2O Emissions: A Meta-Analysis. *Sci. Total Environ.* **2019**, *651*, 2354–2364. [CrossRef]
133. Abbas, F.; Hammad, H.M.; Ishaq, W.; Farooque, A.A.; Bakhat, H.F.; Zia, Z.; Fahad, S.; Farhad, W.; Cerdà, A. A Review of Soil Carbon Dynamics Resulting from Agricultural Practices. *J. Environ. Manag.* **2020**, *268*, 110319. [CrossRef]
134. Cao, Y.; Wang, X.; Bai, Z.; Chadwick, D.; Misselbrook, T.; Sommer, S.G.; Qin, W.; Ma, L. Mitigation of Ammonia, Nitrous Oxide and Methane Emissions during Solid Waste Composting with Different Additives: A Meta-Analysis. *J. Clean. Prod.* **2019**, *235*, 626–635. [CrossRef]
135. Rodrigues, M.Â.; Ladeira, L.C.; Arrobas, M. Azotobacter-Enriched Organic Manures to Increase Nitrogen Fixation and Crop Productivity. *Eur. J. Agron.* **2018**, *93*, 88–94. [CrossRef]
136. Beerling, D.J.; Leake, J.R.; Long, S.P.; Scholes, J.D.; Ton, J.; Nelson, P.N.; Bird, M.; Kantzas, E.; Taylor, L.L.; Sarkar, B. Farming with Crops and Rocks to Address Global Climate, Food and Soil Security. *Nat. Plants* **2018**, *4*, 138–147. [CrossRef]
137. Bossolani, J.W.; Crusciol, C.A.C.; Merloti, L.F.; Moretti, L.G.; Costa, N.R.; Tsai, S.M.; Kuramae, E.E. Long-Term Lime and Gypsum Amendment Increase Nitrogen Fixation and Decrease Nitrification and Denitrification Gene Abundances in the Rhizosphere and Soil in a Tropical No-till Intercropping System. *Geoderma* **2020**, *375*, 114476. [CrossRef]
138. Blanc-Betes, E.; Kantola, I.B.; Gomez-Casanovas, N.; Hartman, M.D.; Parton, W.J.; Lewis, A.L.; Beerling, D.J.; DeLucia, E.H. In Silico Assessment of the Potential of Basalt Amendments to Reduce N_2O Emissions from Bioenergy Crops. *GCB Bioenergy* **2021**, *13*, 224–241. [CrossRef]
139. García-Tejero, I.F.; Carbonell, R.; Ordoñez, R.; Torres, F.P.; Durán Zuazo, V.H. Conservation Agriculture Practices to Improve the Soil Water Management and Soil Carbon Storage in Mediterranean Rainfed Agro-Ecosystems. In *Soil Health Restoration and Management*; Springer: Singapore, 2020; pp. 203–230. [CrossRef]
140. Badgery, W.; Murphy, B.; Cowie, A.; Orgill, S.; Rawson, A.; Simmons, A.; Crean, J. Soil Carbon Market-Based Instrument Pilot–the Sequestration of Soil Organic Carbon for the Purpose of Obtaining Carbon Credits. *Soil Res.* **2020**, *59*, 12–23. [CrossRef]
141. Kimble, J.M.; Follett, R.F.; Stewart, B.A. *Assessment Methods for Soil Carbon*; CRC Press: Boca Raton, FL, USA, 2000; ISBN 1-56670-461-8.
142. Dumbrell, N.P.; Kragt, M.E.; Gibson, F.L. What Carbon Farming Activities Are Farmers Likely to Adopt? A Best–Worst Scaling Survey. *Land Use Policy* **2016**, *54*, 29–37. [CrossRef]
143. Baumber, A.; Metternicht, G.; Cross, R.; Ruoso, L.-E.; Cowie, A.L.; Waters, C. Promoting Co-Benefits of Carbon Farming in Oceania: Applying and Adapting Approaches and Metrics from Existing Market-Based Schemes. *Ecosyst. Serv.* **2019**, *39*, 100982. [CrossRef]
144. Evans, M.C.; Carwardine, J.; Fensham, R.J.; Butler, D.W.; Wilson, K.A.; Possingham, H.P.; Martin, T.G. Carbon Farming via Assisted Natural Regeneration as a Cost-Effective Mechanism for Restoring Biodiversity in Agricultural Landscapes. *Environ. Sci. Policy* **2015**, *50*, 114–129. [CrossRef]
145. Kragt, M.E.; Gibson, F.; Maseyk, F.; Wilson, K.A. Public Willingness to Pay for Carbon Farming and Its Co-Benefits. *Ecol. Econ.* **2016**, *126*, 125–131. [CrossRef]
146. Lin, B.B.; Macfadyen, S.; Renwick, A.R.; Cunningham, S.A.; Schellhorn, N.A. Maximizing the Environmental Benefits of Carbon Farming through Ecosystem Service Delivery. *Bioscience* **2013**, *63*, 793–803. [CrossRef]

147. Mondelaers, K.; Aertsens, J.; Van Huylenbroeck, G. A Meta-analysis of the Differences in Environmental Impacts between Organic and Conventional Farming. *Br. Food J.* **2009**, *111*, 1098–1119. [CrossRef]
148. Brock, C.; Fließbach, A.; Oberholzer, H.; Schulz, F.; Wiesinger, K.; Reinicke, F.; Koch, W.; Pallutt, B.; Dittman, B.; Zimmer, J. Relation between Soil Organic Matter and Yield Levels of Nonlegume Crops in Organic and Conventional Farming Systems. *J. Plant Nutr. Soil Sci.* **2011**, *174*, 568–575. [CrossRef]
149. Liang, Q.; Chen, H.; Gong, Y.; Fan, M.; Yang, H.; Lal, R.; Kuzyakov, Y. Effects of 15 Years of Manure and Inorganic Fertilizers on Soil Organic Carbon Fractions in a Wheat-Maize System in the North China Plain. *Nutr. Cycl. Agroecosyst.* **2012**, *92*, 21–33. [CrossRef]
150. Bengtsson, J.; Ahnström, J.; Weibull, A. The Effects of Organic Agriculture on Biodiversity and Abundance: A Meta-analysis. *J. Appl. Ecol.* **2005**, *42*, 261–269. [CrossRef]
151. Stockdale, E.; Lampkin, N.; Hovi, M.; Keatinge, R.; Lennartsson, E.; Macdonald, D.; Padel, S.; Tattersall, F.; Wolfe, M.; Watson, C. Agronomic and Environmental Implications of Organic Farming Systems. *Adv. Agron.* **2001**, *70*, 2001. [CrossRef]
152. Gustin, G. U.S. Rice Farmers Turn Sustainability into Carbon Credits, with Microsoft as First Buyer. Inside Climate News. 2017. Available online: https://Insideclimatenews.Org/News/26062017/Agriculture-Rice-Methane-Emissions-Carbon-Trading-Microsoft-Climate-Change (accessed on 12 September 2023).
153. Koper, T. *FAQ: Agricultural Carbon Credit Projects*; Clim Trust: Portland, OR, USA, 2014.
154. Zhao, X.; Liu, S.; Pu, C.; Zhang, X.; Xue, J.; Zhang, R.; Wang, Y.; Lal, R.; Zhang, H.; Chen, F. Methane and Nitrous Oxide Emissions under No-till Farming in China: A Meta-analysis. *Glob. Chang. Biol.* **2016**, *22*, 1372–1384. [CrossRef] [PubMed]
155. Kim, D.-G.; Kirschbaum, M.U.; Beedy, T.L. Carbon Sequestration and Net Emissions of CH_4 and N_2O under Agroforestry: Synthesizing Available Data and Suggestions for Future Studies. *Agric. Ecosyst. Environ.* **2016**, *226*, 65–78. [CrossRef]
156. Powlson, D.S.; Whitmore, A.P.; Goulding, K.W. Soil Carbon Sequestration to Mitigate Climate Change: A Critical Re-examination to Identify the True and the False. *Eur. J. Soil Sci.* **2011**, *62*, 42–55. [CrossRef]
157. van der Werf, H.M.; Knudsen, M.T.; Cederberg, C. Towards Better Representation of Organic Agriculture in Life Cycle Assessment. *Nat. Sustain.* **2020**, *3*, 419–425. [CrossRef]
158. Smith, L.G.; Kirk, G.J.; Jones, P.J.; Williams, A.G. The Greenhouse Gas Impacts of Converting Food Production in England and Wales to Organic Methods. *Nat. Commun.* **2019**, *10*, 4641. [CrossRef]
159. Tscharntke, T.; Grass, I.; Wanger, T.C.; Westphal, C.; Batáry, P. Beyond Organic Farming–Harnessing Biodiversity-Friendly Landscapes. *Trends Ecol. Evolut.* **2021**, *36*, 919–930. [CrossRef]
160. Rahmann, G.; Reza Ardakani, M.; Bàrberi, P.; Boehm, H.; Canali, S.; Chander, M.; David, W.; Dengel, L.; Erisman, J.W.; Galvis-Martinez, A.C. Organic Agriculture 3.0 Is Innovation with Research. *Org. Agric.* **2017**, *7*, 169–197. [CrossRef]
161. Zoomers, A.; Van Noorloos, F.; Otsuki, K.; Steel, G.; Van Westen, G. The Rush for Land in an Urbanizing World: From Land Grabbing toward Developing Safe, Resilient, and Sustainable Cities and Landscapes. *World Dev.* **2017**, *92*, 242–252. [CrossRef]
162. Gabriel, D.; Sait, S.M.; Hodgson, J.A.; Schmutz, U.; Kunin, W.E.; Benton, T.G. Scale Matters: The Impact of Organic Farming on Biodiversity at Different Spatial Scales. *Ecol. Lett.* **2010**, *13*, 858–869. [CrossRef] [PubMed]
163. D'Odorico, P.; Rulli, M.C.; Dell'Angelo, J.; Davis, K.F. New Frontiers of Land and Water Commodification: Socio-environmental Controversies of Large-scale Land Acquisitions. *Land Degrad. Dev.* **2017**, *28*, 2234–2244. [CrossRef]
164. Hungate, B.A.; Dukes, J.S.; Shaw, M.R.; Luo, Y.; Field, C.B. Nitrogen and Climate Change. *Science* **2003**, *302*, 1512–1513. [CrossRef]
165. Meier, M.S.; Stoessel, F.; Jungbluth, N.; Juraske, R.; Schader, C.; Stolze, M. Environmental Impacts of Organic and Conventional Agricultural Products—Are the Differences Captured by Life Cycle Assessment? *J. Environ. Manag.* **2015**, *149*, 193–208. [CrossRef] [PubMed]
166. Reganold, J.P.; Wachter, J.M. Organic Agriculture in the Twenty-First Century. *Nat. Plants* **2016**, *2*, 15221. [CrossRef]
167. Larsen, E.; Grossman, J.; Edgell, J.; Hoyt, G.; Osmond, D.; Hu, S. Soil Biological Properties, Soil Losses and Corn Yield in Long-Term Organic and Conventional Farming Systems. *Soil Tillage Res.* **2014**, *139*, 37–45. [CrossRef]
168. Pretty, J.; Farage, P.; Ball, A. Economic Constraints to the Adoption of Carbon Farming. *Can. J. Soil Sci.* **2005**, *85*, 541–547. [CrossRef]
169. Renwick, A.; Ball, A.; Pretty, J. Economic, Biological and Policy Constraints on the Adoption of Carbon Farming in Temperate Regions. *Philos. Trans. R. Soc. Lond. Ser. A Math. Phys. Eng. Sci.* **2002**, *360*, 1721–1740. [CrossRef] [PubMed]
170. Lehmann, J.; Kleber, M. The Contentious Nature of Soil Organic Matter. *Nature* **2015**, *528*, 60–68. [CrossRef] [PubMed]
171. Conant, R.T.; Ogle, S.M.; Paul, E.A.; Paustian, K. Measuring and Monitoring Soil Organic Carbon Stocks in Agricultural Lands for Climate Mitigation. *Front. Ecol. Environ.* **2011**, *9*, 169–173. [CrossRef]
172. Jackson, R.B.; Jobbágy, E.G.; Avissar, R.; Roy, S.B.; Barrett, D.J.; Cook, C.W.; Farley, K.A.; Le Maitre, D.C.; McCarl, B.A.; Murray, B.C. Trading Water for Carbon with Biological Carbon Sequestration. *Science* **2005**, *310*, 1944–1947. [CrossRef]
173. Pattanayak, S.K.; McCarl, B.A.; Sommer, A.J.; Murray, B.C.; Bondelid, T.; Gillig, D.; DeAngelo, B. Water Quality Co-Effects of Greenhouse Gas Mitigation in US Agriculture. *Clim. Chang.* **2005**, *71*, 341–372. [CrossRef]
174. Feng, H.; Kurkalova, L.A.; Kling, C.L.; Gassman, P.W. Transfers and Environmental Co-Benefits of Carbon Sequestration in Agricultural Soils: Retiring Agricultural Land in the Upper Mississippi River Basin. *Clim. Chang.* **2007**, *80*, 91–107. [CrossRef]
175. Plantinga, A.J.; Wu, J. Co-Benefits from Carbon Sequestration in Forests: Evaluating Reductions in Agricultural Externalities from an Afforestation Policy in Wisconsin. *Land Econ.* **2003**, *79*, 74–85. [CrossRef]

176. Smith, P.; Goulding, K.W.; Smith, K.A.; Powlson, D.S.; Smith, J.U.; Falloon, P.; Coleman, K. Enhancing the Carbon Sink in European Agricultural Soils: Including Trace Gas Fluxes in Estimates of Carbon Mitigation Potential. *Nutr. Cycl. Agroecosyst.* **2001**, *60*, 237–252. [CrossRef]
177. Lal, R. Beyond Copenhagen: Mitigating Climate Change and Achieving Food Security through Soil Carbon Sequestration. *Food Secur.* **2010**, *2*, 169–177. [CrossRef]
178. Friel, S.; Dangour, A.D.; Garnett, T.; Lock, K.; Chalabi, Z.; Roberts, I.; Butler, A.; Butler, C.D.; Waage, J.; McMichael, A.J. Public Health Benefits of Strategies to Reduce Greenhouse-Gas Emissions: Food and Agriculture. *Lancet* **2009**, *374*, 2016–2025. [CrossRef] [PubMed]
179. Sun, W.; Canadell, J.G.; Yu, L.; Yu, L.; Zhang, W.; Smith, P.; Fischer, T.; Huang, Y. Climate Drives Global Soil Carbon Sequestration and Crop Yield Changes under Conservation Agriculture. *Glob. Chang. Biol.* **2020**, *26*, 3325–3335. [CrossRef] [PubMed]
180. He, R.; Su, Y.; Leewis, M.-C.; Chu, Y.-X.; Wang, J.; Ma, R.-C.; Wu, D.; Zhan, L.-T.; Herriott, I.C.; Leigh, M.B. Low O_2 Level Enhances CH_4-Derived Carbon Flow into Microbial Communities in Landfill Cover Soils. *Environ. Pollut.* **2020**, *258*, 113676. [CrossRef]
181. Six, J.; Feller, C.; Denef, K.; Ogle, S.; de Moraes Sa, J.C.; Albrecht, A. Soil Organic Matter, Biota and Aggregation in Temperate and Tropical Soils-Effects of No-Tillage. *Agronomie* **2002**, *22*, 755–775. [CrossRef]
182. Smith, K.A.; Ball, T.; Conen, F.; Dobbie, K.; Massheder, J.; Rey, A. Exchange of Greenhouse Gases between Soil and Atmosphere: Interactions of Soil Physical Factors and Biological Processes. *Eur. J. Soil Sci.* **2003**, *54*, 779–791. [CrossRef]
183. Adewale, C.; Reganold, J.P.; Higgins, S.; Evans, R.; Carpenter-Boggs, L. Improving Carbon Footprinting of Agricultural Systems: Boundaries, Tiers, and Organic Farming. *Environ. Impact Assess. Rev.* **2018**, *71*, 41–48. [CrossRef]
184. Salas Castelo, E.M. The Role of Factors That Influence the Adoption of the Australian Carbon Farming Initiative-Emissions Reduction Fund: A Mixed Methods Study. Ph.D. Thesis, James Cook University, Douglas, Australia, 2017. [CrossRef]
185. Ando, A.W.; Coppess, J.; Gramig, B.; Liu, M.; Paulson, N. *The Achilles Heels of Carbon Farming: Operational Constraints on the Next Cash Crop*; SSRN: Rochester, NY, USA, 2022. [CrossRef]
186. Sellars, S.; Schnitkey, G.; Zulauf, C.; Swanson, K.; Paulson, N. What Questions Should Farmers Ask about Selling Carbon Credits? *Farmdoc Dly.* **2021**, *11*, 59.
187. Wongpiyabovorn, O.; Plastina, A.; Crespi, J.M. Challenges to Voluntary Ag Carbon Markets. *Appl. Econ. Perspect. Policy* **2023**, *45*, 1154–1167. [CrossRef]
188. Masuda, Y.J.; Harden, S.C.; Ranjan, P.; Wardropper, C.B.; Weigel, C.; Ferraro, P.J.; Reddy, S.M.; Prokopy, L.S. Rented Farmland: A Missing Piece of the Nutrient Management Puzzle in the Upper Mississippi River Basin? *J. Soil Water Conserv.* **2021**, *76*, 5A–9A. [CrossRef]
189. Olson, K. *Applying for CSP on Rented Land*; Center for Rural Affairs: Hartington, NE, USA, 2021.
190. Schnitkey, G.; Swanson, K.; Paulson, N.; Zulauf, C. Impacts of Rental Arrangements on Cover Crop and Conservation Practice Adoption. *Farmdoc Dly.* **2021**, *11*, 106.
191. Guthrie, G.; Kumareswaran, D. Carbon Subsidies, Taxes and Optimal Forest Management. *Environ. Resour. Econ.* **2009**, *43*, 275–293. [CrossRef]
192. Lubowski, R.N.; Plantinga, A.J.; Stavins, R.N. Land-Use Change and Carbon Sinks: Econometric Estimation of the Carbon Sequestration Supply Function. *J. Environ. Econ. Manag.* **2006**, *51*, 135–152. [CrossRef]
193. Tang, K.; Kragt, M.E.; Hailu, A.; Ma, C. Carbon Farming Economics: What Have We Learned? *J. Environ. Manag.* **2016**, *172*, 49–57. [CrossRef] [PubMed]
194. Nelson, E.; Polasky, S.; Lewis, D.J.; Plantinga, A.J.; Lonsdorf, E.; White, D.; Bael, D.; Lawler, J.J. Efficiency of Incentives to Jointly Increase Carbon Sequestration and Species Conservation on a Landscape. *Proc. Natl. Acad. Sci. USA* **2008**, *105*, 9471–9476. [CrossRef] [PubMed]
195. Ebeling, J.; Yasué, M. Generating Carbon Finance through Avoided Deforestation and Its Potential to Create Climatic, Conservation and Human Development Benefits. *Philos. Trans. R. Soc. B Biol. Sci.* **2008**, *363*, 1917–1924. [CrossRef]
196. Garnaut, R. *The Garnaut Review 2011: Australia in the Global Response to Climate Change*; Cambridge University Press: Cambridge, UK, 2011; ISBN 1-107-69168-0.
197. Eigenbrod, C.; Gruda, N. Urban Vegetable for Food Security in Cities. A Review. *Agron. Sustain. Dev.* **2015**, *35*, 483–498. [CrossRef]
198. Corry, R.C. Global and Local Policy Forces for Landscape Perennialization in Central North American Agriculture. *Geogr. Tidsskr.-Dan. J. Geogr.* **2016**, *116*, 5–13. [CrossRef]
199. Hooper, D.U.; Chapin III, F.S.; Ewel, J.J.; Hector, A.; Inchausti, P.; Lavorel, S.; Lawton, J.H.; Lodge, D.M.; Loreau, M.; Naeem, S. Effects of Biodiversity on Ecosystem Functioning: A Consensus of Current Knowledge. *Ecol. Monogr.* **2005**, *75*, 3–35. [CrossRef]
200. Cornelissen, G.; Martinsen, V.; Shitumbanuma, V.; Alling, V.; Breedveld, G.D.; Rutherford, D.W.; Sparrevik, M.; Hale, S.E.; Obia, A.; Mulder, J. Biochar Effect on Maize Yield and Soil Characteristics in Five Conservation Farming Sites in Zambia. *Agronomy* **2013**, *3*, 256–274. [CrossRef]
201. Garrity, D.P.; Akinnifesi, F.K.; Ajayi, O.C.; Weldesemayat, S.G.; Mowo, J.G.; Kalinganire, A.; Larwanou, M.; Bayala, J. Evergreen Agriculture: A Robust Approach to Sustainable Food Security in Africa. *Food Secur.* **2010**, *2*, 197–214. [CrossRef]
202. Krebs, J.; Bach, S. Permaculture—Scientific Evidence of Principles for the Agroecological Design of Farming Systems. *Sustainability* **2018**, *10*, 3218. [CrossRef]

agronomy

MDPI

Article

Establishing a Soil Health Assessment System for Quaternary Red Soils (Luvisols) under Different Land Use Patterns

Yingying Jiang [1], Zhongxiu Sun [2,*], Yubo Zheng [2], Hongling Wang [1] and Jiaqing Wang [1,*]

[1] College of Life Engineering, Shenyang Institute of Technology, Shenfu Reform and Innovation Demonstration Zone, Shenyang 113122, China; yingying9111@126.com (Y.J.); wanghongling@situ.edu.cn (H.W.)

[2] College of Land and Environment, Shenyang Agricultural University, Shenyang 110866, China; zybwan2023@163.com

* Correspondence: zhongxiusun@syau.edu.cn (Z.S.); jiaqing5212@163.com (J.W.); Tel.: +86-15734005989 (Z.S.)

Abstract: The health status of Quaternary red soil is a comprehensive reflection of the production and ecological service functions, which directly affects agricultural productivity and ecosystem sustainability. Based on the Cornell Soil Health Assessment (CASH) system frame, a health evaluation system for Quaternary red soils was established including the soil's physical, chemical, and biological indicators. The soil's health status under different land use patterns (the buried Quaternary red soil, sparse forest and grassland, grassland, woodland, and arable land) was systematically diagnosed in the low hilly region of western Liaoning Province. The results showed significant differences in the soil health comprehensive index of the Quaternary red soils under different land use patterns (the whole soil), presenting a trend of woodland (0.64) > arable land (0.61) > grassland (0.49) > sparse forest and grassland (0.37) > buried Quaternary red soils (0.33). The woodland and arable land are at a healthy level, the grassland and sparse forest and grassland are at a sub-healthy level, and the buried Quaternary red soil is at an unhealthy level. The health status of the topsoil layer (A) under different land use patterns has a trend of woodland (0.86) > arable land (0.73) > grassland (0.70) > sparse forest and grassland (0.67). This is consistent with the overall health status of the profile, better than that of subsoil layer (B), which presents a trend of arable land (0.41) > grassland (0.40) > woodland (0.38) > sparse forest and grassland (0.34), with relatively poor soil health conditions. Overall, the soil health status of the four land use patterns is better than that of the buried Quaternary red soils, showing an evolution trend towards healthy soil. This indicates that at this stage, human land use activities have to some extent promoted the healthy development of Quaternary red soils. The Quaternary red soils of the woodland have a healthy status, and the land use pattern is suitable and can be scientifically recommended in low mountain and hilly areas.

Keywords: soil properties; soil health evolution; arable land

1. Introduction

Quaternary red soils are mainly distributed in areas south of 30° N and are important soil resources in China [1,2]. Influenced by factors such as the parent material, topography, living organisms, time, climate, and human activities [3], especially climate changes, Quaternary red soils have undergone a certain degree of desilication, iron and aluminum accumulation, and clayification, with the presence of reddish, heavy-textured soils called "red clay" [4]. Due to the strong desilication, iron and aluminum accumulation in Quaternary red soils, and people's unreasonable utilization, they are more prone to issues such as acidification, hardpan formation, erosion, and low fertility, limiting the sustainable use of land resources [5]. Therefore, it is essential to carry out soil health assessments of Quaternary red soils under different land use patterns, systematically and comprehensively diagnose their health status, and provide a theoretical basis for the sustainable management and use of land resources.

Soil health refers to the ability of soil, as a living system, to maintain biological productivity and the environmental quality of air and water and to promote the health of plants, animals, and humans within the scope of ecosystems and land use [6]. Soil is a complex composite formed by the integration of soil physics, soil chemistry, and soil biology; so, a comprehensive selection of multiple indicators should be used to diagnose soil health [7]. Physical indicators include soil texture, bulk density, water content, and aggregate stability, etc. [8–10]. Chemical indicators include soil nitrogen, phosphorus, potassium, and organic carbon, etc. [8–10]. Biological indicators include microbial diversity, total biomass, biomass carbon, and microbial groups, etc. [8–10]. Due to the large quantity, variety, and rapid change of soil organisms, they show dynamic changes in time and space, and it is difficult to form a stable relationship with environments [9–11]. Therefore, there has been some difficulty in conducting quantitative research on them, and the biological indicators have not been widely used [11,12]. However, with the rapid development of bioinformation technology in recent years, scholars have gradually realized the importance of soil biodiversity, and more and more biological indicators have been used to assess soil health conditions [10,13,14]. Soil's physical, chemical, and biological indicators are independent from and related to each other [15]. Scholars rarely use one of the indicators alone in soil health evaluation but synthesize and analyze the evaluation results by combining a variety of evaluation indicators [8,15]. The Cornell Soil Health Assessment (CASH) method involves various forms of soil evaluation indicators and mathematical function models for the graded determination of soil [7], making it a more advanced soil health assessment method.

Based on the research background, the study aims to establish a soil health assessment system for Quaternary red soils under different land use patterns based on the CASH method frame. The results are expected to improve our understanding of the health of Quaternary red soils and provide a theoretical basis for maintaining the multifunctional ecosystem and promote the sustainable utilization and protection of soil resources.

2. Materials and Methods

2.1. The Study Area Descriptions

The northeastern Quaternary red soils are mainly distributed in the western part of Liaoning Province, China. The studied region is mountainous and hilly, located in places from the Inner Mongolia Plateau to the east coastal plain [3,16]. According to several field investigations, Quaternary red soils are mostly found in the middle and lower parts of low hills and gentle slopes, as well as in the high terraces [16]. The Quaternary red soil is classified as Argosols in the Chinese Soil Taxonomy [17], corresponding to Alfisols in the Soil Taxonomy [18] and Luvisols in the World Reference Base for Soil Resources [19]. Combining these and referring to soil survey data, such as the Soil Series of China—Liaoning Volume, along with field investigations, Wujianfang Town in Chaoyang City was selected as the typical research area. The region belongs to the North Temperate Continental Monsoon climate zone, with an average annual temperature of 5.4–8.7 °C and an average annual precipitation of 450–580 mm [20,21].

2.2. Sample Collection

Following several field investigations in the area, a relatively stable region was found to contain an evolution sequence of Quaternary red soils, derived from the same Quaternary red soil stratum, under different land use patterns such as sparse forest and grassland, grassland, woodland, and arable land, and a nearby buried Quaternary red soil underground. The investigated Quaternary red soils were basically consistent in the parent material, topography, climate, and time of soil formation (derived from a stratum), except for the land use pattern. The effects of different land use patterns on the soil under the same soil forming factors could be then discussed. Typical samples of sparse forest and grassland, grassland, wood land, and arable land were collected within the same stratum [3], and a Quaternary red soil profile buried underground was collected as the reference baseline (Figure 1).

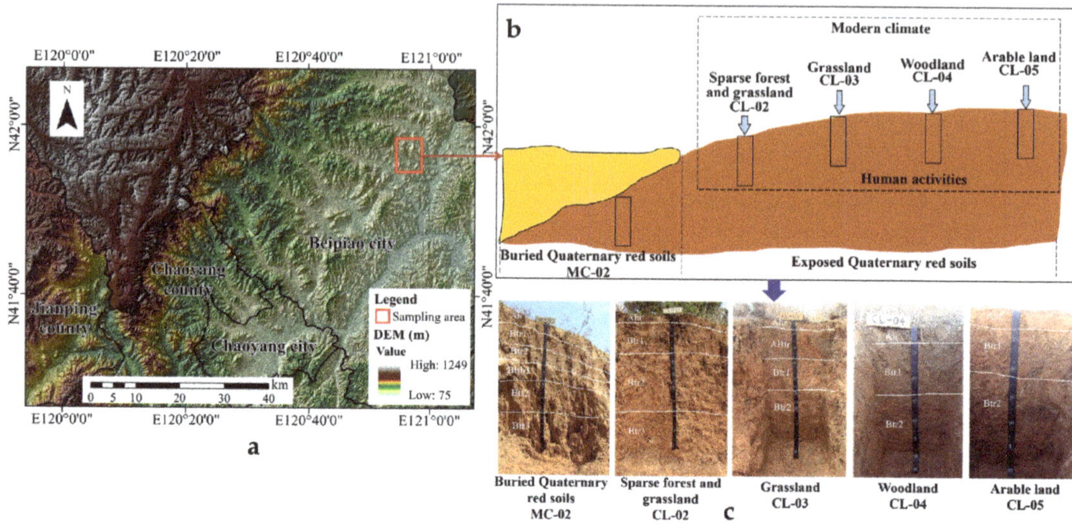

Figure 1. (**a**) The sampling area location of Chaoyang City, Liaoning Province, China; (**b**) the schematic distribution map of the sampling points; (**c**) the profile photos of the Quaternary red soils under different land use patterns. The schematic map was plotted based on the base map of the World Topographic Map (2016) using Arc GIS 10.2.2.

CL-02 represents the sparse forest and grassland. The main herbaceous type of CL-02 is *Themeda triandra Forssk*, and the forest types are *Vitex negundo*, *Ulmus pumila* L., and *Pinus massoniana Lamb.*, with a vegetation coverage of 30%. CL-03 represents the grassland. The main herbaceous type of CL-03 is *Themeda triandra Forssk*, with a vegetation coverage of 30%. CL-04 represents the woodland. The forest types of CL-04 are *Vitex negundo*, *Ulmus pumila*, and shrubbery, with a vegetation coverage of 35%. CL-05 represents the arable land. The main vegetation type of CL-05 in the growing season is maize (*Zea mays* L.). MC-02 represents the buried Quaternary red soil, which is buried under the ground.

Before sowing, around 10,000 kg of organic manure (sheep manure) and 1200 kg of maize fertilizer (the ratio of N-P_2O_5-K_2O was 18-10-12, and the total nutrient content was 40%) were applied per hectare on the arable land. No irrigation was conducted throughout the year. No additional management was given to the other land use patterns during vegetation growth. A typical soil profile was collected for each different land use pattern as the research object. Detailed descriptions of the profiles were conducted according to the Manual of Soil Description and Sampling [22]. The pedogenic horizon samples were collected from bottom to top in the profile. At the same time, the surface samples (0–30 cm) affected by human activities were collected at 10 cm intervals. The collected samples were air-dried indoors and stored for the following analysis.

Soil samples (0–20 cm) under different land use patterns were collected to analyze the microbial diversity. Each sample was collected at 5 random points along the "S" curve using a soil auger (with an inner diameter of 5 cm) and a cutting ring (100 cm^3). The root residues and litter on the soil surface were carefully removed with sterile gloves. The five soil samples under the same land use pattern were thoroughly mixed into a mixed sample. The above sampling process was repeated three times in each land use pattern and packed into sterile polyethylene sealed bags, respectively. The samples were transported in dry ice and stored in the refrigerator at −80 °C immediately upon arrival at the laboratory. The widely used Shannon index was used to show the bacteria diversity, which does not have any units. The higher the Shannon index, the greater the bacteria diversity.

2.3. Laboratory Methods

2.3.1. Basic Soil Physicochemical Properties

Basic physical and chemical properties were determined using conventional laboratory methods. The soil bulk density was determined using the cylindrical core method [23]. For the soil pH, the 10.00 g 10-mesh soil sample was placed in a small beaker. Then, 25 mL of distilled water without CO_2 was added and stirred for 1 min. Then, the beaker stood for 30 min, and the sample was measured using a pH meter (PHSJ-3F, Shanghai, China).

2.3.2. Soil Organic Carbon (SOC) and Soil Total Nitrogen

The 0.04–0.05 g 100-mesh soil sample was placed in a tint boat. After being packed and compacted, the sample was measured by an elemental analyzer produced by the Elementar Analysensysteme (GmbH, Vrrio E1 III) to obtain the contents of SOC and soil total nitrogen [24]:

$$\text{Soil organic matter (SOM)} = \text{SOC} \times 1.724,$$

where 1.724 is the conversion coefficient of soil organic carbon to soil organic matter.

2.3.3. Soil Total Phosphorus and Soil Total Potassium

The 0.25 g 0.149-mesh soil sample was placed in a nickel crucible. Then, 2 g sodium hydroxide was added to the nickel crucible and mixed with the soil sample thoroughly. The nickel crucible was put into a high-temperature electric furnace and melted at 720 °C for 15 min. After cooling, the sample was removed and dissolved and then filled into a 100 mL volumetric bottle for the determination of the total phosphorus [25] and total potassium content [26].

2.3.4. Microbial Diversity

The total DNA in soil was extracted by an Omega M 5635-02 kit. After DNA quantification, the Illumina sequencing library was established with a nucleic acid purifier. MiSeq Reagent Kit V3 (600 cycles) was used to determine 2 × 250 bp double-ended sequencing (Perseno Bio, Shanghai, China, http://www.personalbio.cn, accessed on 1 July 2023). According to the selection of the sequencing region, PCR amplification was performed using specific primers with Barcode. The 338F (5′-barcode + ACTCCTACGGGAGGCAGCA-3′) and 806R (5′-GGACTACHVGGGTWTCTAAT-3′) amplified the V3V4 region of the bacterial 16SrRNA gene. In addition, ITS5 (5′-GGAAGTAAAAGTCGTAACAAGG-3′) and ITS2 (5′-GCTGCGTTCTTCATCGATGC-3′) amplified the ITS1 gene region of the fungus. The Vsearch method was used to remove the low-quality sequences from the sequencing results. According to the sequence structure and primer sequence at both ends of the sequence, the effective sequence was obtained, and the direction of the sequence was corrected. Vsearch software classified the sequences with a similarity threshold above 97% into an OTU (Operation Taxonomic Unit) and generated an OTU table. QIIME2 (April 2019) was used to comparatively analyze the representative sequences of the OTU, the database of 16S rRNA Greengenes, and the UNITE database, and to obtain the species classification information of each OTU. The Alpha diversity index calculated based on the OTU clustering results can reflect the richness and diversity of the microbes [27]. Among them, the Shannon index was used to show the bacteria diversity in the study.

2.4. The CASH System Frame

2.4.1. The Selection of Evaluation Indicators

In the Cornell Soil Health Assessment system, thirty-nine physical, chemical, and biological indicators were selected as the potential evaluation indicators [7]. According to the minimum dataset theory [28], the selected soil health evaluation indicator should be sensitive, dominant, independent, and practical. Indicators that do not have the above characteristics should be removed from the soil health evaluation system. Therefore, the indicators in Table 1 were preliminarily selected to establish the soil health evaluation indicators. In addition, the soil texture has a significant effect on other soil properties and soil functions [7], and a texture indicator can indicate different soil health conditions in soils. Therefore, the soil texture also should be included in the evaluation indicator data set [7].

Table 1. Soil characteristics and functions corresponding to the evaluation indexes in the CASH frame.

Evaluation Indexes	Soil Health Indicators	Soil Function and Properties
Physical Indicators	Soil texture	All functions and properties of soil
	Soil aggregates' stability (0.25–2 mm)	Aeration, infiltration, root growth of superficial layer, soil hardening
	Available water content of soil	Soil water retention
	Soil surface hardness	Root growth of plough layer
	Subsurface hardness of soil	Root growth of deep layer, soil leaching
Biological Indicators	Soil organic matter content	Soil carbon storage, soil water, and fertility conservation
	Soil active carbon content	Soil organic matter provides biological growth ability
	Soil protein content	Nitrogen supply capacity
	Soil respiration	Pressure of soil-borne pests
Chemical Indicators	pH	Toxicity and nutrient availability
	Soil available phosphorus	Phosphorus availability
	Soil available potassium	Potassium availability
	Trace elements	Availability of trace elements, element imbalance, and toxicity

2.4.2. The CASH Level Division

The CASH soil health index (SHI) was calculated based on the scores of the determined indicators. A total score of soil health less than 20% means the level of soil health is very low. A total score of soil health between 20 and 40% means the level of soil health is low. A total score of soil health between 40 and 60% means the level of soil health is medium. A total score of soil health between 60 and 80% means the level of soil health is high. A total score of soil health more than 80% means the level of soil health is very high [7].

2.5. The Selection Method of Soil Health Evaluation Indicators

A preliminary selection of soil health evaluation indicators was conducted using database retrieval and literature analysis. The China National Knowledge Infrastructure (CNKI) database was used as the literature analysis retrieval database, searching for materials related to soil health evaluation from 1 January 2000 to 31 March 2023. In total, 116 soil health evaluation indicators were compiled by reading 132 articles. The indicators were classified into three categories: soil physical evaluation indicators, soil chemical evaluation indicators, and soil biological evaluation indicators.

2.6. Data Processing

We constructed a minimum dataset (MDS) [29] by selecting soil health evaluation indicators through principal component analysis (PCA) [30] and combining normal values (Norm) with correlation analysis to analyze the correlation between the indicators in the MDS. Indicators with smaller main body correlation degrees were retained, and redundant ones were eliminated to the greatest extent. To study the relationship between the soil health indicators and evaluation objects within a certain range, i.e., the relationship between soil indicators and soil health, the fuzzy mathematical membership function was used [30,31]. Based on the CASH method frame, the soil health indicator membership functions were divided into three function types, including the increasing type, decreasing type, and intermediate optimal type [7] (Table 2).

Then, the Bartlett spherical test was used to determine whether the final soil health evaluation indicators selected in the MDS conformed to the normal distribution trend of the initially selected soil indicators [30].

Table 2. Types and calculation formulas of fuzzy mathematics membership functions.

Types of Membership Functions	Computing Formula	Parameter Description
Membership function of incremental type	$y(x) = \begin{cases} 1, & x \geq b \\ \frac{x-a}{b-a}, & a < x < b \\ 0, & x \leq a \end{cases}$	Y(x): membership function; x: the measured value of the index; a: lower limit of the threshold; b: upper limit of the threshold.
Membership function of decreasing type	$y(x) = \begin{cases} 1, & x \leq a \\ \frac{x-b}{a-b}, & a < x < b \\ 0, & x \geq b \end{cases}$	
Membership function of the intermediate optimal form	$y(x) = \begin{cases} 1, & b_2 \geq x \geq b_1 \\ \frac{x-a_1}{b_1-a_1}, & a_1 < x < b_1 \\ \frac{x-a_2}{b_2-a_2}, & a_2 > x > b_2 \\ 0, & x \leq a_1 \text{ or } x \geq a_2 \end{cases}$	Y(x): membership function; X: the measured value of the index; a1: lower limit of the threshold; a2: upper limit of the threshold; b1: lower limit of the optimal value; b2: upper limit of the optimal value.

The CASH method does not take into account the weight of soil attribute indicators, resulting in relatively low accuracy at the regional scale. To further optimize, the weighted comprehensive method was used to determine the comprehensive SHI. Referring to the soil quality index calculation method and according to the weights and membership obtained from PCA, the cumulative method was used to calculate the comprehensive SHI. The formula is as follows [7,32]:

$$SHI = \sum_{i=1}^{n} (H_i \times C_i) \tag{1}$$

where SHI is the comprehensive soil health index, Hi is the membership of the i-th evaluation indicator, Ci is the weight of the i-th evaluation indicator, and n is the number of evaluation indicators [32].

2.7. The Graphic Outline of the Indicator Selection and Soil Health Index Calculation

The graphic outline of the indicator selection and soil health index calculation was showed in the Figure 2.

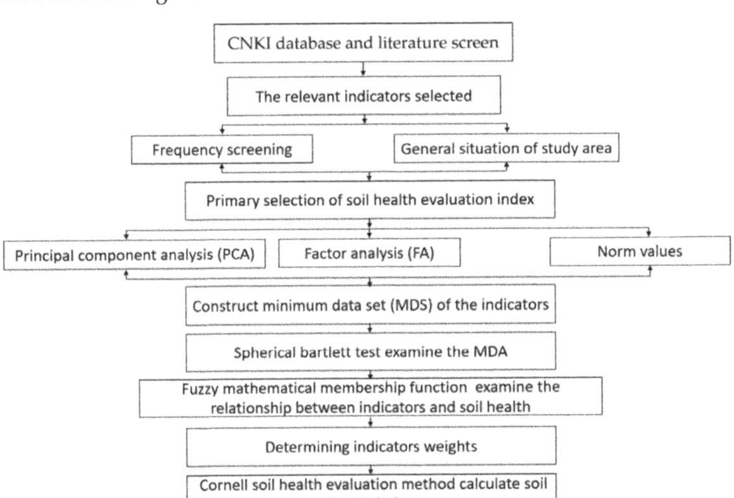

Figure 2. The graphic outline of the indicator selection and soil health index calculation.

3. Results

3.1. Screening Soil Health Evaluation Indicators for Quaternary Red Soils under Different Land Use Patterns

In total, 31 soil physical evaluation indicators were screened (Figure 3). Among them, the usage frequency of soil bulk density was the highest at 41.70%; followed by the soil clay content, the soil layer thickness, and the soil texture, with usage frequencies higher than 20.00%; aggregate stability and soil porosity, as indicators of soil mechanical stability and permeability, had usage frequencies of 16.70% and 15.20%, respectively, in the literature. Considering that soil health evaluation indicators should have simple measurement, strong operability, and a strong correlation with soil while being representative, the infiltration rate, soil saturation, and litter thickness were rarely selected.

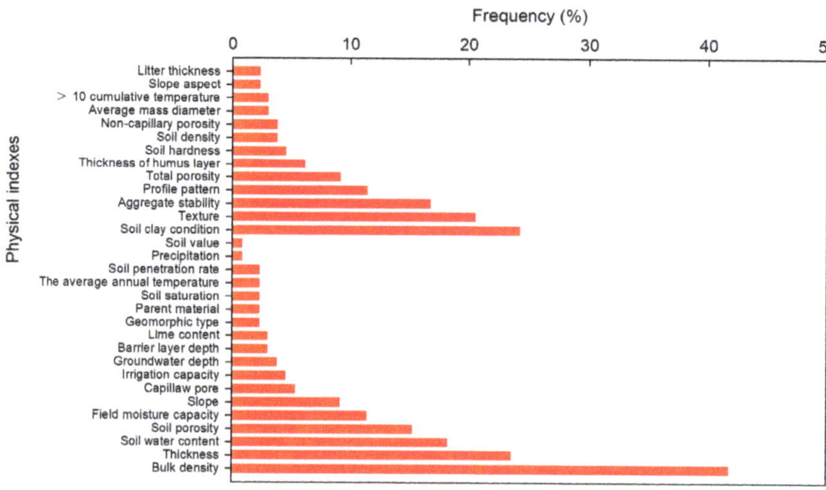

Figure 3. The selection frequencies of soil physical indicators in the literature.

A total of 35 soil chemical indicators were screened (Figure 4). Among them, the usage frequency of organic matter content was the highest at 87.10%; the soil pH value was second at 72.00%; total nitrogen, total phosphorus, and total potassium were the main soil nutrient elements supporting the growth of aboveground crops, participating in crop photosynthesis and other physiological functions, with usage frequencies of 50.00%, 48.50%, and 34.10%, respectively; cation exchange capacity, reflecting the soil buffering performance, had a usage frequency of 19.70%, slightly higher than the conductivity usage frequency of 12.90%; since the measurement methods of soil trace elements were complicated and costly, they were rarely used by researchers.

A total of 29 soil biological indicators were screened (Figure 5). Among them, the usage frequency of soil urease was the highest at 19.00%; phosphatase and catalase, reflecting the soil enzyme activity, had usage frequencies of 16.70% and 11.40%, respectively. With the development of science and technology, the accuracy of soil health evaluation has also increased, and soil microbial diversity can directly reflect the soil fertility status, with a usage frequency of 14.40% through the application of scientific and technological means. However, microbial diversity was rarely selected due to its dynamic changes in time and space; thus, it is difficult to form a stable relationship with environments.

According to the principles that soil health indicators should have representativeness, universality, sensitivity, repeatability, and operability [7], combined with the characteristics of Quaternary red soils [1,2], this study preliminarily selected 12 indicators including the soil layer thickness, the soil bulk density, the soil texture, the clay content, the aggregate stability, the profile configuration, the soil pH value, the organic matter content, the soil nutrient element content (total nitrogen, total phosphorus, and total potassium), and the microbial diversity.

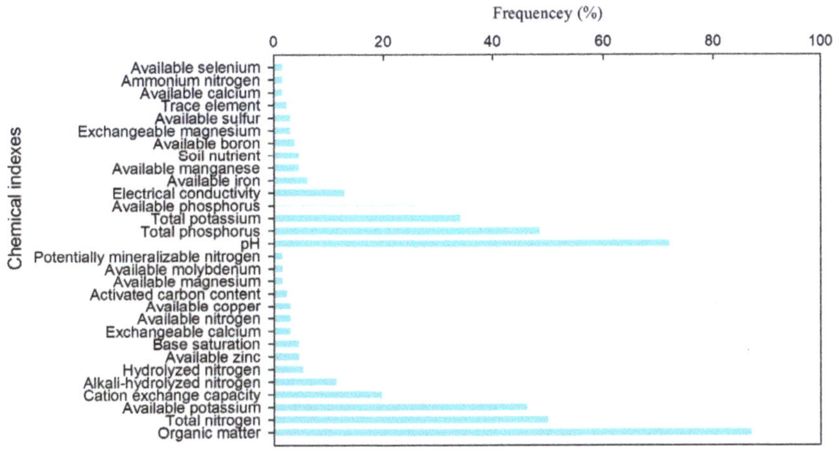

Figure 4. The selection frequencies of soil chemical indicators in the literature.

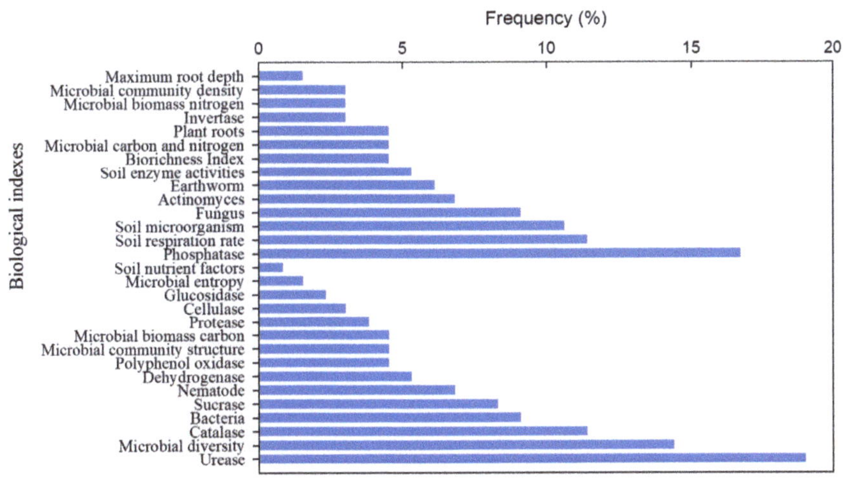

Figure 5. The selection frequencies of soil biological indicators in the literature.

3.2. Final Evaluation Indicator Confirmation

An MDS was constructed, and PCA, Norm values, and correlation analysis were used to screen suitable soil evaluation indicators for the study.

After using SPSSAU for data collation and analysis, the test statistic value for PCA was 0.75, and the eigenvalue was greater than 1. The value after the Bartlett spherical test was lower than the 0.05 significance level, indicating that there were connections between variables and meeting the requirements for PCA [31].

From the loadings of the initially selected 12 soil evaluation indicators on each principal component, the indicators were divided into four key components. The first component matrix included four indicators: profile configuration, soil layer thickness, pH value, and total phosphorus content; the second component matrix contained clay content and organic matter content; the third component matrix contained total potassium content and aggregate stability; the fourth component matrix contained total nitrogen content and microbial diversity. Six soil health evaluation indicators had Norm values greater than 1. The total nitrogen had the highest value for 1.68. The remaining evaluation indicators were ranked by Norm value size in the following order: pH (1.46), organic matter (1.36), bulk density (1.24), total phosphorus (1.23), and total potassium (1.20). One indicator was selected from each of the four component matrices, and three additional indicators were added based on the Norm values. The indicator screening rate reached 41.70%, effectively eliminating the impact of redundant information between the indicators on soil health evaluation and optimizing the soil health evaluation index system. Finally, the MDS included the pH value, the organic matter content, the total potassium content, the microbial diversity, the total nitrogen content, the bulk density, and the total phosphorus content.

3.3. Comparison of the Soil Health Evaluation Indicators of Quaternary Red Soils under Different Land Use Patterns

The soil bulk density of the Quaternary red soils under different land use patterns was in the following order (Table 3): sparse forest and grassland (1.57 g·cm^{-3}) > woodland (1.53 g·cm^{-3}) > arable land (1.49 g·cm^{-3}) > grassland (1.23 g·cm^{-3}) > buried Quaternary red soils (1.12 g·cm^{-3}) (Table 3). The sparse forest and grassland had the greatest bulk density compared to the buried Quaternary red soils. Studies have shown that when the soil pH is between 6.50 and 7.00, crops generally have a better nutrient absorption effect. The pH values of the Quaternary red soils under different land use patterns showed a trend of buried Quaternary red soils (6.09) > arable land (5.91) > woodland (5.84) > grassland (5.81) > sparse forest and grassland (5.76), with no significant difference and overall weak acidity. The organic matter content showed a trend of woodland (12.65 g·kg^{-1}) > arable land (6.26 g·kg^{-1}) > sparse forest and grassland (4.00 g·kg^{-1}) > grassland (3.17 g·kg^{-1}) > buried Quaternary red soils (1.93 g·kg^{-1}), with the woodland having the highest organic matter content. The total nitrogen content showed a trend of woodland (0.70 g·kg^{-1}) > arable land (0.35 g·kg^{-1}) > grassland (0.34 g·kg^{-1}) > sparse forest and grassland (0.30 g·kg^{-1}) > buried Quaternary red soils (0.19 g·kg^{-1}), with the overall soil total nitrogen content being low. The total phosphorus content showed a trend of woodland (0.05%) = arable land (0.05%) = buried Quaternary red soils (0.05%) > grassland (0.04%) = sparse forest and grassland (0.04%), with no significant difference and overall low content. The total potassium content showed a trend of arable land (2.75%) > buried Quaternary red soils (2.72%) > sparse forest and grassland (2.71%) > grassland (2.68%) > woodland (2.64%). The microbial diversity showed a trend of woodland (10.72 mg·g^{-1}) > arable land (10.40 mg·g^{-1}) > sparse forest and grassland (10.15 mg·g^{-1}) > grassland (9.82 mg·g^{-1}) > buried Quaternary red soils (5.96 mg·g^{-1}). Among them, the woodland had a rich microbial content when compared to other Quaternary red soils under different land use patterns, indicating the soil was relatively barren.

Table 3. Comparisons of the Quaternary red soil health evaluation indicators (the whole soil) between different land use patterns.

Index	The Measured Value of the Index				
	Buried Quaternary Red Soil	**Sparse Forest and Grassland**	**Grassland**	**Woodland**	**Arable Land**
pH	6.09	5.76	5.81	5.84	5.91
Bulk density (g·cm^{-3})	1.12	1.57	1.23	1.53	1.49
Organic matter (g·kg^{-1})	1.93	4.00	3.17	12.65	6.26
Total nitrogen (g·kg^{-1})	0.19	0.30	0.34	0.70	0.35
Total phosphorus (%)	0.05	0.04	0.04	0.05	0.05

Table 3. *Cont.*

Index	The Measured Value of the Index				
	Buried Quaternary Red Soil	Sparse Forest and Grassland	Grassland	Woodland	Arable Land
Total potassium (%)	2.72	2.71	2.68	2.64	2.75
Microbial diversity (bacteria)	5.96	10.15	9.82	10.40	10.72

Notes: the numbers are the average of three samples. The microbial diversity does not have any units.

3.4. *Soil Health Evaluation*

3.4.1. Determining the Weights

Soil health indicators refer to indicators that can reflect a certain health attribute of soil [7]. The weights of the indicators are different due to their different impacts on the soil health status [33]. The PCA method can comprehensively balance the related influence of various soil evaluation indicators and is often chosen by researchers [30].

The weight values of various soil health indicators for Quaternary red soils under different land use patterns are shown in Table 4. In the buried Quaternary red soils, the weight proportion of the organic matter was the highest at 16.16%, and the weight proportion of the soil bulk density was the lowest at 11.21%; in the sparse forest and grassland red soils, the weight proportion of the total phosphorus content was the highest at 15.26%, and the weight proportion of the pH value was the lowest at 12.11%; in the grassland red soils, the weight proportion of the total potassium content was the highest at 15.49%, and the weight proportion of the soil bulk density was the lowest at 9.19%; in the woodland red soils, the weight proportion of the organic matter content was the highest at 14.50%, and the weight proportion of the soil bulk density was the lowest at 13.79%; in the arable land red soils, the weight proportion of the pH value was the highest at 16.02%, and the weight proportion of the total phosphorus content was the lowest at 8.46%.

Table 4. Weight values of the Quaternary red soil health evaluation indicators on the MDS under different land use patterns.

Index	Weight Value of Different Land Use Patterns (%)				
	Buried Quaternary Red Soil	Sparse Forest and Grassland	Grassland	Woodland	Arable Land
pH	13.33	12.11	14.90	14.46	16.02
Bulk density	11.21	14.81	9.19	13.79	15.18
Organic matter	16.16	14.24	15.35	14.50	14.86
Total nitrogen	16.13	14.01	15.46	14.47	14.75
Total phosphorus	14.16	15.26	14.97	14.01	8.46
Total potassium	15.85	15.25	15.49	14.48	15.36
Microbial diversity	13.16	14.32	14.63	14.30	15.36

3.4.2. The Coefficient of Variation of the Soil Health Indicators

The smaller the coefficient of variation, the smaller the deviation, the smaller the fluctuation of the measured indicators, and the more stable the indicator. Combined with Pearson correlation analysis, the comparison of the coefficient of variation of various indicators found (Table 5) that the coefficient of variation of the total phosphorus was the highest at 74%, which is a highly sensitive indicator; the coefficient of variation of the soil microbial diversity was between 20% and 50%, which is a moderately sensitive indicator; the coefficients of variation of the pH, bulk density, organic matter, total nitrogen, and total potassium were between 0% and 20%, which are low sensitivity indicators.

Table 5. Coefficients of variation for the seven Quaternary red soil health evaluation indicators under different land use patterns.

Index	pH	Bulk Density	Organic Matter	Total Phosphorus	Total Nitrogen	Total Potassium	Microbial Diversity
Standard deviation (SD)	0.26	0.16	0.44	0.31	0.08	0.01	2.90
Mean	5.94	1.45	6.23	0.42	2.72	0.05	9.51
Coefficient of variation (%)	4.40	11.20	7.00	74.00	2.90	18.80	30.50

3.4.3. The SHI and Its Classification

By running the fuzzy membership function formula, each indicator was calculated separately to obtain the membership of the soil health indicators. Then, combined with the weight values of each indicator obtained from the PCA [30], the SHI values were calculated based on the SHI formula. The soil health index of the minimum dataset (MDS-SHI) (seven final soil indicators) ranged from 0.22 to 0.81, with an average of 0.40.

Referring to the CASH method frame for the soil health level divisions, three scoring functions (increasing, decreasing, and optimal) were used [7,34], combined with the classification standards of the soil nutrient indicators. The upper critical value in the classification standard was set as a score of 10, and the lower critical value was set as a score of 1 [7,34]. Subsequently, the measured values of the selected indicators under different land use patterns were sorted in ascending order, and the score comparison curve was drawn to obtain the scores of each soil health indicator. Finally, the sum of the scores corresponding to the measured values of each indicator was compared with the sum of the scores corresponding to the upper critical value of the overall indicator. In this study, seven soil health evaluation indicators were selected, and the sum of the scores corresponding to the upper critical value of the overall indicators was 70. Finally, three evaluation levels of Quaternary red soils under different land use patterns were determined. According to the results of the SHI calculation, the range of levels was [0–1.00]. The evaluation index SHI $\geq$ 0.6 was classified as healthy soil, the evaluation index 0.4 $\leq$ SHI < 0.6 indicated sub-healthy soil, and the evaluation index SHI < 0.4 indicated unhealthy soil [7,34].

The comprehensive health indexes of the Quaternary red soils under different land use patterns, such as the buried Quaternary red soil, sparse forest and grassland, grassland, woodland, and arable land, were 0.33, 0.37, 0.49, 0.64, and 0.61 (Figure 6), respectively. The health status of the Quaternary red soils under different land use patterns tested showed significant differences with a sequence for woodland > arable land > grassland > sparse forest and grassland > buried Quaternary red soils. The comparison results showed that the soil health value of the woodland was highest, and the buried Quaternary red soils was the lowest.

The MDS-SHIs of the buried Quaternary red soils and sparse forest and grassland soils were 0.33 and 0.37, respectively, less than 0.40, belonging to the unhealthy level; the grassland SHI was 0.49, between 0.40 and 0.60, belonging to the sub-healthy level. The SHI of arable land and woodland was 0.61 and 0.64, respectively, belonging to the healthy soil types. Therefore, the Quaternary red soils of woodland and arable land belong to the healthy level.

In order to further explore the impact of different land use patterns on the health status of Quaternary red soils, the topsoil layer (A) and subsoil layer (B) of each profile were compared.

The pH values of the topsoil layers were not significantly different (Table 6), all samples showed weak acidity. The soil bulk density of the topsoil layer of grassland was the lowest at 1.23g$\cdot$cm^{-3}, and the arable land soil was the highest (1.54 g$\cdot$cm^{-3}). The organic matter and total nitrogen contents of woodland were the highest (30.53g$\cdot$kg^{-1} and 1.58g$\cdot$kg^{-1}, respectively), providing sufficient nitrogen nutrients for plant growth. The total phosphorus content was similar in soils under different land use patterns, all presented low content. The maximum Shannon microbial diversity in arable land was 10.72. The topsoil layers' health status was woodland (0.86) > arable land (0.73) > grassland (0.70) > sparse forest and grassland (0.67) (Figure 6), consistent with the profile soil health evaluation results. Among them, the soil health status of woodland was the best, and that of the sparse forest and grassland was the worst. According to the soil health grading standard, the SHI of the Quaternary red soils under the four land use patterns was greater than 0.60, all belonging to healthy soil types. Overall, the topsoil layers' health status of Quaternary red soils under different land use patterns was relatively good.

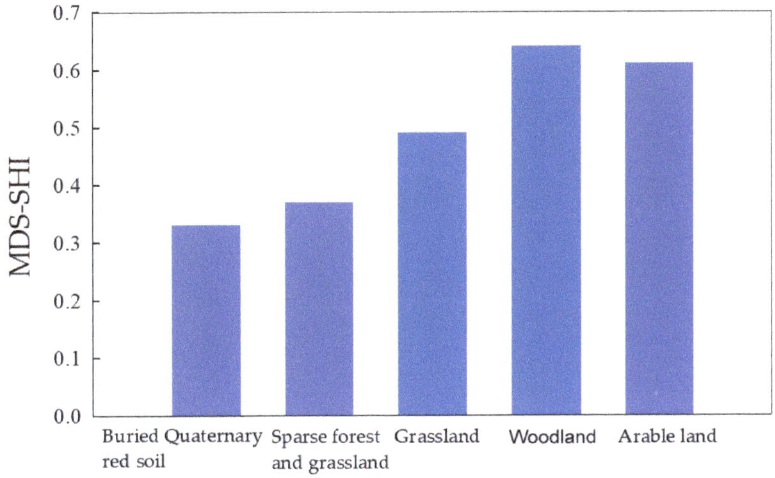

Figure 6. Comparisons of the Quaternary red soil health evaluation indexes under different land use patterns.

Table 6. Contents of the soil health evaluation indexes for A and B horizons of the Quaternary red soils under different land use patterns.

Index	Sparse Forest and Grassland		Grassland		Woodland		Arable Land	
	A	B	A	B	A	B	A	B
pH	5.63	5.8	5.62	5.91	5.67	5.92	5.75	5.99
Bulk density ($g{\cdot}cm^{-3}$)	1.35	1.64	1.23	1.33	1.38	1.61	1.54	1.46
Organic matter ($g{\cdot}kg^{-1}$)	9.33	2.21	7.46	1.36	30.53	3.71	12.93	2.93
Total nitrogen ($g{\cdot}kg^{-1}$)	0.55	0.21	0.78	0.18	1.58	0.27	0.57	0.24
Total phosphorus (%)	0.05	0.04	0.06	0.04	0.07	0.04	0.06	0.05
Total potassium (%)	2.62	2.74	2.61	2.75	2.71	2.60	2.76	2.75
Microbial diversity	10.15	10.15	9.82	9.82	10.4	10.4	10.72	10.72

The differences in the pH values of the subsoil layers in the Quaternary red soil profiles were not significant (Table 6), all showing weak acidity. The grassland had the lowest bulk density (1.23 $g{\cdot}cm^{-3}$), and the sparse forest and grassland had the highest bulk density (1.64 $g{\cdot}cm^{-3}$). Like the topsoil layer analysis results, the woodland had a higher organic matter and total nitrogen content (3.71 $g{\cdot}kg^{-1}$ and 0.27 $g{\cdot}kg^{-1}$, respectively). The differences in the total phosphorus content were not significant, all showing a poor state, but the total potassium content was higher. The microbial diversity of the arable land was the highest. The soil health status of the subsoil layer showed the trend of arable land (0.41) > grassland (0.40) >woodland (0.38) > sparse forest and grassland (0.34) (Figure 7). According to the soil health grading standards, the SHI of the sparse forest and grassland and woodland was less than 0.40, belonging to the unhealthy level; the SHI of the grassland and arable land was between 0.40 and 0.60, belonging to the sub-healthy level; overall, the health status of the subsoil layers under different land use patterns was relatively poor.

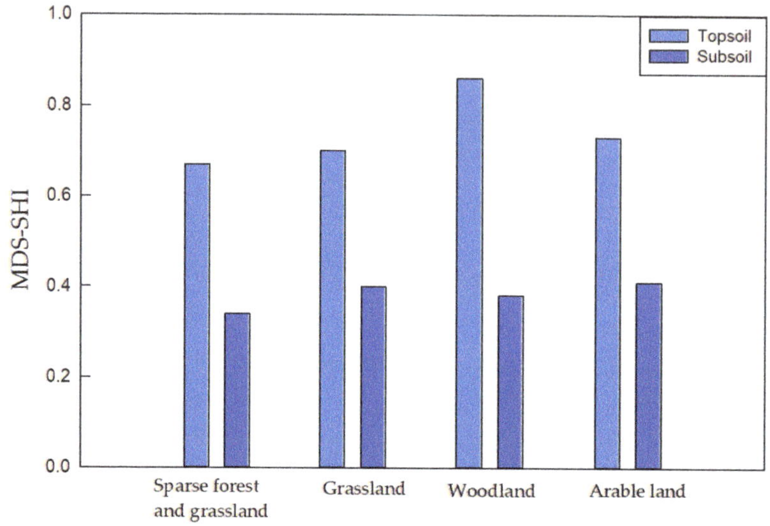

Figure 7. Comparisons of the Quaternary red soil health evaluations for topsoils and subsoils under different land use patterns.

It can be seen that topsoil layers were directly affected by frequent human land use activities, resulting in significant differences in the health status of the topsoil layers and subsoil layers (Figure 7). The health status of the Quaternary red soil topsoil layers (A) under different land use patterns was consistent with the overall profile health status trend, better than the health status of the subsoil layer (B), with the subsoil layer health status showing a trend of arable land > grassland > woodland > sparse forest and grassland. Overall, the health indexes of the sparse forest and grassland, grassland, woodland, and arable land were higher than that of the buried Quaternary red soils (0.33), indicating that certain human land use activities have improved the soil health status, with the woodland having the best health status, making it a suitable land use pattern in low mountain and hilly areas.

3.5. Soil Health Evaluation Result Verifications

In order to evaluate the accuracy of the MDS-SHI, the soil health index of the total dataset (TDS-SHI) (12 initially selected soil indicators) was calculated. The TDS-SHI ranged from 0.20 to 0.71, with an average of 0.31 and a coefficient of variation of 21.30%.

The soil health evaluation results were verified using linear regression, in the case of undetermined independence, to ensure the accuracy of soil health evaluation results. Linear regression analysis was performed based on the relative deviation coefficient. There was a highly significant positive correlation ($p < 0.01$) between the soil health comprehensive index of the TDS-SHI and the MDS-SHI, proving that the MDS-SHI under PCA was reasonable and correct (Figure 8). This indicated that the results of the established health evaluation system of Quaternary red soils under different land use patterns was representative.

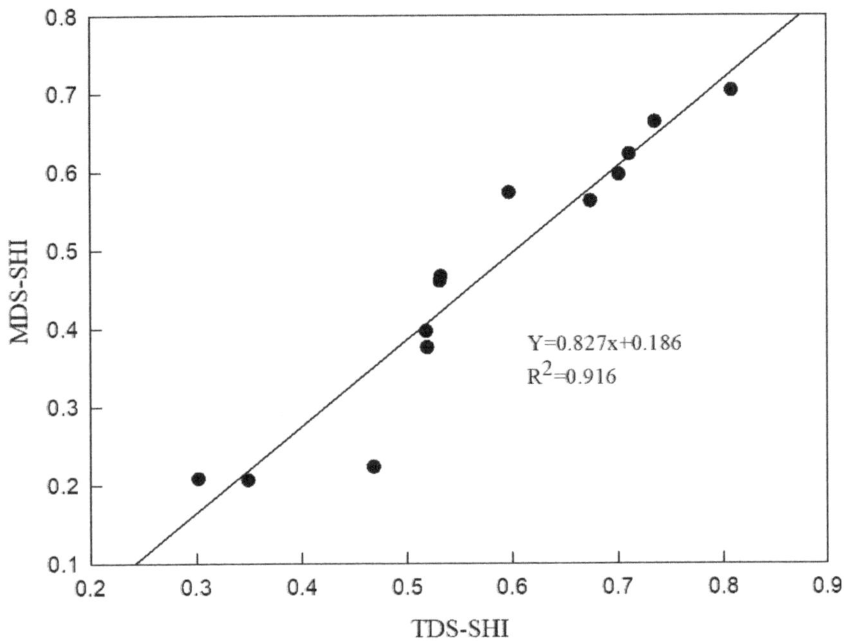

Figure 8. The correlation of the Quaternary red soil health evaluation between the results using the minimum dataset (MDS-SHI) and the total dataset (TDS-SHI).

4. Discussion

Soil health is affected by many factors [33,34]. The established health evaluation system is a preliminary exploration of the study on Quaternary red soils. In order to continuously improve the health evaluation system, more detailed and comprehensive analysis and investigation are needed, such as index optimization, membership function selection, weight value determination, and soil health classification.

4.1. The Selection of the Health Evaluation Indicators for the Quaternary Red Soils under Different Land Use Patterns

The advantage of the SHI method is that it can fully consider the influence of the measured value, weight, and interaction of evaluation indicators on the evaluation results [7,34]. In the selection of the soil health indicator (Figure 2), the representativeness, universality, sensitivity, reproducibility, and the measurement cost, as well as its appropriate range and threshold should be considered [7,35,36]. In the study, 116 potential indicators were firstly selected according to the CNKI. Through frequency screening and combining with factors such as the topography, climate, hydrological conditions, and soil properties, 12 evaluation indexes were further selected. Based on the frame of the Cornell Soil Health Evaluation system [7], seven evaluation indicators including the bulk density, pH, organic matter content, total potassium content, total nitrogen content, total phosphorus content, and microbial diversity were determined using the MDS, PCA, and Norm values. Theoretically, the more evaluation indicators, the closer the evaluation results are to the real condition of the soil health [35,36]. However, the indicators that can be used in the actual operation process are limited. Thus, it is necessary to use limited data to obtain the results closest to reality. Therefore, indicators can be flexibly selected for health evaluation according to the local conditions [34–37]. The significant correlation between the SHI-MDS and SHI-TDS further indicates that the MDS can replace the TDS to accurately evaluate the health status of Quaternary red soils under different land use patterns (Figure 8).

4.2. Changes in the Health Status of Quaternary Red Soils under Different Land Use Patterns

The health index of the Quaternary red soils under different land use patterns was obtained by calculating the SHI-MDS. The reference base of the buried Quaternary red soils was not affected by human activities, maintaining the original characteristics. Long-term compaction resulted in a soil structure that was compact, sticky, and poorly permeable. In addition, with the low precipitation and dry climate, the rate of organic matter mineralization was far higher than the rate of humus. Therefore, the organic matter content, soil nutrient content, and soil microbial diversity were low. The SHI was 0.33, which is in an unhealthy state.

When the buried Quaternary red soils were exposed to the surface, the health status of the Quaternary red soils varied under different human land use activities, showing the trend of woodland (0.64) > arable land (0.61) > grassland (0.49) > sparse forest and grassland (0.37) > buried Quaternary red soils (0.33). According to the classification criteria of soil health [7,34], soils with an index below 0.4 are unhealthy, those between 0.4 and 0.6 are sub-healthy, and those above 0.6 are healthy. Among them, the Quaternary red soils under woodland and arable land use patterns showed a relatively healthy state. Woodland has a larger amount of biomass returned from trees and shrubs each year, and plant residues covering the surface help to conserve soil and water [38]. At the same time, the decomposition of fallen leaves participates in the soil humification process, increasing the soil nutrient content and creating a suitable environment for microbes [38–40]. Higher microbial diversity promotes the accumulation of organic matter and nutrient cycling [39], significantly improving the health status of the red soils. Therefore, the health index of woodland was the highest at 0.64, reaching the standard of healthy soils. The arable land ranked second, with a health index of 0.61. Human activities, such as applying organic fertilizers and chemical fertilizers, increase and maintain soil nutrient content in the arable land [41]. In addition, machine plowing loosens the topsoil layers, improving the structure and permeability of the red soils, making it suitable for microbial survival and enhancing soil health [35]. In the study area, with a semiarid climate and scarce precipitation, plant growth in woodland is greatly restricted due to extensive management. However, the relatively fine management of arable land brings their health status close to that of woodland under the influence of human land use activities. For the grassland (0.49) and sparse forest and grassland (0.37), the aboveground biomass is mainly annual herbaceous plants, and the organic matter returned to the soil through plant residues is limited. Compared to woodland and arable land, these areas have lower economic value and are under extensive management. Their external conditions are insufficient, and their internal conditions are scarce, making their health status lower than that of arable land and woodland and showing a sub-healthy soil state. Overall, the soil health status of the four land use patterns, woodland, arable land, grassland, and sparse forest and grassland, are better than that of the reference base buried red soils (0.33). This indicates that current human land use activities promote the development of red soils towards a better health status.

In order to further explore the impact of different land use patterns on the health status of Quaternary red soils, the topsoil layer (A) and subsoil layer (B) of the red soils were analyzed and studied under different land use patterns. The health index of the topsoil layer ranged from 0.67 to 0.86, showing a trend of woodland (0.86) > arable land (0.73) > grassland (0.70) > sparse forest and grassland (0.67), which was consistent with the overall trend of the soil health status in the profile. The health indexes of the red soils under different land use patterns in the study were greater than 0.6, belonging to the healthy level. In woodland, grasslands, and sparse forest and grasslands, the surface is covered with fallen leaves and branches all year round, which helps to maintain moisture and a suitable temperature [39,40]. At the same time, the decomposition of plant residues returns nutrients to the soil, improving the organic matter content and promoting microbial survival [40]. The increase in microbial diversity also promotes nutrient cycling in the soil, leading to healthier soil development [39]. However, for grasslands and sparse forest and grasslands dominated by annual herbaceous plants, the ground cover is not as high as that of woodland, and the aboveground biomass and coverage rate are lower than those of woodland. These conditions result in a lower humification process and microbial diversity than woodland, leading to a less healthy soil status for the grasslands and sparse forest and grasslands. For arable land, the topsoil layer is directly affected by human plowing and farmyard manure and chemical fertilizers, which improves the fertility status, making it comparable to woodland. Compared to the buried Quaternary red soils (0.33), the health status under different land use patterns has developed to varying degrees towards healthier soil conditions. Human land use activities directly affect the topsoil layer, having the greatest impact on its characteristics [35,41].

As the soil depth changes, the influence of human activities is limited [40], resulting in a weaker impact on the subsoil layer (B). Affected by the continental monsoon climate in the northern temperate zone, the study area has low precipitation and a dry climate [20,21]. Compared to the topsoil layer, which becomes looser and has a higher organic matter content under the influence of human land use activities, the subsoil layer is more compact, with poorer water permeability, and lower nutrient content, which affects the microbial activity, thus influencing the accumulation of humus and the transformation and formation of soil nutrient elements [35,42]. Therefore, the soil health level of the subsoil layer was lower than that of the topsoil layer, and the difference was significant. The SHI ranged from 0.34 to 0.41, presenting a trend of arable land (0.41) > grassland (0.40) > woodland (0.38) > sparse forest and grassland (0.34). Arable land and grasslands exhibited a sub-healthy soil state, while woodland and sparse forest and grasslands exhibited an unhealthy soil state. The reason for the relatively high health status of the arable land is that humans pursue a higher crop yield, applying large amounts of organic and chemical fertilizers [33,43]. Mechanical plowing allows some fertilizers to enter the subsoil layer directly [43]. After plowing, the topsoil layers become loose, and nutrients can easily move to the subsoil layer, making its soil health status better than other land use patterns [43]. Compared with the SHI of the paleosol profile, the soil health status has declined, but compared with the reference buried red soils (0.33), the soil health status has also improved, albeit with a weak degree of change towards a healthy soil state.

Combining the above analysis, it is clear that when buried Quaternary red soils are exposed and used by humans, factors such as topographic features, climatic characteristics, hydrological features, different human land use activities, soil properties, and external environmental characteristics can lead to changes in the soil health. Compared with buried Quaternary red soils, the health status under other land use patterns has improved to varying degrees and is developing towards healthier soil. At this stage, soil health has become an important limiting factor in regional land use in agricultural production practices. It is necessary to plan, manage, and use land scientifically and rationally according to the local conditions. While obtaining crop yields, it is also essential to improve the soil health and gradually achieve the sustainable use of paleosol resources.

5. Conclusions

The buried red soils have not been affected by human land use activities, main-taining their original state. Due to their long-term compacted state, their permeability and aeration are poor, with low organic matter content and low soil microbial diversity. The SHI was 0.33, indicating an unhealthy state. As the buried red soils are exposed to the surface due to erosion and other factors and affected by different land use activities, the soil health improves, transitioning to various degrees of healthier states. This suggests that to a certain extent, land use activities positively impact the health of the Quaternary red soils.

Compared with the buried Quaternary red soils, the health index of soils under different land use patterns showed a trend of woodland (0.64) > arable land (0.61) > grassland (0.49) > sparse forest and grassland (0.37) > buried Quaternary red soils (0.33). Among them, buried Quaternary red soils were classified as unhealthy, grasslands and sparse forest and grass-lands were sub-healthy, while woodlands and arable lands were healthy.

The soil health trend in the topsoil layer (A) of the Quaternary red soils under different land use patterns was consistent with the profile health status, namely woodland (0.86) > arable land (0.73) > grassland (0.70) > sparse forest and grassland (0.67) > buried Quaternary red soils (0.33). Except for buried the Quaternary red soils, the topsoil layer under other land use patterns was in a healthy state.

The health status of the subsoil layer (B) showed a trend of arable land (0.41) > grassland (0.40) > woodland (0.38) > sparse forest and grassland (0.34) > buried Quaternary red soils (0.33) and was in a sub-healthy state.

The soil health conditions under the four types of land use were better than those of the buried Quaternary red soils, showing a trend towards soil health. This suggests that human land use activities in the low mountain and hilly areas have to some extent promoted the healthy development of the Quaternary red soils.

The woodland red soil is in a healthy status and is suitable land use pattern for low mountain and hilly areas. The research results are expected to provide a scientific basis for adjusting the land use structure and a reference for scientific management and use in similar regions, to promote the sustainable use of the Quaternary red soils.

Author Contributions: Conceptualization, Y.J. and Z.S.; Data curation, Y.J., Z.S. and Y.Z.; Formal analysis, Y.J., Z.S. and Y.Z.; Funding acquisition, Y.J., Z.S., H.W. and J.W.; Investigation, Y.J. and Z.S.; Methodology, Y.J., Z.S. and Y.Z.; Project administration, Y.J., Z.S. and Y.Z.; Resources, Y.J. and Z.S.; Software, Y.J., Z.S. and J.W.; Supervision, Y.J. and Z.S.; Validation, Y.J., Z.S., H.W. and J.W.; Visualization, Y.J. and Z.S.; Writing—original draft, Y.J., Z.S. and Y.Z.; Writing—review and editing, Y.J., Z.S. and J.W. All authors have read and agreed to the published version of the manuscript.

Funding: This research was funded by the Applied Basic Research Program of Liaoning Province (No. 2022JH2/101300167), the National Natural Science Foundation of China (No. 42277285), the Basic scientific research project of Liaoning Provincial Department of Education (No. LJKZ1341), and the Applied Basic Research Program of Liaoning Province (No. 2023020244-JH2/1016).

Data Availability Statement: All data are true, valid, and can be made available.

Acknowledgments: The authors sincerely thank all the students and staff who provided input to this study. Our acknowledgements also extend to the anonymous reviewers for their constructive reviews of the manuscript.

Conflicts of Interest: The authors declare no conflict of interest. The funders had no role in the design of the study; in the collection, analyses, or interpretation of data; in the writing of the manuscript, or in the decision to publish the results.

References

1. Duan, S.Y.; Sun, Z.X.; Wang, Q.B.; Jiang, Y.Y.; Han, C.L.; Zhang, Y.W.; Lv, M.F.; Sun, F.J.; Chen, L.M. Characteristics of soil organic carbon distribution of quaternary red soils under different land use patterns. *Chin. J. Soil Sci.* **2021**, *52*, 1078–1084.
2. Sun, Z.X.; Owens, R.P.; Han, C.L.; Chen, H.; Wang, X.L.; Wang, Q.B. A quantitative reconstruction of a loess–paleosol sequence focused on paleosol genesis: An example from a section at Chaoyang, China. *Geoderma* **2016**, *266*, 25–39. [CrossRef]
3. Duan, S.Y.; Sun, Z.X.; Wang, Q.B.; Jiang, Y.Y.; Sun, F.J. Comparison of aggregate composition in quaternary red soil under different land use pattens. *Chin. J. Soil Sci.* **2020**, *51*, 587–596.
4. Liu, L.W.; Gong, Z.T. Development and evolution of red paleosols. *Mar. Geol. Quat. Geol.* **2000**, *20*, 37–42.
5. Kang, L.F.; Rui, L.F.; Hua, W.; Tan, J.A.; Yang, F.W.; Long, H.W. The effects of different land-use types on quality of urban soils. *Ecol. Sci.* **2006**, *25*, 59–63.
6. Doran, J.W.; Sarrantonio, M.; Liebig, M.A. Soil health and sustainability. *Adv. Agron.* **1996**, *56*, 1–54.
7. Moebius-Clune, B.N.; Moebius-Clune, B.K.; Gugino, B.K.; Idowu, O.J.; Schindelbeck, R.R.; Ristow, A.J.; Van Es, H.M.; Thies, J.E.; Shayler, H.A.; McBride, M.B.; et al. *Comprehensive Assessment of Soil Health-The Cornell Framework Manual*; Edition 3.1; Cornell University: Geneva, Switzerland; New York, NY, USA, 2016.
8. Lehmann, J.; Bossio, D.A.; Kögel-Knabner, I.; Rillig, C.M. The concept and future prospects of soil health. *Nat. Rev. Earth Environ.* **2020**, *1*, 544–553. [CrossRef]
9. Fine, A.K.; van Es, H.M.; Schindelbeck, R.R. Statistics, scoring functions, and regional analysis of a comprehensive soil health database. *Soil Sci. Soc. Am. J.* **2017**, *81*, 589–601. [CrossRef]
10. Rinot, O.; Levy, G.J.; Steinberger, Y.; Svoray, T.; Eshel, G. Soil health assessment: A critical review of current methodologies and a proposed new approach. *Sci. Total Environ.* **2019**, *648*, 1484–1491. [CrossRef]
11. Zhu, Y.G.; Peng, J.J.; Wei, Z.; Shen, Q.R.; Zhang, F.S. Linking the soil microbiome to soil health. *Sci. Sin. Vitae* **2021**, *51*, 1–11.
12. Zhang, J.Z.; LI, Y.Z.; Li, Y.; Zhang, J.L.; Zhang, F.S. Advances in the Indicator System and Evaluation Approaches of Soil Health. *Acta Pedol. Sin.* **2022**, *59*, 603–616.

13. Zhu, Y.G.; Shen, R.F.; He, J.Z.; Wang, Y.F.; Han, X.G.; Jia, Z.J. Soil Microflora in China: Progress and Prospect. *Agric. Sci. Eng. China* **2018**, *3*, 6–13.
14. Wall, D.H.; Nielsen, U.N.; Six, J. Soil biodiversity and human health. *Nature* **2015**, *528*, 69–76. [CrossRef] [PubMed]
15. Bünemann, E.K.; Bongiorno, G.; Bai, Z.G. Soil quality-A critical review. *Soil Biol. Biochem.* **2018**, *120*, 105125. [CrossRef]
16. Han, C.L.; Liu, S.H.; Wang, Q.B.; Wang, H.Q. Basic properties and genetic characteristic research of Quaternary paleosol in Chaoyang city of Liaoning. *Chin. J. Soil Sci.* **2009**, *6*, 1233–1239.
17. Chinese Soil Taxonomy Research Group; Institute of Soil Science Chinese Academy of Sciences; Cooperative Research Group on Chinese Soil Taxonomy. *Keys to Chinese Soil Taxonomy*, 3rd ed.; University of Science and Technology of China Press: Hefei, China, 2001. (In Chinese)
18. Soil Survey Staff. *Keys to Soil Taxonomy*, 12th ed.; USDA, NRCS Publications: Washington, DC, USA, 2014.
19. IUSS Working Group WRB. *World Reference Base for Soil Resources 2014, Update 2015 International Soil Classification System for Naming Soils and Creating Legends for Soil Maps*; World Soil Resources Reports No. 106; FAO: Rome, Italy, 2015.
20. Sui, J.Y. Analysis on precipitation features during the flood season of Chaoyang in Liaoning Province in recent 58 Years. *J. Anhui Agric. Sci.* **2011**, *39*, 14931–14932.
21. Zhang, F.M.; Zong, Y.F. The characteristics of temperature change in Chaoyang area of Liaoning Province. *Anhui Agric. Sci. Bull.* **2015**, *21*, 150–160.
22. Zhang, G.L.; Li, D.C. *Manual of Soil Description and Sampling*; Science Press: Beijing, China, 2017. (In Chinese)
23. Grossman, R.B.; Reinsch, T.G. Bulk Density and Linear Extensibility. In *Methods of Soil Analysis: Part 4 Physical Methods*; Dane, J.H., Topp, G.C., Eds.; ASA and SSSA: Madison, WI, USA, 2002; pp. 201–228.
24. Li, J.W.; Jiao, X.G.; Sui, Y.Y.; Cheng, S.J. The Belonging of Dark Brown Soil in the East of Jilin Province in Chinese Soil Taxonomy. *Chin. Agric. Sci. Bull.* **2011**, *27*, 74–79.
25. *GB/T9837-88*; Determination of Soil Total Phosphorus. Standards Press of China: Beijing, China, 1988.
26. *GB/T9836-88*; Determination of Soil Total Potassium. Standards Press of China: Beijing, China, 1988.
27. Frank, S.; Tebbe, C. A new approach to utilize PCR-Single-Strand-Conformation polymorphism for 16S rRNA gene-based. *Appl. Environ. Microbiol.* **1998**, *64*, 4870–4876.
28. Larson, W.E.; Pierce, F.J. *Conservation and Enhancement of Soil Quality: Evaluation for Sustainable Land Management in the Developing Word*; International Board for Soil Research and Management Lnc.: Bangkok, Thailand, 1991; pp. 175–203.
29. Doran, J.W.; Parkin, T.B. *Quantitative Indicators of Soil Quality: A Minimum Data Set*; Soil Science Society of America Special Publication 49; American Society of Agronomy: Madison, WI, USA, 1996; pp. 25–37.
30. Du, Z.F. *Multivariate Statistical Analysis*; Tsinghua University Publishing House Co., ltd.: Beijing, China, 2016.
31. Wang, X.; Jin, L.; Fu, M. Research on Fuzzy Plastic Constitutive Model Based on Membership Function. *Math. Probl. Eng.* **2021**, *2021*, 9901948. [CrossRef]
32. Andrews, S.S.; Karlen, D.L.; Cambardella, C.A. The Soil Management Assessment Framework: A Quantitative Soil Quality Evaluation Method. *Soil Sci. Soc. Am. J.* **2004**, *68*, 1945–1962. [CrossRef]
33. Andrews, S.S.; Karlen, D.L.; Mitchell, J.P. A comparison of soil quality indexing methods for vegetable production systems in Northern California. *Agric. Ecosyst. Environ.* **2002**, *90*, 25–45. [CrossRef]
34. Sheng, F. Introduction and application of Cornell Soil Health Assessment. *Chines J. Soil Sci.* **2014**, *45*, 1289–1296.
35. Wu, K.N.; Yang, Q.J.; Zhao, R. A Discussion on Soil Health Assessment of Arable Land in China. *Acta Pedol. Sin.* **2021**, *58*, 537–544.
36. *NY/T 1634-2008*; Ministry of Agriculture of the People's Republic of China. Rules for Soil Quality Survey and Assessment. Chinese Agriculture Press: Beijing, China, 2008.
37. Raiesi, F. A minimum data set and soil quality index to quantify the effect of land use conversion on soil quality and degradation in native range lands of upland arid and semiarid regions. *Ecol. Indic.* **2017**, *75*, 307–320. [CrossRef]
38. Li, L.; Zhang, Y.; Wang, L.B.; Wang, L.M. Vertical changes of the soil microbial biomass and the correlation analysis in different forests. *J. Cent. South Univ. For. Technol.* **2007**, *2*, 52–60.
39. Wang, Y.; Ding, G.D.; Liu, M.J.; Gao, G.L.; Yu, M.H.; Li, X. Influence of different vegetation types on soil microbial characteristics of typical forest land in Yulin Sandy Area. *Chin. J. Soil Sci.* **2022**, *4*, 97–918.
40. Lauber, C.L.; Ramirez, K.S.; Aanderud, Z.; Lennon, J.; Fierer, N. Temporal variability in soil microbial communities across land use types. *ISME J. Multidiscip. J. Microb. Ecol.* **2013**, *7*, 1641–1650. [CrossRef]
41. Sharma, K.L.; Sharma, S.C.; Bawa, S.; Singh, S.; Chandrika, D.S.; Sharam, V. Combined effect of tillage and organic fertilization on soil quality key indicators and indices in alluvial soils of Indo-Gangetic Plains under rainfed maize-wheat system. *Arch. Agron. Soil Sci.* **2015**, *61*, 313–327. [CrossRef]
42. Yang, Q.J.; Wu, K.N.; Feng, Z.; Zhao, R.; Zhang, X.; Li, X.L.; Ma, N. Soil quality assessment on large spatial scales: Advancement and revelation. *Acta Pedol. Sin.* **2020**, *57*, 565–578.
43. Qi, Y.B.; Darilek, J.L.; Huang, B.; Zhao, Y.; Sun, W.; Gu, Z. Evaluating soil quality indices in an agricultural region of Jiangsu Province, China. *Geoderma* **2009**, *149*, 325–334. [CrossRef]

agronomy

MDPI

Article

Effect of Elevated Air Temperature on the Growth and Yield of Paddy Rice

Dohyeok Oh [1], Jae-Hyun Ryu [2], Hoejeong Jeong [3,4], Hyun-Dong Moon [5,6], Hyunki Kim [5,7], Euni Jo [5,6], Bo-Kyeong Kim [5,6], Subin Choi [5] and Jaeil Cho [5,6,*]

1 Agricultural Resources Research Institute, Gyeonggi Agricultural Research & Extension Services, 61 Yeoncheon-ro, Yeoncheon 11017, Republic of Korea; odh0212@korea.kr
2 Climate Change Assessment Division, National Institute of Agricultural Science, Rural Development Administration, 166 Nongsaengmyeong-ro, Iseo-myeon, Wanju 55365, Republic of Korea; ryujaehyun@korea.kr
3 Crop Production Technology Research Division, National Institute of Crop Science, Rural Development Administration, 20 Jeompiljae-ro, Miryang-si 50424, Republic of Korea; hoejeong@korea.kr
4 Interdisciplinary Program for Agriculture & Life Science, Chonnam National University, 77 Yongbong-ro, Gwangju 61186, Republic of Korea
5 Department of Applied Plant Science, Chonnam National University, 77 Yongbong-ro, Gwangju 61186, Republic of Korea; dignwhgdkz@gmail.com (H.-D.M.); khg650771@gmail.com (H.K.); joeuni2018@gmail.com (E.J.); kbk000417@gmail.com (B.-K.K.); csb7419@gmail.com (S.C.)
6 BK21 Four Center for IT-Bio Convergence System Agriculture, Chonnam National University, 77 Yongbong-ro, Gwangju 61186, Republic of Korea
7 Crop Production and Physiology Division, National Institute of Crop Science, Rural Development Administration, 181 Hyeoksin-ro, Iseo-myeon, Wanju 55365, Republic of Korea
* Correspondence: chojaeil@jnu.ac.kr; Tel.: +82-62-530-2056; Fax: +82-62-530-2059

Abstract: Rice is one of the major food crops, particularly in Asia. However, it is vulnerable to high temperature and has high yield fluctuations. Monitoring crop growth and physiological responses to high temperatures can help us better understand the agricultural impacts of global warming. The aim of this study is to monitor growth, development, and physiological responses to high temperature conditions on paddy rice and to assess their combined effects on yield. In this study, changes to growth, maturity, and senescence in paddy rice throughout the growing season were identified under elevated air temperature conditions created by a temperature gradient field chamber (TGFC). That facility provides a gradient from the ambient air temperature (AT) to 3 °C above AT (AT + 3 °C). To represent crop physiology and productivity, we measured the plant height, chlorophyll, normalized difference vegetation index (NDVI), and maximum photosynthetic rate (A_{max}) to assess growth and physiological processes, and heat stress effects on four yield measurements were assessed using the heating degree day index. Rice height increased more rapidly in the AT + 3 °C treatment from the early growth stage to heading, while SPAD and NDVI decreased more rapidly at AT after heading. The A_{max} of AT and AT + 3 °C was not significantly different in the tillering stage. However, it was higher at AT in the booting stage but higher at AT + 3 °C in the grain filling stage. These results indicate that paddy rice was not affected by heat stress at the tillering stage, but a cumulative effect emerged by the booting stage. Further, photosynthetic capacity was maintained much later into the grain filling stage at AT + 3 °C. These results will be useful for understanding the growth and physiological responses of paddy rice to global warming.

Keywords: paddy rice; crop growth; yield; elevated temperature; spikelet sterility; temperature gradient field chamber (TGFC)

1. Introduction

Rice is a staple crop for more than half the world's population, and the cultivation and consumption of rice are widespread in the Asian region [1]. However, it is vulnerable to high temperature and has high yield fluctuations [2]. The impacts of global warming, associated with climate change, are increasing over time. According to the IPCC's 2021 special report, if the current trend continues, the global mean temperature increase will exceed 1.5 °C above pre-industrial levels between 2030 and 2052, and is projected to reach 3 °C by 2100 [3]. In addition, abnormal weather phenomena, such as extreme high temperatures, are expected to occur more frequently in most areas. Recently, abnormal high temperature phenomena, including heatwaves and tropical nights, have frequently occurred in the Korean peninsula [4,5]. Rice is one of the three major food crops and the second largest food supply in the world after wheat, meeting about 80% of the food calorie requirements of more than half of the world's population. It is a critical crop in the context of food security, but it is vulnerable to high temperatures and has high yield fluctuations. In particular, high temperatures during the reproductive and flowering stages are directly linked to the yield [6–9], and by 2050, 27% of rice-growing areas are expected to experience at least five days of heat stress during these stages [10].

Previous studies have shown that heat stress shortens the growth and development stage of crops (e.g., [11–13]). Moreover, extremely high temperatures reduce the size of crops and decrease the total amount of photosynthesis during the whole growing season [14]. Kim et al. [15] reported that the early termination of grain filling in high temperature conditions was due to a loss of sink capacity rather than leaf senescence, resulting in an increased distribution of dry matter to leaves and stems.

Crop heat stress due to heatwave events will affect growth and physiological processes, ultimately expressed as yield and biomass losses [11]. In particular, photosynthesis and respiration, which are sensitive to temperature conditions, can affect the total biomass of a crop and are often responsible for final yield fluctuations [16]. Many studies have shown that the photosynthetic response to temperature can be explained as a parabolic curve in which photosynthesis is suppressed at low and high temperatures, but the optimal temperature also changes seasonally, with mutable photosynthetic efficiency at varying temperatures [17]. However, it has been reported that light use efficiency at the leaf and canopy scale decreases under stress conditions, such as high and low temperatures, drought, and biotic stresses [18,19].

Increased mean air temperature and abnormal high temperature events, such as extreme heatwaves, affect the growth and physiological conditions of crops throughout the growing season. Moreover, high temperature phenomena during the growing season reduce a crop's capacity to gain any possible beneficial effects of increased accumulated degree days under global warming [20]. The benefits of the expected rise in CO_2 concentrations can also be eliminated due to temperature increases [21]. In addition, even if the temperature is not high throughout the growing period, extreme events such as heatwaves can cause serious damage to rice production because pollen viability in rice exposed to high temperatures is reduced [6,22]. This spikelet sterility will not only reduce yield but will also affect growth and physiological responses during the reproductive growth period.

Monitoring and quantifying the growth and physiological responses of crops to global warming is critical for making informed policy decisions that affect food security. Thus, non-destructive and efficient optical measurement methods have been used in many studies to assess the impacts of climate change on crops. In particular, the SPAD value, which is clearly proportional to the concentration of nitrogen present in the leaves, has been used in many studies to determine the chlorophyll content in the leaves. In addition, many studies have demonstrated that photosynthetic capacity and leaf growth vary with leaf nitrogen content [23–25]. The normalized difference vegetation index (NDVI) is widely used to represent the biomass and health of vegetation, and it is useful for monitoring the growth of crops; moreover, the accumulated effect of the environment on a crop can be expressed as photosynthesis at the leaf level. Therefore, the purpose of this study is to monitor growth, development, and physiological responses to high temperature conditions on paddy rice and to assess their combined effects on yield.

2. Materials and Methods

2.1. Experimental Facility and Conditions

This study was conducted in a sunlit temperature gradient field chamber (TGFC) installed in a paddy field at the Agricultural Practice Training Center of Chonnam National University, Gwangju, Korea (35°10′ N, 126°53′ E; elevation 33 m) in 2016 and 2018. The TGFC is a special-purpose heating facility based on field conditions that consists of one heater and three ventilators: two small fans and one large fan. The entrance is always opened, and the innermost ventilator operates to make the air moving from the entrance to the back smoothly through the chamber. Along the length of the TGFC, four units, labeled T0, T1, T2, and T3, are installed to measure the air temperature and humidity from the entrance (T0) to the back of the chamber (T3). Using the heater and ventilators, the TGFC is designed to maintain the temperature gradually increasing to a 3 °C higher air temperature (AT + 3 °C) at the innermost part of the chamber, compared to the ambient temperature (AT) at the entrance. When the air temperature difference between T0 (AT) and T3 (AT + 3 °C) is lower than 3 °C, the heater is operated to inject heated air, and when it is higher than 3 °C, the speed of the ventilator fans is increased [26]. A detailed depiction of the TGFC is shown in Figure 1.

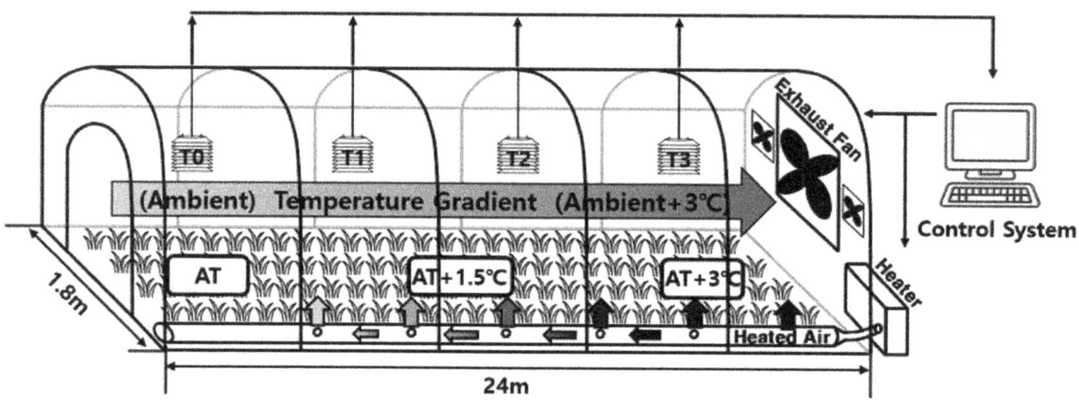

Figure 1. A schematic illustration of the TGFC for exposing paddy rice to a gradient of warmed conditions, from the ambient temperature (AT) at T0 to 3 °C above the ambient temperature (AT + 3 °C) at T3. The mean ambient temperatures during the rice growing season in 2016 and 2018 were 26.5 and 26.9 °C, respectively.

The daily mean, maximum, and minimum air temperatures during the rice growing season (from transplanting to 120 days after transplanting) were 1.2, 1.1, and 1.3 °C greater in 2016 and 1.5, 2.1, and 1.3 °C greater in 2018 relative to the average of 1981–2010 (Figure 2). In particular, there was an unprecedented heatwave event in July and August 2018. In this study, the Ilmi variety of *Oryza sativa* L. *japonica*, one of the leading Korean rice cultivars, was used because it is known to be resistant to high temperature stress [27]. The mid-late maturing Ilmi cultivar was transplanted after sowing the seeds into the TGFC on 3 June 2016 (DOY 155), and 1 June 2018 (DOY 152), and harvested on 13 October 2016 (DOY 287), and 27 September 2018 (DOY 270). The number of transplanted rice plants was 3 per hill, and the plant density was 15 × 30 cm. In this experiment, three TGFCs were used in both years, with 960 (12 × 80) rice plants grown per TGFC in an area of 43.2 m^2 (1.8 × 24 m). Fertilizer treatments of 845.2 g N (nitrogen), 972 g P (phosphorus), and 547.2 g K (potassium) per TGFC, based on an N:P:K of 9:4.5:5.7 kg per 10 acre, were applied at planting. Two more supplemental fertilizers were applied based on a 5:2:3 ratio.

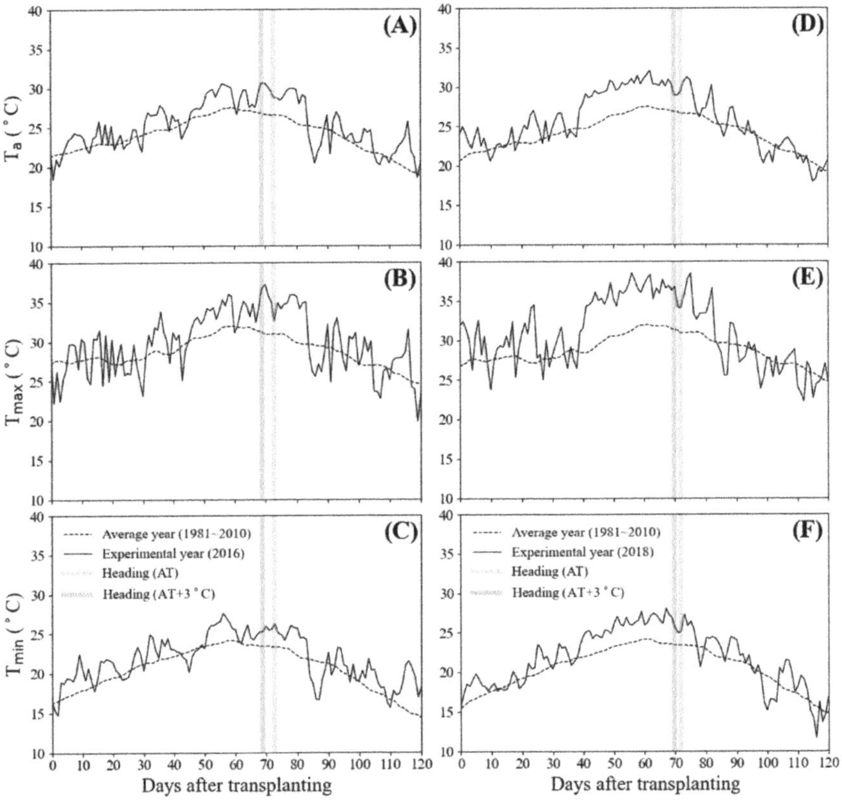

Figure 2. Daily mean air temperature (**A**,**D**), maximum air temperature (**B**,**E**), and minimum air temperature (**C**,**F**) at the experimental site in 2016 (**A**–**C**) and in 2018 (**D**–**F**). The average values for 1981–2010 are shown in each panel. Vertical red and blue lines indicate the heading date at AT and at AT + 3 °C, respectively.

2.2. Meteorological Data Acquisition

Air temperature (T_a) and relative humidity (RH) were measured at T0, T1, T2, and T3, from the entrance to the back of the TGFC. The data were recorded every 5 min in a data logger (CR10X; Campbell Scientific, Logan, UT, USA) and used to automatically maintain the temperature gradient in the TGFC. Other meteorological data, including photosynthetically active radiation (PAR) and solar radiation, were measured inside and outside the TGFC using a quantum sensor (LI-190R; LiCor Inc., Lincoln, NE, USA) and a pyranometer (LI-200R; LiCor Inc., Lincoln, NE, USA), respectively. Temperature data for the experimental years 2016 and 2018, and for the years 1981–2010, in order to contextualize the relative weather conditions of the experimental years, were obtained from the Korea Meteorological Administration (KMA, https://data.kma.go.kr/cmmn/main.do (accessed on 21 November 2023)).

2.3. Heating Degree Day (HDD) Index

The optimum and threshold growth temperature varies according to cultivars and developmental stages, with limits to how much this can be simplified [28,29]. However, in general, the optimum growth temperature of rice is known to be from 20 to 35 °C, and the threshold temperature affecting grain yield is 34–35 °C [6,7,28,30]. The temperature affects various determinants of rice yield and contributes to variation in overall yield [31]. The number of panicles per unit area is directly linked to the number of effective tillers and depends on environmental conditions post transplanting, but it is almost unaffected after the spikelet differentiation stage. The number of spikelets per panicle is determined by the difference between the number of differentiated spikelets and the number of spikelets that degenerate after forming. The differentiation and degeneration of spikelets are most likely to be affected by the environment for 5 to 35 days before heading [32]. The grain filling rate and 1000-grain weight depend on environmental conditions after the heading stage. Heat stress during this period causes spikelet sterility and reduces the accumulation of starch and amylose, negatively affecting the grain filling rate and quality [33].

According to previous studies, the final grain yield shows the most significant relationship with the grain filling rate, and the higher the intensity of heat stress at anthesis, the more the sterility increases, and the more conspicuous the relationship becomes. Also, since the harvest index (HI) depends on the final grain yield, it may change heavily depending on the environmental conditions after the heading stage [29]. Heat stress in plants is a complex function of the intensity and duration of high temperatures. Additionally, in the evaluation of heat stress in crops, not only the instantaneous temperature of the moment but also the accumulated temperature is important. Accumulated heat stress is expressed as growth effects through physiological processes, ultimately affecting the final yield. Thus, in this study, the heating degree day index (HDD) was used as a stress index to comprehensively evaluate the duration and intensity of heat stress. The HDD was calculated for each temperature regime using a modified version of the equations described by Shi et al. [34]:

$$\mathrm{HDD}_{35_{t+5}^{t+50}} = \sum\nolimits_{d_t+5}^{d_t+50} \mathrm{HD}_i \tag{1}$$

$$\mathrm{HDD}_{35_{h-35}^{h-5}} = \sum\nolimits_{d_h-35}^{d_h-5} \mathrm{HD}_i \tag{2}$$

$$\mathrm{HDD}_{35_h^m} = \sum\nolimits_{d_h}^{d_m} \mathrm{HD}_i \tag{3}$$

$$\mathrm{HD}_i = \frac{1}{24}\sum\nolimits_{j=1}^{24} \mathrm{HDD}_j \tag{4}$$

$$\mathrm{HDD}_j = \begin{cases} 0 & \left(T_j \le T_h\right) \\ T_j - T_h & \left(T_j > T_h\right) \end{cases} (\mathrm{j} = 1, 2, \ldots, 24) \tag{5}$$

where d_t, d_h, and dm are the date of transplanting, heading, and maturity; T_j is the hourly temperature, which is based on the temperature recorded every 5 min; and T_h is the threshold temperature, which was set at 35 °C. Equations (1)–(3) represent the HDD for specific factors related to grain yield over a specified period: Equation (1) is for the number of panicles per unit area; (2) is for the number of spikelets per panicle; and (3) is for the yield components, total grain yield, and HI. Equations (4) and (5), respectively, quantify and sum the temperature values above the threshold temperature on day *i*.

2.4. Monitoring of Plant Growth

To identify growth responses and development stages, the plant height and number of tillers were investigated in 10 plants in each temperature regime once a week starting 2 weeks after transplanting. The SPAD value was also measured once a week after the heading stage (69, 75, 84, 89, 96, 110, and 119 days after transplanting (DAT) in 2016 and 75, 83, 90, 97, 105, 112, and 119 DAT in 2018) to monitor changes in the chlorophyll contents of leaves at the leaf level using a portable chlorophyll meter (SPAD-502; Minolta Corp., Osaka, Japan) based on the differential transmission of red and nearinfrared (NIR) radiation. A high SPAD value indicates a large amount of chlorophyll in the leaves. For the SPAD measurement, 20 of the topmost fully expanded leaves were randomly selected, and the average of consecutive values measured 10–15 times at 5 cm away from the leaf tip to the middle of the leaf were used.

A multispectral radiometer (MSR16R; CROPSCAN Inc., Rochester, MN, USA) and spectrometer (AvaSpec-ULS-2048L; Avantes, Apeldoorn, The Netherlands) were used to measure the reflectance of paddy rice in 2016 and 2018, respectively. A multispectral radiometer has 16 wavebands in the 450–1750 nm region and a field of view (FOV) of 28 degrees. The 660 and 800 nm wavelengths were used in this study. A spectrometer has wavebands in the 300–1100 nm range and a 23-degree FOV. All measurements of reflectance were conducted at midday under clear-sky conditions at intervals of 7 to 14 days after the heading stage (69, 78, 82, 89, 97, 109, and 123 DAT in 2016 and 73, 77, 89, 95, 100, 109, and 117 DAT in 2018). The NDVI is useful for monitoring the growth of crops. It was calculated as [35]:

$$\mathrm{NDVI} = \frac{(R_{800} - R_{660})}{(R_{800} + R_{660})} \tag{6}$$

where R_{800} is the reflectance at 800 nm and R_{660} is the reflectance at 660 nm. In this study, interpolated NDVI data were used to monitor time-series variations in the chlorophyll status of leaves, which were related to the photosynthetic capacity at the canopy level.

2.5. Leaf Gas Exchange Measurements

To determine the photosynthetic capacity of leaves at three main phenological stages—tillering (DAT 38 and 41), booting (DAT 41 and 52), and late grain filling (DAT 90) in 2016 and tillering (DAT 27), booting (DAT 53), and late grain filling (DAT 97) in 2018—under two air temperature regimes of AT and AT + 3 °C, light response curves were measured using an LI-6400 portable photosynthesis system incorporating an infrared gas analyzer (LiCor Inc., Lincoln, NE, USA) and equipped with a 2 × 3 cm opaque leaf cuvette with an internal red-blue LED light source (6400-02B LED; LiCor Inc., Lincoln, NE, USA). In addition, as supplementary data, chlorophyll contents were measured for all samples using a portable chlorophyll meter. As samples for the measurement of leaf-level gas exchange, 3 typical middle to topmost fully expanded leaves were selected. All measurement data were recalculated using the actual leaf area in the interior area (6 cm^2) of the leaf cuvette.

All measurements were performed at varying photosynthetically active radiation levels between 09:00 and 12:00 on clear days. Also, in order to reduce the stress to the leaf, each measurement was carried out after adaptation to the environmental conditions inside the leaf cuvette for at least 10–15 min. The measurement entailed starting from the light intensity equivalent to ambient light and measuring from saturated light to weak light, with transitions between light intensity made after the intercellular CO_2 concentration and net photosynthetic rate remained constant for at least 5 min. The conditions inside the leaf cuvette were controlled at a leaf temperature of 25–30 °C and relative humidity (RH) of 40–70%, and a fixed reference CO_2 concentration of 400 μmol mol^{-1} was maintained using a desiccant, soda lime, and 12 g CO_2 cylinders (LiCor Inc., Lincoln, NE, USA). A maximum photosynthetic rate (A_{max}) was calculated for each light response curve using the A/Q curve protocol of the LI-6400 software (https://www.licor.com/env/support/LI-6400/software.html (accessed on 21 November 2023), LiCor Inc., Lincoln, NE, USA). The light-use efficiency at the leaf level (LUE_{leaf}) was calculated as the slope of the linear regression of PAR between 0 and 400 or 500 μmol m^{-2} s^{-1}.

2.6. Yield and Biomass

All yield components and the biomass were measured after harvest. Samples were selected as 10 plants in the center of the experimental plot in each temperature regime, considering border effects. The harvested samples were brought into the laboratory and immediately separated into stem, leaf, and panicles, and then they were dried in a drying oven at 70 °C for 3 days or 80 °C for 2 days, and the above-ground dry matter (AGDM) was measured. Grain yield was estimated using four yield components: the number of panicles per m^2, the number of spikelets per panicle, the grain filling rate (%), and the 1000-grain weight (g). The number of panicles per m^2 was estimated by multiplying the average number of panicles per hill and the number of hills per m^2. HI was calculated by dividing the filled grain weight by the AGDM.

2.7. Statistical Analysis

The growth, physiological responses, and yield of paddy rice to elevated air temperature were statistically analyzed for two temperature regimes, AT and AT + 3 °C. Independent two-sample *t*-tests were conducted to verify the significant differences between temperature regimes for growth and photosynthetic parameters. An analysis of variance (ANOVA) was carried out to assess the statistical significance of the difference of the year and temperature regime and their interaction on yield components, total yield measures, and AGDM. Regression analyses relating yield components and total yield measures to HDD were conducted to characterize the relationships and determine thresholds. All statistical analyses were performed with SPSS (IBM SPSS Statistic for Windows, v23.0, IBM Corp., Armonk, NY, USA).

3. Results and Discussion

3.1. Growth Responses to Elevated Air Temperature

The growth responses to elevated air temperature over the whole growing season were divided into two parts: plant height was observed to compare the physical size changes of rice from the early growth stage to heading. After heading, the changes in SPAD value, indicating the chlorophyll content of leaves [25], and NDVI, indicating the amount of vegetation, greenness, and photosynthetic capacity of vegetation at the canopy level, were also monitored (Figure 3). The plant height during the vegetative growth stage showed a increase under both temperature regimes. Specifically, the plant height increased rapidly in the effective tillering stage but slowed down in the non-productive tillering stage. The plant height then increased again in the reproductive stage, including during panicle initiation. However, a significant resumption of growth during the reproductive stage in the AT + 3 °C treatment did not occur in 2016, with only a slight increase before heading. Some previous studies (e.g., [14,29]) reported that elevated temperature conditions can often reduce the size of plants by growth inhibition due to heat stress and shorten the overall growth period of crops, affecting the accumulative photosynthesis. These discussions are consistent with the result of shortened non-productive tillering stage in the AT + 3 °C treatment. This seems to have contributed to an earlier heading by about 4 and 2 days in 2016 and 2018, respectively. However, in both years, growth in the AT + 3 °C treatment was generally greater from the early growth stage to heading. This implies that growth inhibition by heat stress may not have significantly occurred in the AT + 3 °C treatment. The difference between height growth in the two temperature regimes was greater in 2016 (Figure 3A).

In contrast, SPAD and NDVI decreased more rapidly at AT. The SPAD values were significantly different ($p < 0.01$) between the two temperature regimes after 89 DAT and through the late growth period in 2016 (Figure 3B). The SPAD values in 2018 showed a significant difference ($p < 0.05$) between the two temperature regimes after 90 DAT and through the rest of the experiment, though the difference was slightly reduced at the last measurement (Figure 3D). In 2016 and 2018, the NDVI of AT declined rapidly after 83 DAT and 90 DAT, respectively, while those of the AT + 3 °C treatment declined gradually (Figure 3B,D). These results show that senescence is delayed by the residual assimilation products in the stem or leaves after heading since the higher temperatures at the moment of heading cause spikelet sterility [33]. This means that a higher chlorophyll content is retained for a relatively long period, thus maintaining the photosynthetic capacity. This seems to be related to a previous study's finding that the distribution of dry matter to leaves and stems increased after the end of grain filling as the senescence of the panicle rapidly progressed at high temperatures [27].

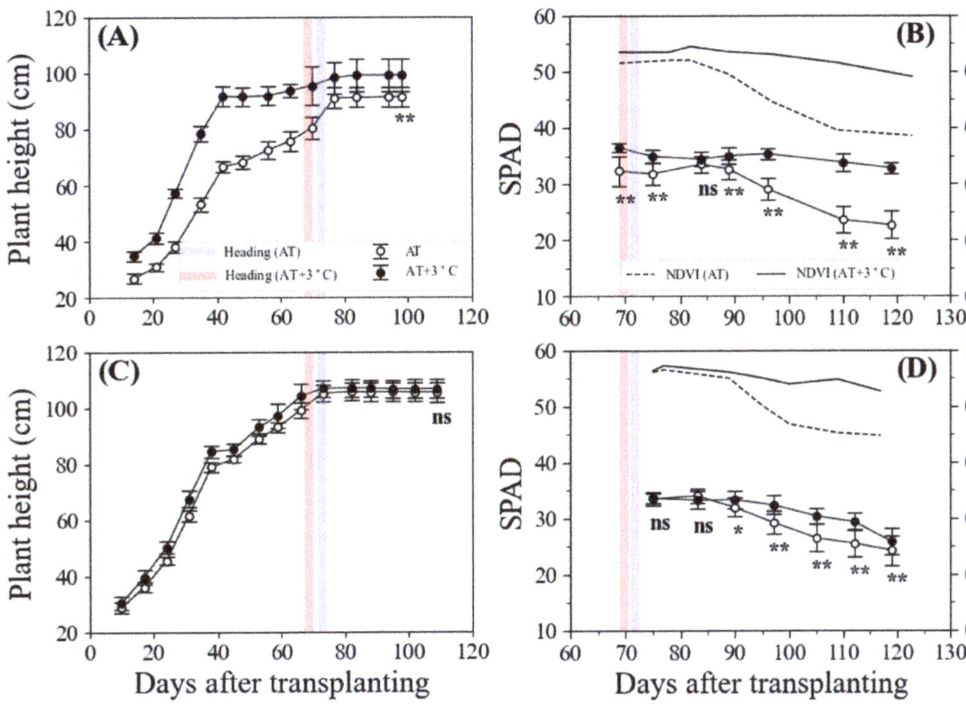

Figure 3. Plant height (cm; (**A**,**C**)), SPAD value, and normalized difference vegetation index (NDVI; (**B**,**D**)) under two temperature regimes in 2016 (**A**,**B**) and 2018 (**C**,**D**). Vertical red and blue lines indicate the heading dates under AT and AT + 3 °C, respectively. Error bars represent standard deviation; $n = 10$ and $n = 20$ for plant height and SPAD, respectively. Significance was assessed using independent two-sample *t*-tests: **, $p < 0.01$; *, $p < 0.05$; ns, not significant ($p \geq 0.05$).

3.2. Leaf Net Photosynthesis

A crop's LUE can be used for spatiotemporal assessment and monitoring plant stress, as LUE_{leaf} decreases in stressed leaves compared to normal leaves [36,37]. There were no significant differences in LUE_{leaf} in both the tillering and booting stages in 2016 ($p \geq 0.05$) (Figure 4A), but in the AT treatment, it was significantly lower at the grain filling stage ($p < 0.01$). In the overall patterns of LUE_{leaf} in 2016, under the AT regime, it showed relatively little difference from the tillering to the booting stage although in 2016 and 2018 it had a decreased and increased pattern, respectively. However, the LUE_{leaf} of AT decreased rapidly in the grain filling stage. Compared to AT, the LUEleaf in the AT + 3 °C regime slightly decreased in the grain filling stage as an extension of the decline from the tillering to the booting stage.

The LUE_{leaf} of the two temperature regimes at the tillering stage in 2018 showed no significant difference ($p \geq 0.05$) (Figure 4C). However, there was a significant difference at the booting stage, when the LUE_{leaf} under AT increased above that of AT + 3 °C, which decreased. Unlike 2016, such a difference of LUE_{leaf} between the two temperature regimes in 2018 may be due to an extreme heatwave event in 2018. Way et al. [17] and Yamori et al. [38] reported that plants adapt to the environment even if the amount of photosynthesis decreases somewhat. Therefore, the relative decrease in photosynthetic capacity at AT + 3 °C at the booting stage of 2018 may have been the result of the small plant height difference between AT and AT + 3 °C. At the grain filling stage, it decreased in both air temperature regimes compared with that of the booting stage, but it decreased more rapidly in the AT treatment, leading to a significant difference ($p < 0.05$). Notably, the variation in the LUE_{leaf} values was quite large at this stage due to the difference of leaf chlorophyll content between the two temperature regimes (see Figure 3B,D).

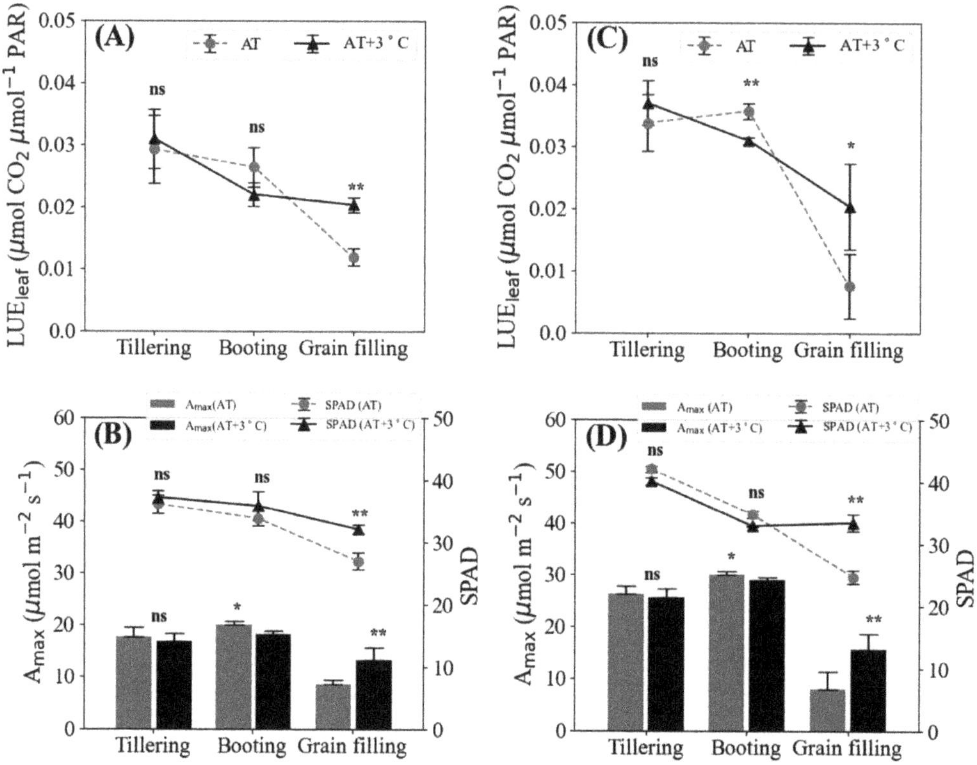

Figure 4. Light-use efficiency (LUE), SPAD values, and maximum photosynthetic rate (A_{max}) under AT and AT + 3 °C at the tillering, booting, and late grain filling stages in 2016 (**A**,**B**) and in 2018 (**C**,**D**). Error bars represent standard deviation; $n = 3$. Significance was assessed using independent two-sample *t*-tests: **, $p < 0.01$; *, $p < 0.05$; ns, not significant ($p \geq 0.05$).

The A_{max} at the booting stage was higher than that of the tillering stage in both years and decreased at the grain filling stage (Figure 4B,D). No difference between the two air temperature regimes was observed at the tillering stage, but A_{max} under AT was significantly higher than that at AT + 3 °C at the booting stage ($p < 0.05$). Specifically, it was about 9% and 4% lower at AT + 3 °C than at AT in 2016 and 2018, respectively. At the grain filling stage, A_{max} in the AT treatment decreased rapidly, whereas in the AT + 3 °C treatment, it decreased only slightly, resulting in a significantly higher A_{max} in the AT + 3 °C treatment ($p < 0.01$).

The changes in SPAD may have significantly related to photosynthetic capacity such as A_{max}. No significant differences in the SPAD values of the photosynthetic samples were found between the temperature treatments at the tillering and booting stages in both years ($p \geq 0.05$). At the grain filling stage, the SPAD value at AT decreased compared to that of the previous stage, but at AT + 3 °C, it either decreased slightly or stayed at a similar level ($p < 0.01$) (see Figure 2B,D and Figure 3B,D). Thus, the A_{max} difference of two temperature regimes at the tillering and booting stages can be interpreted in a similar context to the change in SPAD: the effects of elevated air temperature were not expressed as physiological responses, but these effects seemed to be expressed cumulatively at the booting stage (Figure 4B,D).

Commonly, LUE_{leaf} and A_{max} at AT + 3 °C were higher than those at AT at the grain filling stage. Physiological heat stress and damage may cause lower grain weight and/or more spikelet sterility (see the values of grain filling rate and 1000-grain weight in Table 1). These will bring a loss of sink capacity in the grains, as evidenced by the relatively stable SPAD and NDVI values. Thus, the photosynthetic product stays in the leaves and stems instead of being reallocated to the grains, delaying leaf senescence.

Table 1. Four yield components and total grain yield, above-ground dry matter (AGDM), and the harvest index (HI) under different air temperature regimes (TR) in 2016 and 2018. The values are means ± SE based on n = 3 samples. The ANOVA results at the bottom of the table show the significance levels for the main effects and their interaction.

YEAR	TR	No. of Panicles per m^2	No. of Spikelets per Panicle	Grain Filling Rate (%)	1000-Grain Weight (g)	Grain Yield (g/m^2)	AGDM (g/m^2)	HI
2016	AT	291.21 ± 12.58	90.62 ± 2.70	90.37 ± 0.70	25.23 ± 0.17	596.38 ± 45.65	1388.27 ± 220.83	0.42 ± 0.01
	AT + 3 °C	277.18 ± 7.03	82.36 ± 5.28	0.84 ± 0.77	17.98 ± 0.71	3.48 ± 3.23	1310.72 ± 17.98	0.01 ± 0.00
2018	AT	269.28 ± 14.48	120.53 ± 3.06	87.68 ± 0.57	24.35 ± 0.12	688.50 ± 42.08	1680.29 ± 133.53	0.41 ± 0.01
	AT + 3 °C	267.30 ± 3.78	99.10 ± 7.87	7.31 ± 5.26	20.06 ± 0.37	45.13 ± 34.47	1625.22 ± 95.83	0.03 ± 0.02
ANOVA [z]	YEAR	*	**	*	ns	**	**	ns
	TR	ns	**	**	**	**	ns	**
	YEAR × TR	ns	*	**	**	ns	ns	ns

[z] ANOVA results: **, $p < 0.01$; *, $p < 0.05$; ns, not significant ($p \geq 0.05$).

3.3. Yield Responses to Elevated Air Temperature

Each yield component—the no. of panicles per m^2, no. of spikelets per panicle, grain filling rate, and 1000-grain weight—and the total grain yield, AGDM, and HI were investigated under the two temperature regimes (Table 1). Under the AT regime, all factors were significantly higher ($p < 0.01$) except for the number of panicles per m^2 and AGDM. These results seem to be due to the fact that the influence of water temperature at the tillering stage is larger than the influence of air temperature [28], although there is a relationship between air and water temperature. At AT + 3 °C, the number of panicles per m^2 was 5 and 0.7%, the number of spikelets per panicle was 9 and 18%, the grain filling rate was 99 and 92%, and the 1000-grain weight was 29 and 18% lower than at AT in 2016 and 2018, respectively. These results led to a reduction in grain yield of 99 and 93% and in HI of 98 and 93% in 2016 and 2018, respectively.

The most significant difference was identified in the grain filling rate due to heat stress-induced spikelet sterility. In 2018, when there was a strong heatwave, the grain filling rate dropped by 3% compared to 2016, even in AT. However, the heatwave in 2018 did not reduce grain yield in AT compared to 2016, which appears to be due to the increases in the number of spikelets per panicle and AGDM (See Table 1) with abundant solar radiation and near-full-scale sterility damage (see Figure 5C). On the other hand, photosynthetic capacity (i.e., LUE_{leaf} and A_{max}) at the AT + 3 °C treatment, which did not decrease even during the grain filling stage (see Figure 4B,D), was not of any assistance to grain yield against the damage of spikelet sterility. This is evidenced from the relatively lower values of 1000-grain weight at the AT + 3 °C treatment, compared to AT (Table 1).

3.4. Effects of Heat Stress

As a result of evaluating the effects of heat stress using the HDD index, the number of panicles per m^2 showed a negative relationship with the HDD (Figure 5A), while the number of spikelets per grain showed no statistically significant relationship with the HDD ($p \geq 0.05$) (Figure 5B). The grain filling rate, grain yield, and HI decreased rapidly when the HDD exceeded 5 °C, and were close to 0 when it was above 10 °C (Figure 5C,E,F). This means that when the sum of temperatures greater than or equal to 35 °C per day that rice plants are exposed to from heading to maturity exceeds 5 °C, these factors rapidly decrease. The 1000-grain weight showed a significant negative relationship with the HDD index, with a coefficient of determination (R^2) of 0.70 (Figure 5D). The number of panicles per m^2 and the number of spikelets per panicle, which are determined in the early growth stage, are dominated by water temperature rather than air temperature, while the grain filling rate and 1000-grain weight, determined in the late growth stage, are dominated by air temperature rather than water temperature [28]. Therefore, heat stress due to high temperatures during the early stage of growth may be mitigated by the lower water temperatures relative to air temperatures.

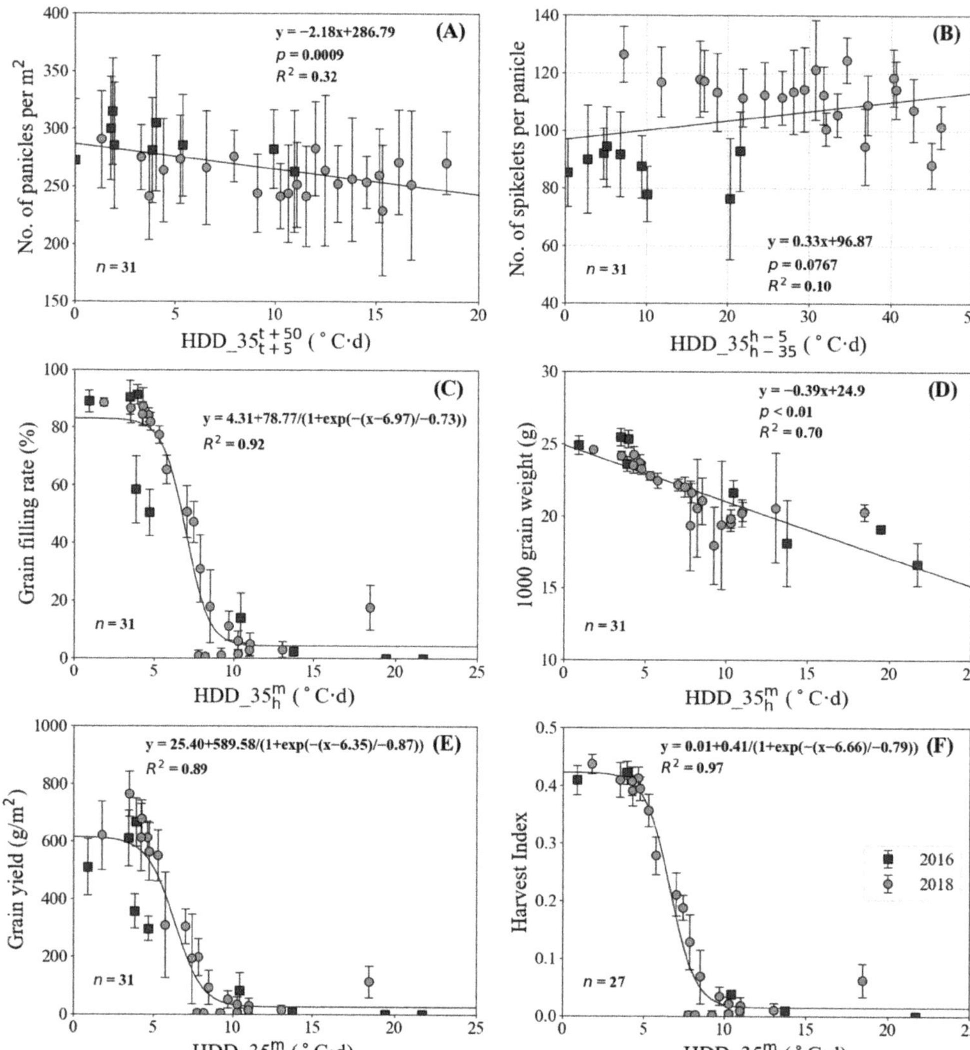

Figure 5. Relationship between the (**A**) no. of panicles per m^2, (**B**) no. of spikelets per panicle, (**C**) grain filling rate, (**D**) 1000-grain weight, (**E**) grain yield, and (**F**) harvest index and the heating degree day (HDD) during specific growing periods: the early growth period ((**A**); from 5 days after to 50 days after transplanting), the late growth period ((**B**); from 35 days to 5 days before heading), and from heading to maturity (**C**–**F**). Error bars represent standard deviation.

Changes in the growth and physiological responses of paddy rice were investigated under two air temperature regimes, AT and AT + 3 °C. During the vegetative growth period, the elevated air temperatures in the TGFC accelerated the expansion of leaves and stems. This is possible to result in increased light capture by the rice canopy, which can cause large amount of canopy photosynthesis. On the other hand, such rapid growth brought the heading date forward. The elevated temperature in the heading stage causes spikelet sterility, which greatly affects the grain filling rate and ultimately largely dominates grain yield.

4. Conclusions

The cumulative amount of photosynthesis can be reduced during the vegetative growth period. Further, the little reduced photosynthetic capacity found at the booting stages under high temperature stress may have a negative effect on grain yield. However, these effects do not seem to be critical on rice grain yield. Because of the sterility-derived delayed aging of leaves, the photosynthetic capacity in the grain filling stage was increased, but it could make up for the loss of grain yield due to spikelet sterility. In 2018, the heatwave event caused some abnormal rate of spikelet sterility even under AT in the TGFC. Given that spikelet sterility increases rapidly with an elevated temperature, the 2018 heatwave can be seen as a critical threshold condition that could damage rice grain yield. Therefore, future prediction studies on rice yield should consider the relationship between high temperature and spikelet sterility rates.

Author Contributions: All authors contributed this paper. D.O., the main author, designed the research, measured and analyzed data, and wrote the manuscript; J.-H.R. and H.J. measured and analyzed data, reviewed the paper, and contributed to the discussion; H.-D.M., H.K., E.J., B.-K.K. and S.C. critically reviewed, edited, and finalized the manuscript; J.C., the corresponding author, designed the research and wrote the manuscript. All authors have read and agreed to the published version of the manuscript.

Funding: This research was funded by the Ministry of Education of the Republic of Korea, grant number NRF-2016R1D1A1B03933218.

Data Availability Statement: Data are contained within the article.

Acknowledgments: The authors would like to thank the Institute for Agricultural Practices Education of Chonnam National University for supporting general field management. The authors would also like to acknowledge the Korean Meteorological Administration for opening the meteorological data portal.

Conflicts of Interest: The authors declare no conflict of interest.

References

1. IRRI. Rice that changed the world: Celebrating 50 years of IR8. In *Rice Today*; International Rice Research Institute: Los Banos, Philippines, 2016.
2. Xu, Y.; Chu, C.; Yao, S. The impact of high-temperature stress on rice: Challenges and solutions. *Crop. J.* **2021**, *9*, 963–976. [CrossRef]
3. IPCC. Climate Change 2021: The Physical Science Basis. In *Contribution of Working Group I to the Sixth Assessment Report of the Intergovernmental Panel on Climate Change*; Masson-Delmotte, V., Zhai, P., Pirani, A., Connors, S.L., Péan, C., Berger, S., Caud, N., Chen, Y., Goldfarb, L., Gomis, M.I., et al., Eds.; Cambridge University Press: Cambridge, UK; New York, NY, USA, 2021.
4. KMA. Korean Climate Change Assessment Report 2020: Summary for Policymakers. In *The Physical Science Basis*; Kim, N.W., Lee, J.H., Cho, G.H., Kim, S.H., Eds.; Korean Meteorology Administration: Seoul, Republic of Korea, 2020; ISBN 978-89-654715-7-9.
5. NIMS. Report of Global Atmosphere Watch 2022: The Greenhouse Gases. In *Atmospheric CO_2 Concentrations*; Lee, S.R., Lee, S.J., Kim, J., Eds.; National Institute of Meteorological Sciences: Seogwipo, Republic of Korea, 2023; pp. 6–13, ISBN 11-1360000-000991-10.4.
6. Satake, T.; Yoshida, S. High temperature-induced sterility in indica rices at flowering. *Jpn. J. Crop Sci.* **1978**, *47*, 6–17. [CrossRef]
7. Jagadish, S.V.K.; Craufurd, P.Q.; Wheeler, T.R. High temperature stress and spikelet fertility in rice (*Oryza sativa* L.). *J. Exp. Bot.* **2007**, *58*, 1627–1635. [CrossRef]

8. Redfern, S.K.; Azzu, N.; Binamira, J.S. Rice in Southeast Asia: Facing risks and vulnerabilities to respond to climate change. *Build Resil. Adapt. Clim. Chang. Agric. Sect.* **2012**, *23*, 295.
9. Jagadish, S.V.K.; Murty, M.V.R.; Quick, W.P. Rice responses to rising temperatures–challenges, perspectives and future directions. *Plant Cell Environ.* **2015**, *38*, 1686–1698. [CrossRef]
10. Gourdji, S.M.; Sibley, A.M.; Lobell, D.B. Global crop exposure to critical high temperatures in the reproductive period: Historical trends and future projections. *Environ. Res. Lett.* **2013**, *8*, 024041. [CrossRef]
11. Krishnan, P.; Ramakrishnan, B.; Reddy, K.R.; Reddy, V.R. High-temperature effects on rice growth, yield, and grain quality. *Adv. Agron.* **2011**, *111*, 87–206.
12. Xiong, D.; Ling, X.; Huang, J.; Peng, S. Meta-analysis and dose-response analysis of high temperature effects on rice yield and quality. *Environ. Exp. Bot.* **2017**, *141*, 1–9. [CrossRef]
13. Zhao, C.; Liu, B.; Piao, S.; Wang, X.; Lobell, D.B.; Huang, Y.; Huang, M.; Yao, Y.; Bassu, S.; Ciais, P.; et al. Temperature increase reduces global yields of major crops in four independent estimates. *Proc. Natl. Acad. Sci. USA* **2017**, *114*, 9326–9331. [CrossRef] [PubMed]
14. Stone, P. The effects of heat stress on cereal yield and quality. In *Crop Responses and Adaptations to Temperature Stress: New Insights and Approaches*; Basra, A.S., Ed.; CRC Press: Boca Raton, FL, USA, 2023; pp. 243–252.
15. Kim, J.; Shon, J.; Lee, C.K.; Yang, W.; Yoon, Y.; Yang, W.H.; Kim, Y.G.; Lee, B.W. Relationship between grain filling duration and leaf senescence of temperate rice under high temperature. *Field Crops Res.* **2011**, *122*, 207–213. [CrossRef]
16. Peng, S.; Huang, J.; Sheehy, J.E.; Laza, R.C.; Visperas, R.M.; Zhong, X.; Centeno, G.S.; Khush, G.S.; Cassman, K.G. Rice yields decline with higher night temperature from global warming. *Proc. Natl. Acad. Sci. USA* **2004**, *101*, 9971–9975. [CrossRef] [PubMed]
17. Way, D.A.; Yamori, W. Thermal acclimation of photosynthesis: On the importance of adjusting our definitions and accounting for thermal acclimation of respiration. *Photosynth. Res.* **2014**, *119*, 89–100. [CrossRef]
18. Rascher, U.; Pieruschka, R. Spatio-temporal variations of photosynthesis: The potential of optical remote sensing to better understand and scale light use efficiency and stresses of plant ecosystems. *Precis. Agric.* **2008**, *9*, 355–366. [CrossRef]
19. Goerner, A.; Reichstein, M.; Rambal, S. Tracking seasonal drought effects on ecosystem light use efficiency with satellite-based PRI in a Mediterranean forest. *Remote Sens. Environ.* **2009**, *113*, 1101–1111. [CrossRef]
20. Zhang, S.; Tao, F.; Zhang, Z. Changes in extreme temperatures and their impacts on rice yields in southern China from 1981 to 2009. *Field Crops Res.* **2016**, *189*, 43–50. [CrossRef]
21. Krishnan, P.; Swain, D.K.; Bhaskar, B.C.; Nayak, S.K.; Dash, R.N. Impact of elevated CO_2 and temperature on rice yield and methods of adaptation as evaluated by crop simulation studies. *Agric. Ecosyst. Environ.* **2007**, *122*, 233–242. [CrossRef]
22. Endo, M.; Tsuchiya, T.; Hamada, K.; Kawamura, S.; Yano, K.; Ohshima, M.; Higashitani, A.; Watanabe, M.; Kawagishi-Kobayashi, M. High temperatures cause male sterility in rice plants with transcriptional alterations during pollen development. *Plant Cell Physiol.* **2009**, *50*, 1911–1922. [CrossRef]
23. Sinclair, T.R.; Horie, T. Leaf nitrogen, photosynthesis, and crop radiation use efficiency: A review. *Crop Sci.* **1989**, *29*, 90–98. [CrossRef]
24. Gulías, J.; Flexas, J.; Mus, M.; Cifre, J.; Lefi, E.; Medrano, H. Relationship between maximum leaf photosynthesis, nitrogen content and specific leaf area in Balearic endemic and non-endemic Mediterranean species. *Ann. Bot.* **2003**, *92*, 215–222. [CrossRef]
25. Uddling, J.; Gelang-Alfredsson, J.; Piikki, K.; Pleijel, H. Evaluating the relationship between leaf chlorophyll concentration and SPAD-502 chlorophyll meter readings. *Photosynth. Res.* **2007**, *91*, 37–46. [CrossRef]
26. Horie, T.; Nakagawa, H.; Nakano, J.; Hamotani, K.; Kim, H.Y. Temperature gradient chambers for research on global environment change. III. A system designed for rice in Kyoto, Japan. *Plant Cell Environ.* **1995**, *18*, 1064–1069. [CrossRef]
27. Kim, Y.S.; Kim, I.S.; Shin, S.Y.; Park, T.H.; Park, H.M.; Kim, Y.H.; Lee, G.S.; Kang, H.G.; Lee, S.H.; Yoon, H.S. Overexpression of dehydroascorbate reductase confers enhanced tolerance to salt stress in rice plants (*Oryza sativa* L. japonica). *J. Agron. Crop Sci.* **2014**, *200*, 444–456. [CrossRef]
28. Yoshida, S. *Fundamentals of Rice Crop Science*; International Rice Research Institute: Los Baños, Philippines, 1981.
29. Wahid, A.; Gelani, S.; Ashraf, M.; Foolad, M.R. Heat tolerance in plants: An overview. *Environ. Exp. Bot.* **2007**, *61*, 199–223. [CrossRef]
30. Morita, S.; Siratsuchi, H.; Takanashi, J.; Fujita, K. Effect of high temperature on ripening in rice plant. Analysis of the effect of high night and high day temperature applied to the panicle in other parts of the plant. *Jpn. J. Crop Sci.* **2004**, *73*, 77–83. [CrossRef]
31. Chae, J.C. *Science of Rice Production*; Hyangmunsa Press: Seoul, Republic of Korea, 2005.
32. Lee, J.H. *Science of Rice Cultivation*; Sunjinmunhwasa Press: Seoul, Republic of Korea, 2001.
33. Chen, J.; Tang, L.; Shi, P.; Yang, B.; Sun, T.; Cao, W.; Zhu, Y. Effects of short-term high temperature on grain quality and starch granules of rice (*Oryza sativa* L.) at post-anthesis stage. *Protoplasma* **2017**, *254*, 935–943. [CrossRef]
34. Shi, P.; Zhu, Y.; Tang, L.; Chen, J.; Sun, T.; Cao, W.; Tian, Y. Differential effects of temperature and duration of heat stress during anthesis and grain filling stages in rice. *Environ. Exp. Bot.* **2016**, *132*, 28–41. [CrossRef]
35. Rouse, J.W.; Haas, R.H.; Schell, J.A.; Deering, D.W. Monitoring vegetation systems in the Great Plains with ERTS. In Proceedings of the Third ERTS Symposium, NASA SP-351, Washington, DC, USA, 10–14 December 1974; pp. 309–317.
36. Barton, C.V.M.; North, P.R.J. Remote sensing of canopy light use efficiency using the photochemical reflectance index: Model and sensitivity analysis. *Remote Sens. Environ.* **2001**, *78*, 264–273. [CrossRef]

37. Gamon, J.A. Reviews and syntheses: Optical sampling of the flux tower footprint. *Biogeosciences* **2015**, *12*, 4509–4523. [CrossRef]
38. Yamori, W. Photosynthetic response to fluctuating environments and photoprotective strategies under abiotic stress. *J. Plant Res.* **2016**, *129*, 379–395. [CrossRef]

agronomy

MDPI

Article

Influence of Interannual Climate Conditions on the Composition of Olive Oil from Centenarian Olive Trees

Leticia Bortoluzzi [1,2,3], Susana Casal [4], Rebeca Cruz [4], António M. Peres [1,2], Paula Baptista [1,2] and Nuno Rodrigues [1,2,*]

[1] Centro de Investigação de Montanha (CIMO), Instituto Politécnico de Bragança, Campus de Santa Apolónia, 5300-253 Bragança, Portugal; leticia.bortoluzzi@ipb.pt (L.B.); peres@ipb.pt (A.M.P.); pbaptista@ipb.pt (P.B.)
[2] Laboratório Associado para a Sustentabilidade e Tecnologia em Regiões de Montanha (SusTEC), Instituto Politécnico de Bragança, Campus de Santa Apolónia, 5300-253 Bragança, Portugal
[3] Departamento de Ingeniería Agrária, Universidad de Léon, Av. Portugal, n° 41, 24071 Léon, Spain
[4] LAQV/REQUIMTE, Laboratory of Bromatology and Hydrology, Faculty of Pharmacy, University of Porto, Rua de Jorge Viterbo Ferreira, 228, 4050-313 Porto, Portugal; sucasal@ff.up.pt (S.C.); rcruz@ff.up.pt (R.C.)
* Correspondence: nunorodrigues@ipb.pt; Tel.: +351-273-303-277

Abstract: In recent years, occurrences of heat waves and drought have become increasingly frequent, highlighting the undeniable impact of climate change. The rise in temperatures and decline in rainfall have had severe repercussions on olive trees' behavior and olive oil production. This study aims to evaluate the effects of two-year climate variations on olive oils from centenarian olive trees situated in the Côa Valley region of Northern Portugal. A selection of 25 centenarian plants was made, and the climate influence on fatty acid content, tocopherols, individual phenols, oxidative stability, and antioxidant activity was assessed over two consecutive years. During the second year of the study, a significant variation (p-value < 0.05) in the proportion of palmitic acid was observed, which increased from 12.9% to 13.6%. Conversely, stearic and arachidic acids exhibited a decrease from 2.7% to 2.3% and from 0.37% to 0.35%, respectively. Analysis of the oils revealed a noteworthy difference (p-value < 0.05) in the concentration of β-tocopherol. The concentration of oils derived from hydroxytyrosol and tyrosol significantly decreases (p-value < 0.005) during the second year. Additionally, significant differences (p-value < 0.005) were observed in the total phenol content and the percentage of ABTS inhibition, both of which decreased in the second year. These findings reinforce the notion that climatic conditions play a key role in shaping the composition of olive oils.

Keywords: climate change; olive oil; composition; interannual variations

1. Introduction

In recent years, agriculture has been significantly affected by climate change, resulting in negative consequences for crop yield, product composition, and overall food security [1–4]. These global changes have had a noticeable impact on various regions, including the Mediterranean, where frequent occurrences of higher temperatures, erratic precipitation patterns, and extreme winter and summer droughts have been observed [5,6]. These events have raised concerns about the ability of traditional crops, particularly the olive tree (*Olea europaea* L.), which thrives in both hot and dry summers and moderately cold and wet winters, to cope with the adverse effects of climate change [7–11].

Numerous studies have investigated the influence of climate conditions on the chemical properties of olive oils [12–17]. The composition of olive oil primarily consists of triglycerides, other glycerol esters, and some free fatty acids, accounting for approximately 99% of the oil's content. It also contains high amounts of phenols [18,19], sterols, fat-soluble vitamins, chlorophylls and carotenoids, which are synthesized through various biosynthetic pathways, triglycerides [20], sterols [21], tocopherols, and fatty acids [20,22,23], among other compounds. The concentrations and types of compounds present in olive oil are influenced by genetic factors, as well as geographic and climatic conditions, cultivation practices, tree age, olive maturation at harvest, extraction methods, and storage [24,25]. Generally, olive oils obtained from olives grown in colder regions tend to have higher levels of antioxidants and monounsaturated fatty acids (MUFA, such as oleic acid and the oleic/linoleic acid ratio), while oils produced in warmer regions are richer in saturated (SFA) and polyunsaturated acids (PUFA) like palmitic and linolenic acids, respectively [26–29]. The relationship between rainfall and the inhibition of phenol and tocopherol synthesis has also been explored, revealing an increase in their contents under stressful conditions. There have been several works that have studied the oils obtained from centenary olive trees in order to enhance this heritage [30–33]. Furthermore, several authors have reported that centenarian olive trees produce fruits that originate oils with exceptional nutritional and sensory characteristics [29,33,34].

Considering the constant climatological situations, it is crucial to assess the production performance of different olive tree cultivars under diverse climate conditions. Therefore, the objective of this study is to evaluate the effects of climate conditions in the Vale do Côa region on the composition of olive oil obtained from centenarian olive trees. The study spanned two consecutive years and recorded changes in vital bioactive compounds closely linked to the oils' nutritional quality and oxidative stability. The observed variations in the examined physicochemical parameters can be primarily attributed to climatic conditions, as each olive oil sample was extracted using the same pilot scale extraction facility and processing conditions (e.g., temperature and malaxation time). Additionally, the olives were manually harvested from a selection of 25 centenarian trees at the same maturation stage.

2. Materials and Methods

2.1. Selection, Harvest, and Extraction

A total of twenty-five specimens of centenarian olive trees were selected and georeferenced in eight dispersed plots in the Côa Valley region (Portugal). The selection was based on several parameters that indicate its advanced age, such as structural integrity, general appearance, and information provided by the producers. Each plant was selected, marked, and georeferenced which allowed their harvest in the in the different years of study. The plants integrate the traditional olive groves of the region, which are rainfed and follow similar agronomic practices and low vegetative development. In general, plants were only slightly pruned every year, with slight impacts on production, and no major annual fluctuations were observed. The selected plants are uncharacterized centenarian specimens, and no information is available about the variety. The soil is a schist leptosol, with a very thin and poor layer, with an approximate slope of 20 to 30%; the plants are planted at 10 × 10 m, with 100 plants per hectare. In the years under study, no phytosanitary treatment was carried out, the soil was mobilized in spring and autumn to control weeds, and in April, fertilization was applied using a compound fertilizer (7:14:14), consisting of 7% N (5% ammoniacal N and 2% urea N), 14% P_2O_5 (11% water soluble), and 14% K_2O. The plants were analyzed over two consecutive seasons (2020 and 2021). For a comprehensive analysis, approximately 4–5 kg of olives per specimen were collected to ensure a representative sample. The harvest period occurred from 25th to 31st of October in 2020, and from 2nd to 10th of November in 2021, approximately 150 to 160 days after the flowering. The fruits were collected when they reached a maturity index between M2 and M3, as per the guidelines set by the International Olive Council [35]. These indices are characterized by fruit skins with red spots covering less than half (M2) or more than half (M3) of the olive. Within 24 h of harvesting, oils were extracted using an Abencor pilot plant (Comercial Abengoa S.A., Seville, Spain).

The milling process employed the MC2 MM100 hammer mill (Seville, Spain), equipped with a 5.5 mm diameter sieve and a 1.5 kW single-phase motor. Subsequently, the resulting olive paste underwent temperature-controlled malaxation in a MC2 Thermo-Mixer TB-100 (Seville, Spain). This machine offered eight work stations and eight mixing flasks, each with temperature control and individual blades to ensure thorough mixing. Finally, the olive paste underwent centrifugation using the MC2 Centrifugal Machine CF-100 (Seville, Spain). This machine featured a stainless-steel drum rotating at 3500 rpm, driven by a 1.5 kW three-phase motor, and included an automatic timer. To eliminate solid particles and residual water, the samples were filtered using Whatman no. 4 paper over anhydrous sodium sulfate. Finally, the extracted oil was stored in 125 mL dark glass bottles and kept in a dark place at room temperature situated between 20 and 25 °C for subsequent analysis.

2.2. Meteriological Data

In order to assess the impact of yearly climate conditions on the quality of olive oil, parameters like the temperature (minimum—TMIN, average—TAVE, and maximum—TMAX), and precipitation for the two harvest years (2020 and 2021) were collected for the months of fruit development and fruit maturation. The meteorological data used in this study were obtained from the IPMA—Portuguese Institute for the Sea and the Atmosphere. These data were collected from weather stations located closest to the study areas, specifically the Moncorvo station (Station No.: 0654; Latitude: 41.1899°; Longitude: −7.0185°; Altitude: 539 m). The collected data covers the months from June to November (2020 and 2021), as shown in Figure 1.

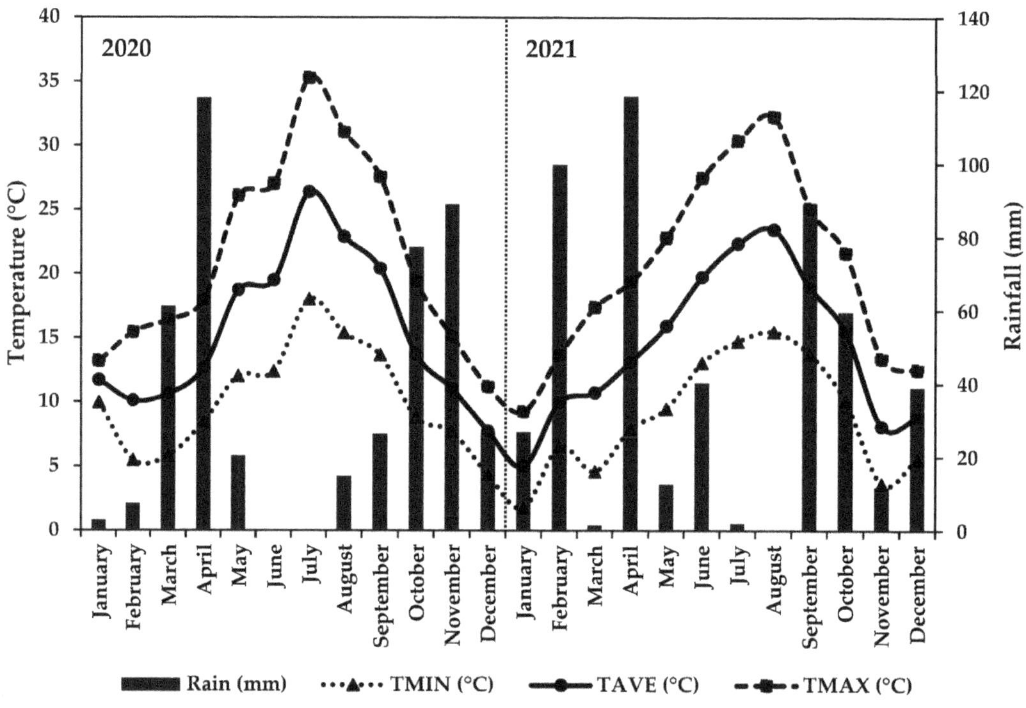

Figure 1. Meteorological data of average rainfall values and the minimum, average, and maximum temperatures (TMIN, TAVE, and TMAX, respectively) collected at the Moncorvo station for the years of 2020 and 2021.

2.3. Fatty Acid Profile

The relative percentages of fatty acid levels were determined using chromatography, following the guidelines outlined in document COI/T.20/Doc. No 28/Rev. 3, dated November 2022 [36]. The analysis was conducted using a Chrompack CP 9001 (Burladingen, Germany) chromatograph equipped with a split-splitless injector set at 230 °C, an FID detector at 250 °C, and an automatic sampler Chrompack CP-9050 (Burladingen, Germany).

For the analysis, a 1 µL injection (at a 1:50 proportion) was made, and separation was achieved using a fused silica capillary column (Select FAME: 50 m × 0.25 mm i.d. Varian, Palo Alto, CA, USA) manufactured by Agilent. Helium was employed as the carrier gas, with an internal pressure maintained at 140 kPa. The determination of relative percentages was carried out by internally normalizing the chromatographic peak area between myristic and lignoceric methyl esters. To identify and quantify the peaks, a standard mixture of certified fatty acid methyl esters obtained from Sigma (Barcelona, Spain) was used.

2.4. Tocopherol Contents

The determination of α-, β-, and γ-tocopherol levels was conducted utilizing high-performance liquid chromatography (HPLC) following the methodology outlined in ISO 9936:2016 [37], with certain modifications. To begin, 50 mg of filtered and diluted olive oil in n-hexane (Sigma-Aldrich, Germany) was combined with the internal standard tocol (2-methyl-2-(4,8,12-trimethyltridecyl) chroman-6-ol from Matreya Inc. (Pleasant Gap, PA, USA). Subsequently, the mixture underwent centrifugation at room temperature, with a speed of 13,000 rpm for 5 min, and the resulting supernatant was collected for HPLC analysis. The HPLC equipment employed a normal phase silica column (Supelcosil^{TM} LC-SI; dimensions: 7.5 cm × 3 mm; particle size: 3 µm, from Supelco, Bellefonte, PA, USA) maintained at a temperature of 23 °C. Elution was performed using a mobile phase consisting of 1,4-dioxane (Sigma-Aldrich—Allentown, PA, USA) in n-hexane (2.5%, *v*/*v*) at a flow rate of 0.75 mL/min. Analysis of the data was carried out utilizing the ChromNAV Control Center program—JASCO Chromatography Data Station (Jasco, Tokyo, Japan), and the identification of compounds was achieved by comparing their retention times with authentic standards obtained from Sigma (Barcelona, Spain). The final results were expressed as milligrams per kilogram (mg/kg) of olive oil and the content of vitamin E was calculated as the sum of concentrations of α-, β-, and γ-tocopherols.

2.5. Olive Oils Total Content of Hydroxytyrosol and Tyrosol Derivatives after Acid Hydrolysis of Secoiridoids

The phenolic profile of oils obtained from centenary olive trees in different years was evaluated following the methodology described by Marx et al. [38]. For this, an HPLC-DAD system from Jasco (Tokyo, Japan) was used with a data transmitter (LC–NetII/ADC), two integrated pumps (PU–4180), an automatic sampler (AS–4050), a column oven (ECOM Eco2000, Zlin, Czech Republic), and the DAD (MD–4010). The separation of compounds was performed on a C18 reversed phase column (Kinetex C18; particle size: 2.6 µm; pore size: 100 Å; LC length: 100 mm; internal diameter: 3.00 mm, Phenomenex, Madrid, Spain), at 35 °C, using a mobile phase composed of water and acetonitrile, both with 0.1% formic acid, at a flow rate of 0.8 mL/min for 20 min. All samples were injected in duplicates. The total contents of hydroxytyrosol or tyrosol after hydrolysis were expressed as the individual sum in mg of hydroxytyrosol or tyrosol, respectively, per kg of oil. Calibration curves of hydroxytyrosol and tyrosol (R^2 = 0.9992 and 0.9990, respectively) were prepared in methanol/water (80:20, *v*/*v*) in a concentration range of 0.00031 to 0.0160 mg/mL.

2.6. Oxidative Stability, Total Reducing Capacity, and Antioxidant Activity

The Rancimat 743 equipment (Metrohm CH, Herisau, Switzerland) was used to assess the oxidative stability (OS) of the samples. Three grams of olive oil were heated at a temperature of 120.0 ± 1.6 °C, while clean, filtered, and dry air was supplied at a flow rate of 20 L/h. Volatile compounds released during heating were collected in water, and the water conductivity (µS/cm) was continuously monitored. The point at which the conductivity curve exhibited an inflection corresponded with the OS value (in h).

To determine the total phenolic compounds, extracts of the oils were obtained through methanol–water microextraction. The evaluation was performed using a UV-Vis spectrophotometer (VIS/UV-1280 Shimadzu) at 765 nm, as previously described [39]. For the microextraction, 1 mL of methanol–water and 0.5 g of oil were combined in an Eppendorf tube, vigorously mixed on a vortex for 1 min, and then centrifuged for 5 min at 13,200 rpm. The resulting supernatant was collected and transferred to a 10 mL vial. This centrifugation process was repeated two more times, with an additional 1 mL of methanol–water added before each repetition. The supernatants were collected and combined in the same vial, ensuring a final volume of 5 mL with methanol–water. Next, a solution was prepared by adding 1500 µL of ultrapure water, 0.1 µL of Folin–Ciocalteu reagent, and 100 µL of the extract. The mixture was vortexed and allowed to stand in the dark for 3 min. Afterward, 300 µL of 20% (w/v) sodium carbonate solution was added, vortexed again, and kept in the dark for 1 h at room temperature (20–22 °C). Finally, the absorbance of the sample was measured at 765 nm, and the quantification was made using a calibration curve correlating the absorbance with the gallic acid concentration of a methanolic solution ($R^2 \geq 0.9999$). The results were expressed as milligrams of gallic acid equivalents [GAE] per kilogram of oil.

The antioxidant activity of the samples was evaluated by measuring their total reducing power and ABTS (2,2′-azinobis (3-ethylbenzothiazoline-6-sulfonic acid)) scavenging ability. The total reducing power was determined by assessing the elimination of the DPPH (2,2-diphenyl-1-picrylhydrazyl) radical using a spectrophotometer (UV-VIS/UV-1280 Shimadzu spectrophotometer) at 517 nm and 20 °C. For this analysis, sample extraction was conducted using methanol–water microextraction. A DPPH control solution was prepared by mixing 0.1 mL of methanol with 3.9 mL of DPPH (0.06 mM) and allowing it to stand in the dark for 30 min before measuring the absorbance. For the olive oil samples, 0.1 mL of the methanolic extract was added to 3.9 mL of DPPH (0.06 mM), and the solution was left in the dark for 30 min before measuring the absorbance. The results were expressed as the percentage reduction in DPPH activity. The ability of the samples to inhibit the ABTS radical, in comparison to a standard antioxidant reference (trolox), was determined following the methodology previously described [40]. To generate the ABTS radical, 440 µL of $K_2S_2O_8$ (140 mM) was added to 25 mL of ABTS (7 mM), and the solution was kept in the dark for 12–16 h, at room temperature. After the specified time, the solution was diluted in ethanol until the absorbance, at 734 nm, reached 0.70 ± 0.02. Finally, 100 µL of olive oil was added to 2 mL of the adjusted ABTS solution, and the average absorbance was measured at 734 nm. The results were expressed as the percentage of inhibition.

2.7. Statistical Analysis

The statistical differences in the composition and characteristics of the oils between two consecutive harvest years (2020 and 2021) were assessed using the t-Student test. The application of the Welch's correction depended on whether equal or unequal variances between the groups could be assumed, as determined by the F-test. PCA was employed as an unsupervised pattern recognition tool to determine if the evaluated composition and characteristics of the olive oils (such as OS, antioxidant activity, tocopherols and phenolics contents, and fatty acids profile) could differentiate the oils based on the crop year. Additionally, LDA was used, in conjunction with the SA algorithm to identify the most influential non-redundant discriminative parameters. These parameters were selected based on the best correct classification performances (i.e., sensitivities) achieved during training and LOO-CV internal validation. The statistical analysis was conducted using the R statistical program, specifically the open-source packages available in RStudio version 2021.09.0, also known as the "Ghost Orchid" Release (077589bc, 20 September 2021); the significance level was set at 5%.

3. Results and Discussion

3.1. Meteorological Data

The meteorological data uncovered variations in both the volume and temporal distribution of rainfall and temperature across the time-periods considered for both years (Figure 1). Specifically, in 2020, the distribution of rainfall followed a more consistent pattern, characterized by a dry period in June and a gradual increase in precipitation starting from August and culminating in a peak in November (88.9 mm). Conversely, 2021 exhibited significant rainfall levels in June (40.30 mm), followed by a dry phase and a subsequent surge in precipitation, with the highest amounts observed in September (89.30 mm).

Additionally, differences in temperature were observed between the two years. In 2020, the warmest period occurred in July, registering maximum temperatures of 35.9 °C, followed by a gradual decline in subsequent months, reaching the lowest minimum in November (7 °C). Conversely, in 2021, temperatures remained elevated, with the highest value recorded in August (32.26 °C) and notable minimums in November (3.6 °C).

By examining the maximum temperatures within the time periods, it became evident that the peak temperature in the second year was slightly delayed compared to the first year. In 2020, July stood out as the hottest month, while in 2021, the maximum temperatures were observed in August. Overall, temperatures were generally higher in 2020.

3.2. Fatty Acid Composition

This study aimed to investigate the impact of climatic conditions on the concentration of fatty acids in olive oil. To achieve this, olive oil samples were obtained from olive trees aged over 100 years during two consecutive harvest seasons, namely 2020 and 2021. All samples met the concentration standards set by the International Olive Council [36]. The fatty acid composition, expressed as relative percentages, exhibited typical proportions of major fatty acids found in olive oils. The most abundant fatty acid was oleic acid (59–79%), followed by palmitic acid (9–16%) and linoleic acid (2–18%). On the other hand, stearic acid (1–3%), palmitoleic acid (0.5–2%), linolenic acid (0.6–1.2%), arachidic acid (0.2–0.4%), and eicosenoic acid (0.2–0.3%) were present in the lowest concentrations (Table 1).

The results of the study indicated that the composition of the oils was influenced by the year of harvest. Specifically, a significant variation (p-value < 0.05) was observed in the proportion of palmitic acid, which increased from 12.9% to 13.6%. Individually, it was observed that 44% of the studied plants increased their palmitic acid content in the second year of harvest. It was found that 8% of the plants had a lower content of this compound, and 48% of the plants did not observe any effects during the year (Table S1 of the Supplementary Materials). In contrast, stearic acid concentrations decreased by 2.7–2.3%, i.e., 84% of the studied plants dropped their content in the second year. The same was observed for arachidonic acid, which decreased by 0.37–0.35%; that is, 44% of the plants under study (Table S1). No significant variations were observed in the remaining fatty acids or proportion of total saturated (SFA), monounsaturated (MUFA), and polyunsaturated (PUFA) fatty acids (Table 1). These findings are consistent with other studies conducted in different regions [15,31,41]. Rey-Giménez et al. [15] reported similar observations, noting that warmer years, particularly during the fruit development period, led to higher concentrations of palmitic acid, reaching up to 15.2%. In colder climates, concentrations were found to be around 13.5%.

These studies, along with others, suggest that such conditions may increase the levels of SFA and PUFA, while decreasing MUFA [28]. In extreme temperatures and arid climates, such as in desert regions, palmitic acid levels may exceed the permissible 20% threshold, while oleic acid values may fall below 50% [26]. The region's climatic data were analyzed over two years, revealing consistent patterns in temperature and precipitation during 2020. Rainfall began in August with a total of 14.8 mm and reached its peak in November at 88.9 mm. Concurrently, temperatures gradually decreased from 35 °C in July to 19.3 °C in October and 15.1 °C in November (Figure 1). In contrast, 2021 exhibited irregularities, with rainfall occurring early in June at 40.3 mm and reaching its maximum level in September at 89.9 mm. The rainfall values for July, August, and September were abnormal for this time, resulting from a storm that caused heavy rains. Excessive water from storms was not infiltrated into the soil, depriving the crops of the benefits of precipitation during this period. This phenomenon of heavy rainfall induced by storms is a consequence of climate change experienced in recent years. In August, high temperatures and dry conditions were observed, with a peak temperature of 32.2 °C. Compared to the previous year, October and November saw a temperature increase of 2 °C, measuring 21.6 °C and 17.6 °C, respectively (Figure 1). Additionally, the 2 °C temperature rise and significant decrease in precipitation during the harvest period in October and November could have affected the synthesis of palmitic acid. The reduction in the concentration of stearic acid in 2021 might be attributed to its association with the elongation of palmitic acid [42]. Conversely, the concentrations of arachidonic acid, which were already low, were easily influenced by the erratic climatic conditions experienced throughout the year. Some authors state that genetic factors determine the composition of fatty acids in olive oil, while others state that climatic conditions also have a significant impact. Therefore, it is essential to study the oil content of centenarian olive trees in different regions with varying climates to understand their resilience and unique composition. For example, a study conducted by Rodrigues et al. [33] investigated centenarian olive trees in Portugal during a period from August to September, characterized by intense droughts and less than 30 mm of precipitation. They observed significant variations in the concentration of palmitic acid in the olive oils, depending on the specific olive cultivar. The cv. Redondal exhibited an increase in palmitic acid concentration of 12–13%, while the cv. Madural showed a decrease by 11–12%. The levels of stearic acid and arachidic acid found in this study were consistent with previous findings, ranging from 1–3% and 0.3–0.5%, respectively. Similarly, Hijawi [41] examined centenarian olive trees in Palestine and discovered fluctuations in oils' fatty acid concentration based on the year. Palmitic acid levels ranged from 11.9–14.6% in one year and 10.9–14.9% in another, while stearic acid concentrations ranged from 2.2–4.5% and 3.7–4.4%, respectively. The levels of arachidic acid were 0.3–0.7% in one year and 0.6–0.9% in another, indicating high levels of this compound.

The fruit growth phase occurs from June to August, followed by oil formation and fruit maturation from September to November. In 2021, unusual rainfall patterns were observed during these months. While the normal fruit formation period experienced 40.3 mm of rainfall in June, the maturation months saw a shift in rainfall to September (89.9 mm), October (59.6 mm), and November (11.8 mm). These rainfall values in August and September deviate from the typical pattern and were the result of two stormy days with heavy rains. It is known that droughts during the maturation phase, which usually starts in September, negatively affect the fat content of the fruits [28,43]. The increased rainfall in October and November, combined with a 2 °C temperature rise compared to 2020, may have influenced the higher level of palmitic acid. However, due to the high intensity of rainfall, the water ran off without infiltrating the soil, resulting in the plants being unable to utilize this water effectively or create a water reservoir for their needs. Temperature conditions between flowering and harvesting also impact fruit characteristics [44]. Elevated temperatures during the maturation period generally lead to higher levels of SFA and PUFA but lower levels of MUFA [26,45,46]. However, in the present study, for these parameters, no significant differences were observed between years. However, a more detailed analysis of the results, genotype by genotype, can be found in the tables included in the Supplementary Materials. These tables show that there are some significant inter-annual differences on the oils' compositions, being the observed changes depending on the genotype, i.e., increasing/decreasing trends were observed and differ from genotype to genotype. However, for the main compounds, similar trends (increasing/decreasing) for the majority of the studied genotypes were observed. For example, regarding the fatty acid's relative abundance, for SFA, MUFA, and PUFA, the year had a significant effect in the related contents found in 20%, 48%, and 76% of the oils/genotypes. Concerning tocopherols, the year showed a significant effect on the vitamin E content of the oils extracted in 68% of the studied genotypes, with an increasing content observed in 60% of those genotypes. Also, although the annual variation of the total content of hydroxytyrosol + tyrosol after acids hydrolysis was genotype-dependent, a decreasing trend was observed in 64% of the oils extracted from each single genotype. On the other hand, the oxidative stability and the total phenolic content showed an increasing trend with the year in the oils extracted from 60%

and 88% of the studied genotypes, but the DPPH and ABTS antioxidant capacities showed a decreasing trend for 88% and 68% of the oils extracted from the studied genotypes.

Table 1. Mean ± standard deviation (%) of the relative abundance of fatty acids determined in olive oils extracted from olives of centenarian olive trees located in the Côa Valley, Portugal, for two consecutive crop years (2020 and 2021), and the corresponding percentage change with the crop year (Δ, in %).

Fatty Acid Relative Abundance (%)	Crop Year		*p*-Value	Δ (%)
	2020	2021		
Palmitic acid ($C_{16:0}$)	12.9 ± 1.6 [b] (9.7–17.3)	13.6 ± 1.2 [a] (10.7–16.4)	0.0114 *	+5.4
Palmitoleic acid ($C_{16:1}$)	0.9 ± 0.3 [a] (0.5–2.0)	0.9 ± 0.3 [a] (0.6–1.4)	0.9339 **	+1.1
Stearic acid ($C_{18:0}$)	2.7 ± 0.4 [a] (1.7–3.6)	2.3 ± 0.3 [b] (1.6–2.9)	<0.0001 *	−14.8
Oleic acid ($C_{18:1}$)	71.6 ± 6.1 [a] (59.4–79.1)	72.3 ± 5.3 [a] (64.3–79.8)	0.4910 **	+1.0
Linoleic acid ($C_{18:2}$)	9.8 ± 5.1 [a] (3.4–18.6)	8.8 ± 4.6 [a] (2.9–16.4)	0.3221 **	−10.2
Linolenic acid ($C_{18:3}$)	0.9 ± 0.2 [a] (0.7–1.2)	0.9 ± 0.2 [a] (0.6–1.2)	0.9250 **	0.0
Arachidic acid ($C_{20:0}$)	0.4 ± 0.0 [a] (0.3–0.5)	0.4 ± 0.0 [b] (0.3–0.5)	0.0138 *	−5.4
Eicosenoic acid ($C_{20:1}$)	0.3 ± 0.0 [a] (0.2–0.4)	0.3 ± 0.0 [a] (0.2–0.4)	0.3072 **	+3.6
SFA	16.3 ± 4.0 [a] (13.0–20.5)	16.4 ± 4.0 [a] (13.3–19.5)	0.3958 **	+0.6
MUFA	73.0 ± 6.1 [a] (61.1–80.6)	74.4 ± 5.5 [a] (67.2–81.4)	0.2195 **	+2.0
PUFA	10.7 ± 5.2 [a] (4.1–19.5)	9.4 ± 4.6 [a] (3.7–16.0)	0.2041 **	−11.9

* *t*-Student test with Welch's correction (unequal variances according to the F-test). ** *t*-Student test without correction (equal variances according to the F-test). *p*-values for the *t*-Student test. Different letters in the same row show statistically differences from the given mean ($p < 0.05$).

3.3. *Tocopherols Content*

The total amounts of vitamin E are variable, clearly depending on several factors, such as the cultivar, fruit maturation, environmental conditions, agronomic factors (for example, irrigation, fertilization, incidence of pests and diseases), and the conditions used during production of oil extraction, storage, etc. [13]. The analysis of the olive oils revealed the presence of α-, β-, and γ- vitamers, which were quantified. Among these vitamers, α-tocopherol was the predominant form in olive oils, constituting over 90% of the total amount as expected in olive oil [10]. Analysis of data from 2020 and 2021 (Table 2) showed that α-tocopherol levels ranged from 121 to 362 mg/kg in 2020 and from 134 to 323 mg/kg in 2021, with no significant differences between the two crop years. However, a significant difference related to the crop year (*p*-value < 0.05) was observed for β-tocopherol, which increased from 1.3 to 2.0 mg/kg; that is, 68% of the plants under study increased the content of this compound in the second year (Table S2). Although no significant differences in α-tocopherol were observed in the present study, there are several authors who report that there are certain factors that affect the amount of this vitamer, namely the maturation of the fruit influences the amount, decreasing during ripening [17], the cultivar [13], low temperatures, and precipitation also influence the amount of vitamer (less precipitation, higher content) [47].

Table 2. Tocopherols (α-, β-, and γ-) and total vitamin E mass contents (mean ± standard deviation, mg/kg of olive oil) for olive oils extracted from olives of centenarian olive trees located in the Côa Valley (Portugal), for two consecutive crop years (2020 and 2021), and respective variation with crop-year (Δ, in %).

Tocopherol Concentration (mg/kg)	Crop Year		*p*-Value	Δ (%)
	2020	2021		
α-tocopherol	254.3 ± 70.1 [a] (121.0–362.3)	252.3 ± 63.1 [a] (134.0–323.3)	0.8820 **	−0.8
β-tocopherol	1.3 ± 0.6 [b] (0.1–2.1)	2.0 ± 0.6 [a] (1.0–2.9)	<0.0001 *	+54.2
γ-tocopherol	4.5 ± 3.3 [a] (0.4–14.6)	5.0 ± 2.2 [a] (2.0–9.6)	0.3650 **	+11.4
Vitamin E	260.1 ± 72.5 [a] (122.7–369.3)	259.3 ± 64.8 [a] (138.0–342.8)	0.9552 **	−0.3

* *t*-Student test with Welch's correction (unequal variances according to the F-test). ** *t*-Student test without correction (equal variances according to the F-test). *p*-values for the *t*-Student test. Different letters in the same row show statistically differences from the given mean ($p < 0.05$).

The average concentration of vitamin E in the oils was 260 mg/kg, with a maximum value of 369 mg/kg (Table 2). These findings align with the existing literature on olive oils, where vitamin E concentrations typically range from 80 to 500 mg/kg [17], and β-tocopherol concentrations typically vary from 1 to 2 mg/kg, consistent with previous studies [34].

Examining climate data for the two crop years, special attention is given to 2021, characterized by higher temperatures in compared to July (Figure 1). These rainfall values are unusually high for this period, which typically experiences no rainfall. These fluctuations during the months of oil accumulation and fruit ripening may have contributed to a significant decrease in β-tocopherol levels. It is known that β-tocopherol declines during ripening and can exhibit significant variations due to external conditions, particularly considering its naturally low concentration [48]. Previous studies have emphasized the influence of climate on tocopherol composition, suggesting that regions with lower rainfall levels tend to have higher levels of α-tocopherol, due to the plant's synthesis of this compound in response to water stress. Additionally, low temperatures during the harvest period can negatively impact tocopherol concentrations [25].

In non-Mediterranean climates, olive oils have been found to have higher concentrations of β-tocopherol compared to the present study, ranging from 2 to 6 mg/kg, while alpha-tocopherol concentrations were below 250 mg/kg. These variations were attributed to different climatic and extraction conditions [45,49]. In locations characterized by high temperatures, such as olive groves in Egypt, the α-tocopherol concentrations in oils can reach nearly 800 mg/kg [50].

In previous studies, the concentration of vitamin E in olive oils extracted from centenarian olive trees in Portugal exhibited a range of 140 to 354 mg/kg, depending on the specific cultivar and harvest year. Notably, the cvs. Lentisca and Madural closely matched the values obtained in this study, averaging at 224–335 mg/kg [29].

Although tocopherols are present in olive oils as minor compounds, they play a crucial role in protecting the oil against autoxidation and possess biological activity when consumed in the human diet. Besides their stability-enhancing function, it is vital to explore the potential health benefits associated with tocopherols to enhance the value of olive oils derived from centenarian trees. Since the composition of tocopherols is influenced by various factors, documenting the variations in tocopherol composition among these specimens under different climatic conditions is of utmost importance.

3.4. Olive Oils Total Content of Hydroxytyrosol and Tyrosol Derivatives

This study aimed to analyze the effect of the year on the levels of two families of phenolic acids, hydroxytyrosol- and tyrosol-based in two different years. The research revealed a noteworthy decline (p-value < 0.05) in the concentrations of these compounds from the first year to the second year. Hydroxytyrosol compounds' values ranged from 290 to 220 mg/kg, while tyrosol ones exhibited a decrease from 170 to 135 mg/kg. The concentration range for these compounds was found to be 130–400 mg/kg and 36.5–427 mg/kg, respectively (Table 3). It was observed that this decrease occurred in 68% of the plants for hydroxytyrosol and 56% of the plants for tyrosol (Table S3).

Table 3. Hydroxytyrosol, tyrosol, and total hydroxytyrosol + tyrosol based compounds (mean ± standard deviation, mg/kg of olive oil) for olive oils extracted from olives of centenarian olive trees located in the Côa Valley (Portugal), for two consecutive crop years (2020 and 2021), and respective variations with crop-year (Δ in %).

Phenolic Acids Concentration (mg/kg)	Crop Year		p-Value	Δ (%)
	2020	2021		
Hydroxytyrosol	290.2 ± 68.0 ^{a} (130.5–400.1)	220.50 ± 62.22 ^{b} (130.80–380.30)	<0.0001 *	−24.0
Tyrosol	170.3 ± 92.0 ^{a} (36.5–427.3)	135.93 ± 59.19 ^{b} (41.80–338.40)	0.0169 *	−20.2
Hydroxytyrosol + Tyrosol	460.6 ± 142.4 ^{a} (167.4–772.7)	356.4 ± 98.7 ^{b} (179.5–532.3)	<0.0001 *	−3.3

* t-Student test with Welch's correction (unequal variances according to the F-test). p-values for the t-Student test. Different letters in the same row show statistically differences from the given mean (p < 0.05).

Jiménez-Herrera et al. [51] conducted an investigation focusing on the impact of water stress, specifically drought, on the levels of phenolic acids in traditional olive groves. The motivation behind this study stemmed from water scarcity issues in Mediterranean countries, which led to losses in the olive industry. The researchers observed that under drought conditions and high temperatures, particularly during fruit growth, the concentration of hydroxytyrosol and tyrosol decreased, while the total phenol content benefited from these conditions. In the present study, the climatic conditions during fruit development were analyzed. It was noted that in 2020, the initial month's experienced dryness, with rainfall, gradually increasing until harvest. Conversely, in 2021, droughts occurred in June and August, accompanied by high temperatures during harvesting months and high precipitation in October (59.6 mm) and November (11.8 mm) (Figure 1). These conditions likely contributed to a higher degradation of these compounds rather than their synthesis, potentially explaining the negative impact on phenolic acid concentrations in 2021.

Anticipating the effects of climate change on water availability, Faghim et al. [52] investigated the response of olive trees in Tunisian olive groves, located in arid and semi-arid regions, under different irrigation conditions. They found lower levels of hydroxytyrosol and tyrosol in oils obtained from non-irrigated olive groves compared to oils from irrigated olive groves. The difference was statistically significant (p-value < 0.05), with a reduction of the contents for both compounds being observed. Specifically, hydroxytyrosol levels decreased from 290.2 to 220.5 mg/kg, while tyrosol levels decreased from 170.3 to 135.9 mg/kg. Among the olive tree varieties, cv. Chemlali exhibited higher levels of tyrosol compared to hydroxytyrosol, aligning with findings reported by other researchers [45]. However, Uslu et al. [53], who evaluated the influence of irrigation over a two-year period, found high concentrations of these compounds in Mediterranean varieties. They observed a wide range of hydroxytyrosol levels, ranging from 307.3 to 1449.1 mg/kg. Additionally, they noticed that irrigated varieties exhibited the greatest reduction in tyrosol levels from one year to the next. The demand for high-quality olive oil with distinct phenolic compound concentrations is increasing [54]. The presence of phenolic acids in olive oil can vary due to various factors, including the genotype, ripening stage of the fruit, agroclimatic conditions, production year, and geographical origin [55]. Given the significant variation in hydroxytyrosol and tyrosol concentrations in olive oils, it is crucial to characterize them. Furthermore, there is a possibility of discovering previously unknown centenarian olive trees with favorable phenolic acid values and observing their potential adaptation to climate change.

3.5. Oxidative Stability, Antioxidant Activity, and Total Reducing Capacity

The antioxidant capacity of the olive oils was evaluated by measuring oxidative stability, total phenolic compounds, and antioxidant activity using the DPPH and ABTS methods. Significant differences (p-value < 0.005) were observed only for total phenols and the ABTS inhibition percentage with the crop year. Over the period from 2020 to 2021, the average concentration of total phenols increased from 375 to 569 mg/kg and reached a maximum value of 965 mg/kg. It was found that this increase occurred in 72% of the plants under study (Table S4). These findings align with previous studies that reported levels ranging from 50 to 1000 mg/kg [56]. The maximum value of ABTS inhibition was 54.7%, but there was a decline from 40.8% to 35.0% between the first and second years (Table 4). This decrease was observed for 60% of the plants under study (Table S4). In terms of weather conditions, it is observed that the levels of total phenols generally tend to rise, while the concentrations of phenolic acids and other compounds with antioxidant capacity may decrease. The increase in phenolic compounds can be attributed to higher concentrations of oleuropein, a phenolic compound responsible for the characteristic bitter taste in olive oils [51,57]. The ripening process of the fruit involves various chemical and enzymatic reactions, which lead to the production of free phenols and changes in the quantity of different phenolic compounds [58]. Interestingly, unlike hydroxytyrosol and tyrosol compounds, the total phenols exhibited an increase in 2021 under the analyzed conditions. This particular year was characterized by irregularities in temperature and precipitation. Stress-inducing factors, such as early rains in June and September, potentially accelerated fruit ripening, while low precipitation and higher temperatures in August, October, and November may have influenced the synthesis of certain phenolic compounds.

Table 4. Oxidative stability (hours), DPPH and ABTS (inhibition %) and total reduction capacity (mg GAE per kg of oil) (mean ± standard deviation) for the olive oils extracted from olives of centenarian olive trees located in the Côa Valley (Portugal), for two consecutive crop years (2020 and 2021), and respective variation with crop-year (Δ, in %).

Antioxidant Activity	Crop Year		p-Value	Δ (%)
	2020	2021		
Oxidative stability	18.2 ± 8.3 [a] (6.6–31.6)	19.0 ± 9.1 [a] (8.0–44.5)	0.6472 **	+4.7
Total reduction capacity	375.2 ± 160.6 [b] (61.3–710.7)	569.5 ± 189.5 [a] (234.1–965.7)	2.32×10^{-7} *	+51.8
DPPH	57.1 ± 17.7 [a] (12.6–84.1)	53.2 ± 18.0 [a] (17.1–88.6)	0.2922 **	−6.8
ABTS	40.8 ± 8.5 [a] (21.1–54.7)	35.0 ± 5.6 [b] (23.9–45.8)	9.07×10^{-5} *	−14.2

* t-Student test with Welch's correction (unequal variances according to the F-test). ** t-Student test without correction (equal variances according to the F-test). p-values for the t-Student test. Different letters in the same row show statistically differences from the given mean ($p < 0.05$).

In 2021, there was a decline observed in both the radical inhibition capacity and the presence of phenolic acids, which are closely associated with the antioxidant properties of the olive oil. Moroccan olive oils also exhibited significant variations in total phenolic content across different crop years, primarily influenced by climatic factors, particularly precipitation [20]. A similar trend was noticed in the oils of cv. Koroneiki, with phenolic contents ranging from 493.7 to 566.3 mg/kg [20]. Once phenolic compounds could act as a defense against trees or plants, the cumulative rainfall can play a crucial role in shaping the phenolic composition of olives, as water deficit creates a stressful environment for the olive trees that stimulates the production of these beneficial compounds [25,58]. Phenolic compounds play a vital role in preventing oxidation and are directly linked to the oxidative stability values [59]. No notable differences were observed in the present study regarding oils from Portuguese centenarian olive trees, with the average oxidative stability ranging from 18 to 19 h and a maximum value of 44.5 h. Similar findings have been reported for olive oils produced in other geographical regions [13,20,40]. On the other hand, the DPPH antioxidant method consistently yielded values between the years, with the average inhibition ranged from 53 to 57%. The presence of different phenolic compounds in varying concentrations contributes to the unique characteristics of an olive oil, including its flavor, aroma, quality, as well as its relevance in terms of nutrition and shelf-life [60]. Exploring the antioxidant activity offers promising prospects in characterizing centenarian olive trees.

3.6. Discrimination of Olive Oils by Crop Year

The observed climate variations in 2020 and 2021 had a noticeable impact on the composition of the studied olive oils. Specifically, the fatty acids, tocopherols, phenolic acids, and antioxidant activity of the oils were affected. These effects provide an opportunity to differentiate the oils based on their crop year using principal component analysis (PCA). The PCA analysis, which considered 17 independent variables (OS; DPPH; ABTS; hydroxytyrosol; tyrosol; α-, β-, and γ-tocopherol; TPC; $C_{16:0}$, $C_{16:1}$, $C_{18:0}$, $C_{18:1}$, $C_{18:2}$, $C_{18:3}$, $C_{20:0}$, and $C_{20:1}$), revealed three main components: PC1, PC2, and PC3. These components accounted for 37.0%, 16.5%, and 12.3% of the data variability, respectively. Consequently, it became possible to distinguish different batches of olive oil based on the harvest years, as depicted in Figure 2. Olive oils harvested in 2020 were positioned in the positive quadrant of PC1 and negatively in terms of PC2 and PC3. Conversely, olive oils from the 2021 harvest exhibited an opposite pattern, with positive values for PC2 and PC3, and negative values for PC1.

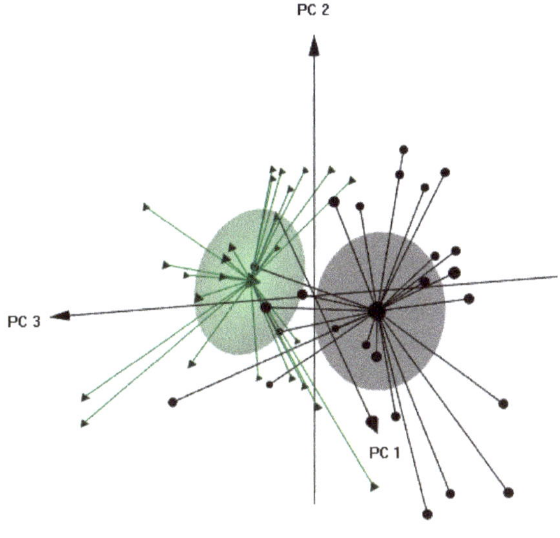

Figure 2. 3D PCA plot (PC1: 37%, PC2: 16.5%, and PC3: 1.2%): olive oils extracted from olives of centenarian olive trees located in the Côa Valley (Portugal) unsupervised differentiation according to the harvest year (2020: black •; 2021: green ▲) based on 17 independent variables (OS; DPPH; ABTS; hydroxytyrosol; tyrosol; α-, β-, and γ-tocopherol; TPC; $C_{16:0}$, $C_{16:1}$, $C_{18:0}$, $C_{18:1}$, $C_{18:2}$, $C_{18:3}$, $C_{20:0}$, and $C_{20:1}$).

To further validate the effectiveness of the unsupervised classification, linear discriminant analysis (LDA) was employed in combination with the simulated annealing (SA) variable selection algorithm. This approach aimed to identify the most relevant variables among the initial 17 used in PCA, focusing on those with the highest discriminatory power. By utilizing data from seven selected variables (ABTS; α-, β-, and γ-tocopherol; TPC; $C_{16:0}$, and $C_{18:0}$), a classification LDA-SA model was developed.

This model successfully discriminated the analyzed oils based on their respective harvest year (2020 or 2021), achieving sensitivities of 100% and 98% for the training and leave-one-out cross-validation (LOO-CV) procedures, respectively. Notably, in the LOO-CV validation, only one oil from 2021 was misclassified as belonging to 2020.

4. Conclusions

The primary objective of this study was to examine how interannual climatic conditions influence the composition of olive oil derived from centenarian olive trees. These trees have endured diverse environmental conditions throughout the years and possess genetic potential to enhance agroecosystem resilience. Over a two-year period, the impact of yearly variations on the oils extracted from 25 centenarian olive trees of the Côa Valley region was analysed. The findings revealed significant effects of climatic variations observed during the years 2020 and 2021, particularly on fatty acids, tocopherols, phenolic acids, and antioxidant activity in the oils under investigation. During warmer years, especially during the fruit development period, the concentration of palmitic acid, the most variable compound, reached 15.22%, while in colder climates, it reached 13.53%. Regarding tocopherols, only the influence of the year was observed for β-tocopherol. Furthermore, abnormal rainfall caused by storms was found to potentially alter the composition of phenolic compounds. Water scarcity creates a state of stress that increases the production of phenolics in olive oils. Excess water will lead to a decrease in phenolic content. Although these results are preliminary, they provide a foundation for future studies exploring the response of centenarian olive trees to different scenarios. Understanding these changes is crucial for informed decision making.

Supplementary Materials: The following supporting information can be downloaded at: https://www.mdpi.com/article/10.3390/agronomy13122884/s1, Table S1: Mean ± standard deviation (%) of the relative abundance of fatty acids determined in olive oils extracted from olives of centenarian olive trees located in the Côa Valley, Portugal, for two consecutive crop years (2020 and 2021); Table S2: Tocopherols (α-, β-, and γ-) and total vitamin E mass contents (mean ± standard deviation, mg/kg of olive oil) for olive oils extracted from olives of centenarian olive trees located in the Côa Valley (Portugal), for two consecutive crop years (2020 and 2021); Table S3: Hydroxytyrosol, tyrosol and total hydroxytyrosol + tyrosol based compounds (mean ± standard deviation, mg/kg of olive oil) for olive oils extracted from olives of centenarian olive trees located in the Côa Valley (Portugal), for two consecutive crop years (2020 and 2021); Table S4: Oxidative stability (hours). DPPH and ABTS (inhibition %) and total reduction capacity (mg GAE per kg of oil) (mean ± standard deviation) for the olive oils extracted from olives of centenarian olive trees located in the Côa Valley (Portugal) for two consecutive crop years (2020 and 2021).

Author Contributions: Conceptualization, N.R. and P.B.; methodology, L.B., R.C. and N.R.; software, A.M.P.; formal analysis, L.B., R.C., S.C. and N.R.; investigation, L.B., S.C., R.C., A.M.P., P.B. and N.R.; resources, N.R.; writing—original draft preparation, L.B.; writing—review and editing, N.R., S.C., A.M.P. and P.B.; supervision, N.R. and P.B.; project administration, N.R. and P.B.; funding acquisition, N.R. All authors have read and agreed to the published version of the manuscript.

Funding: This research was funded by Foundation for Science and Technology (FCT, Portugal) financial support from national funds FCT/MCTES to CIMO (UIDB/00690/2020 and UIDP/00690/2020), SusTEC (LA/P/0007/2020) and to LAQV/Requimte (UIDB/50006/2020 and UIDP/50006/2020); to the Project "OLIVECOA—Centennial Olive Trees in the Côa Valley Region: rediscovering the past to enhance the future" (ref.COA/BRB/0035/2019). Nuno Rodrigues and Rebeca Cruz thanks the FCT- Foundation for Science and Technology, P.I., for the National funding through the institutional program contract for scientific employment.

Data Availability Statement: Data available on request. The data are not publicly available due to [Data are contained within the article].

Acknowledgments: The authors acknowledge the collaboration of the Instituto Português do Mar e da Atmosfera (IPMA) for providing the meteorological data.

Conflicts of Interest: The authors declare no conflict of interest. The funders had no role in the design of the study; in the collection, analyses, or interpretation of data; in the writing of the manuscript; or in the decision to publish the results.

References

1. Langsdorf, S.; Löschke, S.; Möller, V.; Okem, A.; Officer, S.; Rama, B.; Belling, D.; Dieck, W.; Götze, S.; Kersher, T.; et al. Terrestrial and Freshwater Ecosystems and Their Services. In *Climate Change 2022: Impacts, Adaptation and Vulnerability. In Contribution of Working Group II to the Sixth Assessment Report of the Intergovernmental Panel on Climate Change*; IPCC: Cambridge, UK, 2022; pp. 197–377. [CrossRef]
2. Rickards, L.; Howden, S.M. Transformational Adaptation: Agriculture and Climate Change. *Crop Pasture Sci.* **2012**, *63*, 240–250. [CrossRef]
3. Campbell, B.M.; Vermeulen, S.J.; Aggarwal, P.K.; Corner-Dolloff, C.; Girvetz, E.; Loboguerrero, A.M.; Ramirez-Villegas, J.; Rosenstock, T.; Sebastian, L.; Thornton, P.K.; et al. Reducing Risks to Food Security from Climate Change. *Glob. Food Sec.* **2016**, *11*, 34–43. [CrossRef]
4. Nugroho, A.D.; Prasada, I.Y.; Lakner, Z. Comparing the Effect of Climate Change on Agricultural Competitiveness in Developing and Developed Countries. *J. Clean. Prod.* **2023**, *406*, 137139. [CrossRef]
5. Amassaghrou, A.; Barkaoui, K.; Bouaziz, A.; Alaoui, S.B.; Fatemi, Z.E.A.; Daoui, K. Yield and Related Traits of Three Legume Crops Grown in Olive-Based Agroforestry under an Intense Drought in the South Mediterranean. *Saudi J. Biol. Sci.* **2023**, *30*, 103597. [CrossRef]
6. Temani, F.; Bouaziz, A.; Daoui, K.; Wery, J.; Barkaoui, K. Olive Agroforestry Can Improve Land Productivity Even under Low Water Availability in the South Mediterranean. *Agric. Ecosyst. Environ.* **2021**, *307*, 107234. [CrossRef]
7. Arenas-Castro, S.; Gonçalves, J.F.; Moreno, M.; Villar, R. Projected Climate Changes Are Expected to Decrease the Suitability and Production of Olive Varieties in Southern Spain. *Sci. Total Environ.* **2020**, *709*, 136161. [CrossRef]
8. Fraga, H.; Moriondo, M.; Leolini, L.; Santos, J.A. Mediterranean Olive Orchards under Climate Change: A Review of Future Impacts and Adaptation Strategies. *Agronomy* **2020**, *11*, 56. [CrossRef]
9. Arias Montevechio, E.; Crispin Cunya, M.; Fernández Jorquera, F.; Rendon, E.; Vásquez-Lavin, F.; Stehr, A.; Ponce Oliva, R.D. Traditional Crops and Climate Change Adaptation: Insights from the Andean Agricultural Sector. *Clim. Dev.* **2023**, *15*, 723–737. [CrossRef]
10. Deiana, P.; Motroni, A.; Filigheddu, M.R.; Dettori, S.; Nieddu, G.; Mercenaro, L.; Alfei, B.; Culeddu, N.; Santona, M. Effect of Pedoclimatic Variables on Analytical and Organoleptic Characteristics in Olive Fruit and Virgin Olive Oil. *Eur. J. Agron.* **2023**, *148*, 126856. [CrossRef]
11. Fraga, H.; Guimarães, N.; Freitas, T.R.; Malheiro, A.C.; Santos, J.A. Future Scenarios for Olive Tree and Grapevine Potential Yields in the World Heritage Côa Region, Portugal. *Agronomy* **2022**, *12*, 350. [CrossRef]
12. Valente, S.; Machado, B.; Pinto, D.C.G.A.; Santos, C.; Silva, A.M.S.; Dias, M.C. Modulation of Phenolic and Lipophilic Compounds of Olive Fruits in Response to Combined Drought and Heat. *Food Chem.* **2020**, *329*, 127191. [CrossRef] [PubMed]
13. Rodrigues, N.; Casal, S.; Cruz, R.; Peres, A.M.; Baptista, P.; Pereira, J.A. GxE Effects on Tocopherol Composition of Oils from Very Old and Genetically Diverse Olive Trees. *JAOCS J. Am. Oil Chem. Soc.* **2020**, *97*, 497–507. [CrossRef]
14. Lodolini, E.M.; Ali, S.; Mutawea, M.; Qutub, M.; Arabasi, T.; Pierini, F.; Neri, D. Complementary Irrigation for Sustainable Production in Olive Groves in Palestine. *Agric. Water Manag.* **2014**, *134*, 104–109. [CrossRef]
15. Rey-Giménez, R.; Sánchez-Gimeno, A.C. Crop Year, Harvest Date and Clone Effects on Fruit Characteristics, Chemical Composition and Olive Oil Stability from an Empeltre Clonal Selection Grown in Aragon. *J. Sci. Food Agric.* **2022**, *102*, 5778–5786. [CrossRef]
16. Aparicio, R.; Ferreiro, L.; Alonso, V. Effect of Climate on the Chemical Composition of Virgin Olive Oil. *Anal. Chim. Acta* **1994**, *292*, 235–241. [CrossRef]
17. Beltrán, G.; Jiménez, A.; del Rio, C.; Sánchez, S.; Martínez, L.; Uceda, M.; Aguilera, M.P. Variability of Vitamin E in Virgin Olive Oil by Agronomical and Genetic Factors. *J. Food Compos. Anal.* **2010**, *23*, 633–639. [CrossRef]
18. Bengana, M.; Bakhouche, A.; Lozano-Sánchez, J.; Amir, Y.; Youyou, A.; Segura-Carretero, A.; Fernández-Gutiérrez, A. Influence of Olive Ripeness on Chemical Properties and Phenolic Composition of Chemlal Extra-Virgin Olive Oil. *Food Res. Int.* **2013**, *54*, 1868–1875. [CrossRef]
19. Bongiorno, D.; Di Stefano, V.; Indelicato, S.; Avellone, G.; Ceraulo, L. Bio-Phenols Determination in Olive Oils: Recent Mass Spectrometry Approaches. *Mass Spectrom. Rev.* **2021**, *42*, 1462–1502. [CrossRef]
20. Mansouri, F.; Moumen, A.B.; Belhaj, K.; Richard, G.; Marie-Laure Fauconnier, G.R.; Sindic, M.; Ahmed Elamrani, H.S.C. Effect of Crop Season on the Quality and Composition of Extra Virgin Olive Oils from Greek and Spanish Varieties Grown in the Oriental Region of Morocco. *Emir. J. Food Agric.* **2018**, *30*, 549–562. [CrossRef]

21. Rey-Giménez, R.; Sánchez-Gimeno, A.C. Authenticity in Olive Oils from an Empeltre Clonal Selection in Aragon (Spain): How Environmental, Agronomic, and Genetic Factors Affect Sterol Composition. *Foods* **2022**, *11*, 2587. [CrossRef]
22. Romero, M.P.; Tovar, M.J.; Ramo, T.; Motilva, M.J. Effect of Crop Season on the Composition of Virgin Olive Oil with Protected Designation of Origin "Les Garrigues". *J. Am. Oil Chem. Soc.* **2003**, *80*, 423–430. [CrossRef]
23. Giuffrè, A.M. The Effects of Cultivar and Harvest Year on the Fatty Alcohol Composition of Olive Oils from Southwest Calabria (Italy). *Grasas Aceites* **2014**, *65*, e011. [CrossRef]
24. Mele, M.A.; Islam, M.Z.; Kang, H.M.; Giuffre, A.M. Pre-and Post-Harvest Factors and Their Impact on Oil Composition and Quality of Olive Fruit. *Emir. J. Food Agric.* **2018**, *30*, 592. [CrossRef]
25. Lechhab, T.; Lechhab, W.; Cacciola, F.; Salmoun, F. Sets of Internal and External Factors Influencing Olive Oil (*Olea Europaea* L.) Composition: A Review. *Eur. Food Res. Technol.* **2022**, *248*, 1069–1088. [CrossRef]
26. Rouina, Y.B.; Zouari, M.; Zouari, N.; Rouina, B.B.; Bouaziz, M. Olive Tree (*Olea europaea* L. Cv. Zelmati) Grown in Hot Desert Climate: Physio-Biochemical Responses and Olive Oil Quality. *Sci. Hortic.* **2020**, *261*, 108915. [CrossRef]
27. Oğraş, Ş.Ş.; Kaban, G.; Kaya, M. The Effects of Geographic Region, Cultivar and Harvest Year on Fatty Acid Composition of Olive Oil. *J. Oleo Sci.* **2016**, *65*, 889–895. [CrossRef] [PubMed]
28. Beltrán, G.; Del Rio, C.; Sánchez, S.; Martínez, L. Influence of Harvest Date and Crop Yield on the Fatty Acid Composition of Virgin Olive Oils from Cv. Picual. *J. Agric. Food Chem.* **2004**, *52*, 3434–3440. [CrossRef]
29. Squeo, G.; Silletti, R.; Mangini, G.; Summo, C.; Caponio, F. The Potential of Apulian Olive Biodiversity: The Case of Oliva Rossa Virgin Olive Oil. *Foods* **2021**, *10*, 369. [CrossRef]
30. Rodrigues, N.; Casal, S.; Peres, A.M.; Baptista, P.; Pereira, J.A. Seeking for Sensory Differentiated Olive Oils? The Urge to Preserve Old Autochthonous Olive Cultivars. *Food Res. Int.* **2020**, *128*, 108759. [CrossRef]
31. Alowaiesh, B.; Singh, Z.; Fang, Z.; Kailis, S.G. Harvest Time Impacts the Fatty Acid Compositions, Phenolic Compounds and Sensory Attributes of Frantoio and Manzanilla Olive Oil. *Sci. Hortic.* **2018**, *234*, 74–80. [CrossRef]
32. Petruccelli, R.; Giordano, C.; Salvatici, M.C.; Beghè, D.; Rodolfi, M.; Fabbri, A.; Benelli, C. Characterization and Conservation of "Olivo Della Strega": An Ancient Olive Tree, Precious Resource for Natural and Cultural Heritage. *Rend. Lincei* **2021**, *32*, 311–324. [CrossRef]
33. Rodrigues, N.; Casal, S.; Pinho, T.; Cruz, R.; Peres, A.M.; Baptista, P.; Pereira, J.A. Fatty Acid Composition from Olive Oils of Portuguese Centenarian Trees Is Highly Dependent on Olive Cultivar and Crop Year. *Foods* **2021**, *10*, 496. [CrossRef] [PubMed]
34. Sabetta, W.; Mascio, I.; Squeo, G.; Gadaleta, S.; Flamminii, F.; Conte, P.; Di Mattia, C.D.; Piga, A.; Caponio, F.; Montemurro, C. Bioactive Potential of Minor Italian Olive Genotypes from Apulia, Sardinia and Abruzzo. *Foods* **2021**, *10*, 1371. [CrossRef] [PubMed]
35. International Olive Council. *Guide for the Determination of the Characteristics of Oil-Olives;* COI/OH/Doc. No 1; International Olive Council: Madrid, Spain, 2011; p. 39.
36. International Olive Council. *Determination of the Content of Waxes and Fatty Acid Ethyl Esters by Capillary Gas Chromatography;* COI/T.20/Doc. No 28/Rev. 3 November 2022; International Olive Council: Madrid, Spain, 2022; pp. 1–21.
37. *ISO 9936:2016;* Animal and Vegetable Fats and Oils—Determination of Tocopherol and Tocotrienol Contents by High-Performance Liquid Chromatography. ISO: Geneva, Switzerland, 2016.
38. Marx, Í.M.G.; Casal, S.; Rodrigues, N.; Veloso, A.C.A.; Pereira, J.A.; Peres, A.M. Estimating Hydroxytyrosol-Tyrosol Derivatives Amounts in Cv. Cobrançosa Olive Oils Based on the Electronic Tongue Analysis of Olive Paste Extracts. *LWT* **2021**, *147*, 111542. [CrossRef]
39. Cherif, M.; Rodrigues, N.; Veloso, A.C.A.; Pereira, J.A.; Peres, A.M. Kinetic Study of the Microwave-induced Thermal Degradation of Cv. Arbequina Olive Oils Flavored with Lemon Verbena Essential Oil. *J. Am. Oil Chem. Soc.* **2021**, *98*, 1021–1032. [CrossRef]
40. Samaniego Sánchez, C.; Troncoso González, A.M.; García-Parrilla, M.C.; Quesada Granados, J.J.; López García de la Serrana, H.; López Martínez, M.C. Different Radical Scavenging Tests in Virgin Olive Oil and Their Relation to the Total Phenol Content. *Anal. Chim. Acta* **2007**, *593*, 103–107. [CrossRef] [PubMed]
41. Hijawi, T. Characterizing of Oil Quality and Fatty Acid Profiles of Old Olive Trees in Palestine. *J. Oleo Sci.* **2021**, *70*, 1585–1606. [CrossRef]
42. Portarena, S.; Farinelli, D.; Lauteri, M.; Famiani, F.; Esti, M.; Brugnoli, E. Stable Isotope and Fatty Acid Compositions of Monovarietal Olive Oils: Implications of Ripening Stage and Climate Effects as Determinants in Traceability Studies. *Food Control* **2015**, *57*, 129–135. [CrossRef]
43. Ângelo, M.; Carlos, R.; Correia, M. *Manual da Safra e Contra Safra Olival;* Instituto Politécnico: Bragança, Portugal, 2009.
44. Borges, T.H.; López, L.C.; Pereira, J.A.; Cabrera-Vique, C.; Seiquer, I. Comparative Analysis of Minor Bioactive Constituents (CoQ10, Tocopherols and Phenolic Compounds) in Arbequina Extra Virgin Olive Oils from Brazil and Spain. *J. Food Compos. Anal.* **2017**, *63*, 47–54. [CrossRef]
45. García-Inza, G.P.; Castro, D.N.; Hall, A.J.; Rousseaux, M.C. Responses to Temperature of Fruit Dry Weight, Oil Concentration, and Oil Fatty Acid Composition in Olive (*Olea europaea* L. Var. 'Arauco'). *Eur. J. Agron.* **2014**, *54*, 107–115. [CrossRef]
46. Ayton, J.; Mailer, R.J.; Haigh, A.; Tronson, D.; Conlan, D. Quality and oxidative stability of australian olive oil according to harvest date and irrigation. *J. Food Lipids* **2007**, *14*, 138–156. [CrossRef]

47. Ben Hmida, R.; Gargouri, B.; Chtourou, F.; Abichou, M.; Sevim, D.; Bouaziz, M. Study on the Effect of Climate Changes on the Composition and Quality Parameters of Virgin Olive Oil "Zalmati" Harvested at Three Consecutive Crop Seasons: Chemometric Discrimination. *ACS Omega* **2022**, *7*, 40078–40090. [CrossRef] [PubMed]
48. Giuffrè, A.M. The Evolution of Free Acidity and Oxidation Related Parameters in Olive Oil during Olive Ripening from Cultivars Grown in the Region of Calabria, South Italy. *Emir. J. Food Agric.* **2018**, *30*, 539–548.
49. Yu, L.; Wang, Y.; Wu, G.; Jin, J.; Jin, Q.; Wang, X. Chemical and Volatile Characteristics of Olive Oils Extracted from Four Varieties Grown in Southwest of China. *Food Res. Int.* **2021**, *140*, 109987. [CrossRef] [PubMed]
50. Benincasa, C.; Russo, A.; Romano, E.; Elsorady, M.E.; Perri, E.; Muzzalupo, I. Chemical and sensory analysis of some Egyptian virgin olive oils. *J. Nutr. Food Sci.* **2011**, *5*, 118. [CrossRef]
51. Jiménez-Herrera, R.; Pacheco-López, B.; Peragón, J. Water Stress, Irrigation and Concentrations of Pentacyclic Triterpenes and Phenols in *Olea Europaea* L. Cv. Picual Olive Trees. *Antioxidants* **2019**, *8*, 294. [CrossRef] [PubMed]
52. Faghim, J.; Mohamed, M.B.; Bagues, M.; Nagaz, K.; Triki, T.; Guasmi, F. Irrigation Effects on Phenolic Profile and Extra Virgin Olive Oil Quality of "Chemlali" Variety Grown in South Tunisia. *S. Afr. J. Bot.* **2021**, *141*, 322–329. [CrossRef]
53. Uslu, N.; Özcan, M.M. The Effect of Irrigation and Harvest Time on Bioactive Properties of Olive Fruits Issued from Some Olive Varieties Grown in Mediterranean Region. *Eur. Food Res. Technol.* **2020**, *246*, 2587–2599. [CrossRef]
54. Wani, T.A.; Masoodi, F.A.; Gani, A.; Baba, W.N.; Rahmanian, N.; Akhter, R.; Wani, I.A.; Ahmad, M. Olive Oil and Its Principal Bioactive Compound: Hydroxytyrosol—A Review of the Recent Literature. *Trends Food Sci. Technol.* **2018**, *77*, 77–90. [CrossRef]
55. Famiani, F.; Farinelli, D.; Rollo, S.; Camposeo, S.; Di Vaio, C.; Inglese, P. Evaluation of Different Mechanical Fruit Harvesting Systems and Oil Quality in Very Large Size Olive Trees. *Span. J. Agric. Res.* **2014**, *12*, 960–972. [CrossRef]
56. Baccouri, B.; Sieren, T.; Mohamed, S.N.; Willenberg, I. Fingerprinting of Tocopherol, Phenolic Compounds and Oxidative Properties of Unstudied Minor and Rare Tunisian Olive Oils. *S. Afr. J. Bot.* **2023**, *156*, 54–64. [CrossRef]
57. García-Garví, J.M.; Sánchez-Bravo, P.; Hernández, F.; Sendra, E.; Corell, M.; Moriana, A.; Burgos-Hernández, A.; Carbonell-Barrachina, Á.A. Effect of Regulated Deficit Irrigation on the Quality of "Arbequina"; Extra Virgin Olive Oil Produced on a Super-High-Intensive Orchard. *Agronomy* **2022**, *12*, 1892. [CrossRef]
58. Ben Brahim, S.; Kelebek, H.; Ammar, S.; Abichou, M.; Bouaziz, M. LC–MS Phenolic Profiling Combined with Multivariate Analysis as an Approach for the Characterization of Extra Virgin Olive Oils of Four Rare Tunisian Cultivars during Ripening. *Food Chem.* **2017**, *229*, 9–19. [CrossRef] [PubMed]
59. Reboredo-Rodríguez, P.; González-Barreiro, C.; Cancho-Grande, B.; Valli, E.; Bendini, A.; Gallina Toschi, T.; Simal-Gandara, J. Characterization of Virgin Olive Oils Produced with Autochthonous Galician Varieties. *Food Chem.* **2016**, *212*, 162–171. [CrossRef] [PubMed]
60. Nenadis, N.; Mastralexi, A.; Tsimidou, M.Z. Physicochemical Characteristics and Antioxidant Potential of the Greek PDO and PGI Virgin Olive Oils (VOOs). *Eur. J. Lipid Sci. Technol.* **2019**, *121*, 1800172. [CrossRef]

agronomy

MDPI

Article

Effect of Low Temperature on Photosynthetic Characteristics, Senescence Characteristics, and Endogenous Hormones of Winter Wheat "Ji Mai 22" during the Jointing Stage

Fengyin Zhang [1,2], Nan Jiang [1], Hanqi Zhang [1], Zhiguo Huo [1,2,*] and Zaiqiang Yang [1,*]

1 Collaborative Innovation Center of Forecast and Evaluation of Meteorological, Nanjing University of Information Science & Technology, Nanjing 210044, China; 20211208020@nuist.edu.cn (F.Z.); 202211080002@nuist.edu.cn (N.J.); 202212080025@nuist.edu.cn (H.Z.)

2 State Key Laboratory of Severe Weather (LASW), Chinese Academy of Meteorological Sciences, No.46, Zhongguancun South Street, Haidian District, Beijing 100081, China

* Correspondence: huozg@cma.gov.cn (Z.H.); yzq@nuist.edu.cn (Z.Y.); Tel.: +86-13-60106-4240 (Z.H.); +86-13-40196-3706 (Z.Y.)

Abstract: To investigate the effects of low-temperature (LT) stress on photosynthetic properties and senescence characteristics of winter wheat leaves during the jointing stage, an environmental temperature control experiment was designed at Nanjing University of Information Science and Technology in 2023, using *Triticum aestivum* L. *cv.* "Ji Mai 22" as the test material. Four different temperature levels were set: 18 °C/8 °C (daily maximum/daily minimum temperature; CK), 13 °C/3 °C, 10 °C/0 °C, and 7 °C/3 °C. The duration of each treatment was 2, 4, and 6 days, respectively. The experimental findings reveal that the changes in physiological parameters of winter wheat leaves under low-temperature stress treatments are nonlinear. Under the 3 °C LT treatment, the photosynthetic parameters and endogenous hormone levels of wheat leaves significantly decrease after 6 days of stress. Under the 0 °C LT treatment, the photosynthetic parameters, leaf pigment content, and endogenous hormones of wheat decrease significantly, while under the −3 °C LT treatment, all the parameters of winter wheat leaves show a significant decline. Generally, the "Ji Mai22" wheat cultivar has a lower growth temperature limit of −3 °C during the jointing stage.

Keywords: wheat (*Triticum astivum* L. *cv.*); net photosynthetic rate; stomatal conductance; transpiration rate; chlorophyll content; protective enzyme activity; endogenous hormone levels

1. Introduction

Wheat (*Triticum astivum* L. *cv.*) belongs to the genus *Triticum* in the Poaceae family, which is classified into two types based on sowing time: spring wheat and winter wheat. Winter wheat is sown in autumn and is known for its strong cold resistance. Wheat is the world's largest cereal crop and is widely distributed globally due to its strong adaptability [1]. Winter wheat accounts for about 75% of the total global wheat area. In particular, China is the largest producer of wheat in the world, with winter wheat accounting for over 80% of the total wheat area and over 85% of the total wheat production [2]. Among them, the Huang-Huai-Hai wheat-growing region in China is the largest wheat-producing area, with a wheat planting area of 240 million acres, accounting for 68% of China's wheat area. Therefore, safe production of winter wheat plays a vital role in ensuring China's—and global—food security [3]. The experimental materials in this study were selected from the largest winter wheat variety, "Ji Mai 22," in the Huang-Huai-Hai region of China. With its strong tillering ability and high yield, "Ji Mai 22" has an average annual expansion area of 15 million acres, with a total planting area exceeding 2.35 billion acres, ranking first in China.

Low-temperature (LT) stress is one of the primary meteorological disasters that affect the growth and development of winter wheat [4]. In recent years, with the backdrop of global climate instability, extreme temperature events have become more frequent. The Huang-Huai-Hai region in China has witnessed frequent occurrences of extreme LT disasters, such as hail and snow, during spring. These events have inflicted substantial losses on local agriculture and have led to significant reductions in wheat production in some areas [5]. After the jointing stage, the cold resistance of winter wheat decreases, making it highly sensitive to LT stress [6,7]. The frequent occurrence of low-temperature stress during the jointing stage has become a major threat to the safe production of winter wheat in China. Therefore, by studying the changes in various physiological parameters of the predominant variety "Ji Mai 22" leaves under different LT stresses in the Huang-Huai-Hai region, which is prone to frequent LT disasters, we can not only separate the low-temperature disaster indicators during the jointing stage of "Ji Mai 22" but also provide a basis for scientific cultivation management of wheat in China and global winter wheat production [8].

Photosynthesis is an important physiological process in crop metabolism, and it is also one of the first physiological functions of plants to be affected under LT stress [7]. The slightly lower temperatures can activate subcellular antioxidant systems and enhance the tolerance of winter wheat to subsequent LT stress, while excessive LT stress can inhibit wheat growth by suppressing oxidative bursts in the photosynthetic apparatus [9]. Scholars have found that under LT stresses, the net photosynthetic rate, transpiration rate, and stomatal conductance of winter wheat leaves decrease with decreasing stress temperature, while the intercellular carbon dioxide concentration shows a trend of initially decreasing and then increasing with decreasing stress temperature [10]. Photosynthesis in wheat leaves is weakened at 0 °C, and photosynthetic organs are damaged at temperatures below −2 °C [11]. Under LT stress, the photosynthetic activity of "Ji Mai 22" winter wheat decreased more significantly with the increasing severity of the stress, reaching its limit at −7 °C [12]. LT stress also limited the degree of stomatal opening in wheat, resulting in an increase and then a decrease in photosynthetic rate, while photosynthetic pigment content tended to decrease [13,14].

The senescence characteristics of winter wheat leaves have a significant impact on yield formation, and many scholars have conducted extensive research on chlorophyll, protective enzyme activity, and endogenous hormones [15–17]. LT stress significantly reduces the activity of pigment synthesis enzymes in winter wheat, leading to a decrease in chlorophyll content [18]. Using spectral monitoring techniques for diagnosing leaf chlorophyll content under LT stress, it was observed that with increasing severity of LT stress, the chlorophyll content in winter wheat leaves showed a trend of initially increasing and then decreasing [19]. LT can cause the accumulation of a large amount of reactive oxygen species in winter wheat leaves during the jointing stage, leading to lipid peroxidation of cell membranes [20]. Research has shown that the activity of superoxide dismutase (SOD) and peroxidase (POD) in the leaves of cold-resistant winter wheat varieties significantly increases with the increasing severity of LT stress [21]. Under three LT stresses, the activities of SOD, POD, and CAT protective enzymes in winter wheat leaves significantly increased, while the number and weight of grains per year decreased significantly, and the changes in the number of wheat plants and thousand-grain weight were not significant [22].

Endogenous hormones play a critical regulatory role in plant growth, development, and response to stress. Previous studies on the relationship between endogenous hormones and cold resistance in winter wheat have shown that under stress conditions, the levels of indole-3-acetic acid (IAA) and gibberellins (GA) in the roots and floral organs during the booting stage decrease [23], while abscisic acid (ABA) significantly accumulates [24], exerting a significant impact on wheat yield. Cold stress leads to a significant accumulation of ABA in the leaves during the flowering stage of winter wheat, and with increasing severity of LT stress, the levels of ABA, IAA, and GA in the leaves show a trend of initially increasing and then decreasing [25].

The adaptability of plants to LT stress has always been a global research topic, and LT stress is a challenge faced by agricultural development in China [26]. Currently, research on the growth and development of winter wheat mainly focuses on the economic yield formation stage after heading and flowering, while there is less research on the sensitive jointing stage. Studies on endogenous hormones in winter wheat are mainly related to tillering and growth, with limited research on leaf stress resistance. We hypothesize that the physiological parameters of winter wheat leaves do not continuously decrease as the severity of LT stress increases. Within a certain range of LT, leaves may trigger their self-protection mechanisms, leading to an increase in some physiological parameters. However, when the temperature stress reaches the limits that the leaves can withstand, growth ceases and the leaves begin to senesce. Therefore, this study aims to (1) explore the effects of LT stress on the photosynthetic properties of winter wheat leaves; (2) study the changes in protective enzyme activity in winter wheat leaves under LT stress; (3) investigate the variations of endogenous hormones (ABA, IAA, CTK, GA) in winter wheat leaves under LT stress; (4) use correlation to analyze the effects of different LT stresses on the photosynthetic and senescence characteristics of "Ji Mai 22" winter wheat, and propose the minimum temperature suitable for its growth, so as to provide a scientific basis for the prevention and control of LT disasters during the wheat jointing stage.

2. Materials and Methods

2.1. Experimental Designs

The winter wheat variety "Ji Mai 22" (semi-winter variety) from China was selected as the experimental material. Winter wheat seeds were procured from greenhouse growers in Nanjing, China, and were planted in mid-October 2022 in a Venlo-type greenhouse at Nanjing University of Information Science and Technology (NUIST). The dimensions of the greenhouse were 5.0 m (height) × 9.6 m (width) × 30.0 m (length), with a north–south orientation. Scholars have conducted extensive research on the suitable temperature, humidity, soil environment, planting density, and other factors in the Huang-Huai-Hai wheat-growing region [27,28]. In this study, we have taken into consideration these indicators to establish our parameters. The planting density was 300 plants·m^2. The soil medium consisted of peat soil: perlite: vermiculite 2:1:1 ($v/v/v$), with pH ranging from 7.5 to 8. At the time of sowing, compound fertilizer (N-P-K: 30%-10%-30%) as a basal fertilizer was applied at 2500 g/m^2, diluted with water (fertilizer: water = 1:100). The winter wheat was irrigated once every 7 days, ensuring that each plant was irrigated to 70% of the field capacity (monitored by a soil moisture meter) to avoid water deficiency.

According to the suitable growth conditions for winter wheat [29], the temperature inside the greenhouse was set at 8 °C (minimum)/18 °C (maximum). The humidity was set at 75% ± 5, and the light intensity was set at 800 μmol·m^{-2}·s^{-1}. The photoperiod was set at 12/12 h (daytime from 6 a.m. to 6 p.m.). When the base internode of the wheat started elongating and was exposed 2 cm above the soil surface (jointing stage), the winter wheat was transplanted into nutrient pots with a height of 25 cm, upper diameter of 20 cm, and lower diameter of 18 cm. In each nutrient pot, five holes were dug near the middle position, and three plants were transplanted in each hole. The soil medium in the pots consisted of peat soil: perlite: vermiculite in a ratio of 2:1:1 ($v/v/v$). Two days after transplantation, a foliar fertilizer was applied, using a compound fertilizer (N-P-K: 15%-15%-15%) and urea. The ratio of compound fertilizer to urea was 4:1, and the diluted fertilizer was applied at a rate of 2000 g per square meter, diluted with water at a ratio of 1:600 (fertilizer: water). During the experiment, the plants were watered at least every two days (most of the time, watering was done once or twice a day). All plants were irrigated to 80% of the field capacity (monitored using a soil moisture meter) to avoid water deficiency. The potted plants were placed in a controlled environment chamber (TPG1260, Australia) for the environmental control experiment (Table 1).

Table 1. Experimental design.

Marker	Treatments	Temperature [°C]	Duration [d]
T1	CK-2d	18/8	2
T2	CK-4d	18/8	4
T3	CK-6d	18/8	6
T4	3–2d	13/3	2
T5	3–4d	13/3	4
T6	3–6d	13/3	6
T7	0–2d	10/0	2
T8	0–4d	10/0	4
T9	0–6d	10/0	6
T10	−3–2d	7/-3	2
T11	−3–4d	7/-3	4
T12	−3–6d	7/-3	6

(1) To make it easier to analyze the study's results, "minimum temperature-stress days" represent each treatment in the subsequent expressions. (2) To make the experimental results easy to observe, the T1, T2, and T3 treatments at the control temperature were collectively referred to as CK.

The environment chamber temperatures were designed according to the range of low temperatures that occur in northern China during the post-spring winter wheat jointing stage [30,31], as shown in Figure 1. Four temperature levels (daily maximum temperature/daily minimum temperature) were set: 7 °C/−3 °C, 10 °C/0 °C, 13 °C/3 °C, and 18 °C/8 °C (CK). The light intensity was set at 800 μmol·m^{-2}·s^{-1}, and the duration was 2, 4, and 6 days, respectively. During the experiment, the relative humidity was set at 70 ± 5%, and the photoperiod was set at 12/12 h (daytime from 6:00 a.m. to 6:00 p.m.).

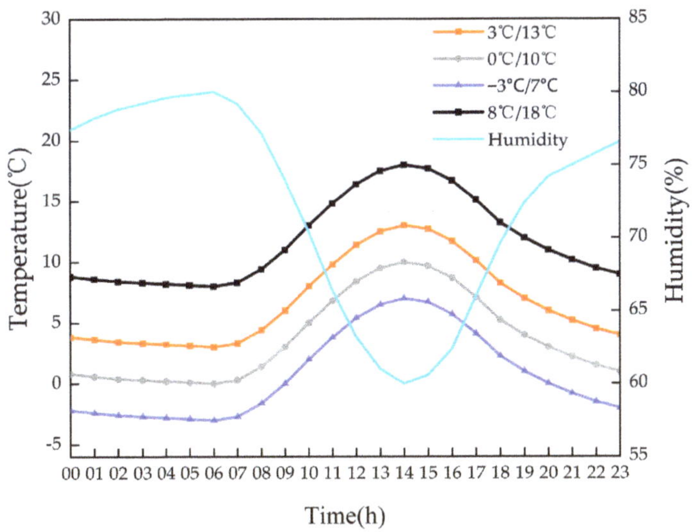

Figure 1. The setting of dynamic temperature and humidity of an artificial climate chamber.

2.2. Experimental Treatments and Leaf Sampling

On the morning of the day when the LT experiments commenced at 8 a.m., winter wheat at the jointing stage, exhibiting relatively uniform growth, was randomly selected and subjected to LT treatments within an artificial climate chamber. There was a total of 12 LT treatments. On the one hand, the recommended planting density for wheat is 200–300 plants·m^2. On the other hand, it was necessary to include a sufficient number of wheat plants in each LT treatment to measure various parameters. Therefore, for each LT treatment, 11 pots of winter wheat were placed (2 pots for measuring photosynthetic parameters,

2 pots for measuring protective enzyme activity, 2 pots for measuring pigment content, 2 pots for measuring endogenous hormones, and 3 spare pots), with each environmental chamber having an area of 2.25 m^2 and a total of 12 environmental chambers simultaneously conducting treatments.

At the end of the LT stress periods on day 2 (48 h), day 4 (96 h), and day 6 (144 h), winter wheat at the jointing stage, exhibiting relatively uniform growth, was selected for the measurements of photosynthetic parameters and pigment content. Additionally, fresh samples of the selected leaves were wiped clean and placed in zipper bags, then rapidly frozen in liquid nitrogen, and stored at −80 °C to measure protective enzyme activity and endogenous hormone content. Each of the measured indicators was sampled three times.

2.3. *Testing Content and Methods of Measurement*

2.3.1. Determination of Light Response Parameters

Light curve data of winter wheat leaves were measured using the LI-6400 Portable Photosynthesis System (LI-COR Biosciences Inc., Lincoln, NE, USA) between 9:00 a.m. and 11:00 a.m. on sunny days. Measurements were taken at the same blade position of the fully expanded penultimate leaf. During the measurements, the leaf chamber temperature in the LI-6400 was set at 24 °C, the CO_2 concentration was set at 400 $\mu mol \cdot mol^{-1}$, and photosynthetically active radiation (PAR) was set at 1800, 1600, 1400, 1200, 1000, 800, 600, 500, 400, 300, 200, 100, 50, and 0 $\mu mol \cdot m^{-2} \cdot s^{-1}$ [32]. The measured parameters included Pn (net photosynthetic rate), Gs (stomatal conductance to water vapor), Tr (transpiration rate), etc. When measuring the photosynthetic characteristics, three LI-6400 instruments were simultaneously used to measure three wheat leaves and three replicates were used for the determination of photosynthetic characteristics. The values are presented as "mean ± standard deviation (SD)".

The light response curve data of winter wheat leaf were fitted with a rectangular hyperbolic model after measurement [33]. The expression of the model is as follows:

$$P_n(i) = \alpha \frac{1 - \beta I}{1 + \gamma I} I - R_d \tag{1}$$

where α represents the slope of the light response curve of plant photosynthesis at I = 0, that is, the initial slope of the light response curve, also known as the initial quantum efficiency. β represents the correction coefficient, γ is a coefficient independent of light intensity, and Rd represents dark respiration.

2.3.2. Determination of Chlorophyll Content

The determination of chlorophyll content was conducted following the method proposed by Li [34] as follows. Select healthy and mature leaf samples from the top of winter wheat plants. Gently wipe the leaf surfaces to remove dust and remove the leaf veins. Weigh 0.2 g of leaf and then crush it into small pieces. Place the crushed leaf in 25 mL of 95% ethanol solution, seal it, and protect it from light for 48 h until the complete extraction of chlorophyll from the leaf. Measure the absorbance at 665 nm, 649 nm, and 470 nm wavelengths using a spectrophotometer (UV1800 Shimadzu) for colorimetric determination, with three replicates for each treatment. Calculate the chlorophyll content using the following formula:

$$\text{chla} = 13.95D_{665} - 6.88D_{649} \tag{2}$$

$$\text{chlb} = 24.96D_{665} - 7.32D_{665} \tag{3}$$

$$\text{chl}(a + b) = \text{chla} + \text{chlb} \tag{4}$$

$$\text{car} = (1000D_{470} - 2.05\text{Chla} - 114.8\text{Chlb})/245 \tag{5}$$

where chla, chlb, chl(a + b), and car represent chlorophyll a ($mg \cdot g^{-1}$), chlorophyll b ($mg \cdot g^{-1}$), total chlorophyll ($mg \cdot g^{-1}$), and carotenoids ($mg \cdot g^{-1}$), respectively. D665, D649, and D470 represent the absorbance values of the extracting solution at 665 nm, 649 nm, and 470 nm, respectively.

2.3.3. Determination of Protective Enzyme Activity

In each wheat plant, weigh 0.5 g of single-leaf samples from the same position at the top leaf and place them into a mortar. Add 5 mL of phosphate buffer solution with a pH of 7.8. Grind the mixture in an ice bath and then transfer the homogenate into a centrifuge tube. Freeze the tube and centrifuge for 20 min. Finally, transfer the supernatant (enzyme solution) into a test tube and store it at 0–4 °C.

The activity of superoxide dismutase (SOD), peroxidase (POD), and catalase (CAT) is determined by the nitroblue tetrazolium (NBT) colorimetric method [35,36], the guaiacol method [37], and the UV spectroscopic method [38], respectively.

2.3.4. Determination of Endogenous Hormone Content

First, prepare 0.1 g/zeatin (ZT), 0.01 g/abscisic acid (ABA), 0.1 g/indole-3-acetic acid (IAA), and 5 g/gibberellin (GA) mother solutions separately. Mix these reagents to prepare a series of standard mixed solutions with different concentrations and store them in brown volumetric flasks at 4 °C for further use.

The second step involves referring to the method of Yang et al., with some modifications [39–41] to analyze the four endogenous hormones, zeaxanthin (ZT), gibberellin (GA), indole-3-acetic acid (IAA), and abscisic acid (ABA), using high-performance liquid chromatography (HPLC). Specifically, use 10 mL of 80% cold methanol for overnight extraction of frozen plant samples. Filter the extract and then perform two additional extractions using 10 mL of 80% cold methanol. Combine the clear extracts from each extraction. Deodorize the residue with an equal volume of petroleum ether. Discard the ether phase and retain the aqueous phase. Repeat this step three times. Evaporate the combined aqueous phase under reduced pressure at 37 °C until the volume is reduced to one-fourth of the original volume. Adjust the pH of the solution to 2.8 and then extract with an equal volume of ethyl acetate. Discard the aqueous phase, and combine the ester phase. Repeat this step three times. Evaporate the ester phase under reduced pressure at 37 °C until it reaches a volume of 1 mL. Adjust the volume to 2 mL with methanol, and after filtration through a 0.45 μm microporous membrane, the solution will be ready for analysis.

The chromatographic column used in the experiment is Agilent 5 HC-C18 (150 × 4.6 mm, 5 μm). The mobile phase used is a mixture of methanol and 0.075% acetic acid aqueous solution in a ratio of 45:55 with a flow rate of 0.7 mL/min. The column temperature is set at 35 °C, and each injection volume is 20 μL. The wavelength range for detection is 210 nanometers.

2.4. Statistical Analysis

For statistical analyses in this investigation, all data were the mean ± standard deviation (SD) of 3 biological replications. In order to investigate the effects of temperature, duration, and their interaction on various physiological parameters of leaf tissue, SPSS 24.0 (SPSS Inc., Chicago, IL, USA) was used for two-way ANOVA, Duncan's multiple comparison test ($p = 0.05$), and correlation analysis.

3. Results

3.1. Effect of LT on the Photosynthetic Characteristics of Wheat Leaves

3.1.1. Effect of LT Stress on Photosynthetic Rate (Pn) of Leaves

Figure 2 shows changes in the light response curve of winter wheat leaves under different LT stresses. From Figure 2a, it can be observed that except for the −3 °C treatment, the photosynthetic rate (Pn) of winter wheat leaves showed no significant difference during the first 400 $\mu mol \cdot m^{-2} \cdot s^{-1}$ of PAR (photosynthetically active radiation) and then slightly decreased after reaching the light saturation point. Under the CK treatments, the Pn of wheat leaves reached a maximum of 29.16 $\mu mol \cdot m^{-2} \cdot s^{-1}$ at a PAR of 1600 $\mu mol \cdot m^{-2} \cdot s^{-1}$.

Under the 3 °C treatments, the Pn of leaves reached a maximum of 25.67 μmol·m^{-2}·s^{-1} at a PAR of 1200 μmol·m^{-2}·s^{-1}, which was 88.03% of the CK value. Under the 0 °C treatment, the Pn of leaves reached a maximum of 18.15 μmol·m^{-2}·s^{-1} at a PAR of 1000 μmol·m^{-2}·s^{-1}, which was 62.24% of the CK value. Under the −3 °C treatment, the Pn of leaves reached a maximum of 2.24 μmol·m^{-2}·s^{-1} at a PAR of 400 μmol·m^{-2}·s^{-1}, which was only 7.70% of the CK value and showed an unclear changing trend. From Figure 2b,c, it can be observed that the trend of Pn in winter wheat leaves remained consistent with the 2d LT stress when subjected to 4d and 6d stress. However, the Pn at the light saturation point is lower in the prolonged stress conditions compared with the 2d stress. Furthermore, under −3 °C treatment, the maximum net photosynthetic rate of winter wheat tended toward 0.

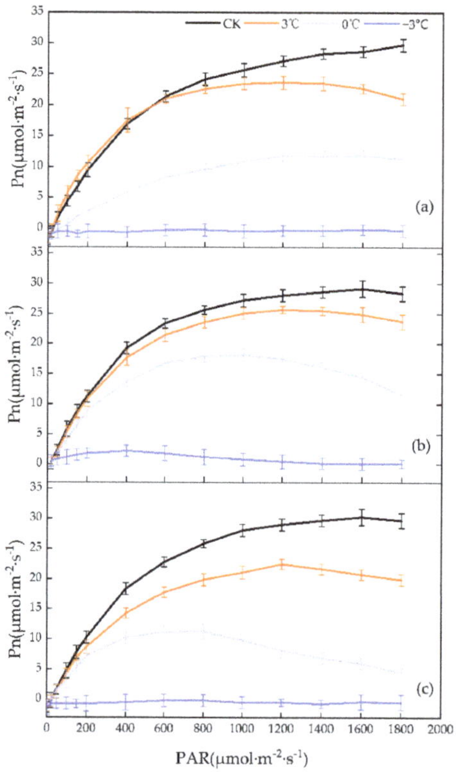

Figure 2. Changes in the light response curves of winter wheat leaves under different LT (low-temperature) stresses. (**a**) Light response curves under various LT stresses after 2 days of treatment. (**b**) Light response curves under various LT stresses after 4 days of treatment. (**c**) Light response curves under various LT stresses after 6 days of treatment. Three replicates were used for determination, and each value is presented as "mean ± standard deviation (SD)".

In conclusion, the photosynthetic rate of winter wheat leaves under different treatments increased to a saturation point and then slightly declined. As the severity of LT stress increased, the net photosynthetic rate of winter wheat leaves reached the light saturation point more quickly, but with a smaller maximum net photosynthetic rate. The −3 °C LT stress had a significant impact on the photosynthetic rate of winter wheat leaves, almost causing a complete cessation of photosynthesis in the leaves.

3.1.2. Effect of LT Stress on Stomatal Conductance (Gs) of Leaves

Stomata are the main channels through which plants exchange gases with the external environment. They play a crucial role in balancing water loss and gaining carbon for biomass production. Stomatal conductance (Gs) represents the degree of stomatal opening and is a major factor influencing plant photosynthesis, respiration, and transpiration [42]. Figure 3 shows changes in Gs of winter wheat leaves under different LT stresses. From Figure 3a, it can be observed that for the same duration of stress, Gs of winter wheat leaves decreased significantly with lower stress temperatures. Among them, Gs under T4 treatment showed little change compared with T1, while Gs of winter wheat leaves under T10 treatment reached a maximum of 0.03 mmol·m^{-2}·s^{-1} at a PAR of 1800 μmol·m^{-2}·s^{-1}, only 33.04% of the maximum value of T1. From Figure 3b, it can be seen that when PAR was between 100–200 μmol·m^{-2}·s^{-1}, Gs of leaves under T2, T5, T8, and T11 treatments rapidly increased, followed by a slow overall increase. Gs of leaves under the T8 treatment was significantly lower than that under the T5 treatment. From Figure 3c, it can be seen that after 6 days of stress, Gs of leaves under T10 treatment could still maintain a relatively high level, while Gs of leaves under T6 treatment decreased significantly compared with 2d and 4d stresses. Gs of leaves under T9 and T12 treatments remained at a lower level.

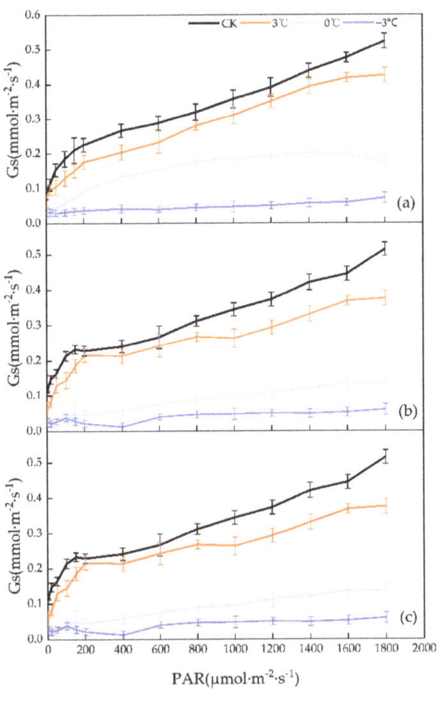

Figure 3. Changes in Gs (stomatal conductance to water vapor) of winter wheat leaves under different LT (low-temperature) stresses. (**a**) Gs variation under various LT stresses after 2 days of treatment. (**b**) Gs variation under various LT stresses after 4 days of treatment. (c) Gs variation under various LT stresses after 6 days of treatment. Three replicates were used for determination, and each value is presented as "mean ± standard deviation (SD)".

In summary, as the severity of LT stress increases, Gs of winter wheat leaves tended to decrease. Specifically, a significant decrease in Gs under 3 °C LT stress occurred after 6 days of stress, while a noticeable decrease in Gs under 0 °C LT stress occurred after 4 days of stress. Under −3 °C LT stress, Gs of leaves remained at a consistently lower level throughout the stress period.

3.1.3. Effect of LT Stress on Leaf Transpiration Rate (Tr)

Transpiration of winter wheat leaves can help regulate temperature [43]. Figure 4 shows changes in the leaf transpiration rate (Tr) of winter wheat leaves under different LT stresses. From Figure 4a, it can be observed that except for the T10 treatment, Tr of winter wheat leaves under the other LT treatments rapidly increased with increasing PAR and then stabilized. As the severity of LT stress increased, leaf Tr decreased. Under the T10 treatment, Tr of leaves remained constant at a level of 0.30 mmol·m^{-2}·s^{-1}, which was only 8.62% of the highest value under the T1 treatment. From Figure 4b, it can be seen that leaf Tr under the T2 treatment remained at a relatively high level, while Tr under the T5 treatment reached its peak and decreased significantly earlier than under the T4 treatment. Tr of leaves under T8 and T11 treatments remained at a lower level. From Figure 4c, it can be seen that leaf Tr under the T3 treatment remained at a higher level. Tr of leaves under the T6 treatment reached its peak earlier than under the T5 treatment and was higher than under the T3 treatment when PAR was between 100–600 μmol·m^{-2}· s^{-1}, then rapidly decreased. Tr of leaves under T9 and T12 treatments remained at a lower level.

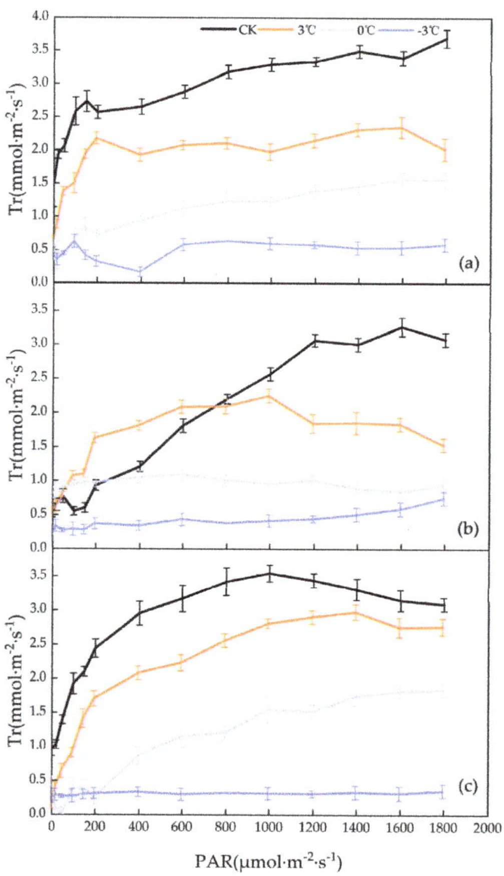

Figure 4. Changes in leaf Tr (transpiration rate) of winter wheat leaves under different LT (low-temperature) stresses. (**a**) Variation of Tr under various LT stresses after 2 days of treatment. (**b**) Variation of Tr under various LT stresses after 4 days of treatment. (**c**) Variation of Tr under various LT stresses after 6 days of treatment. Three replicates were used for determination, and each value is presented as "mean ± standard deviation (SD)".

In summary, as the severity of LT stress increases, the transpiration rate (Tr) of winter wheat leaves decreases significantly. Under CK treatment, the Tr of winter wheat leaves steadily increases. Under 3 °C treatments, the Tr of winter wheat leaves declines notably after 6 days of stress. Under 0 °C and −3 °C treatments, the Tr of winter wheat leaves remains consistently at a lower level.

3.2. *Effect of LT Stress on the Pigment Content of Winter Wheat Leaves*

The temperature stress duration and the interaction between temperature and stress duration have a highly significant impact ($p < 0.05$) on the content of chlorophyll a (chla), chlorophyll b (chlb), total chlorophyll content (chl(a + b)), and carotenoids (car) (shown in Table S1).

From Figure 5a, it can be observed that under T1, T2, and T3 treatments, the chla content in wheat leaves showed an increasing trend, indicating that the leaves were in the growth phase. In the early stage of stress (2d), there was not much difference in chla content among the different LT treatments compared with T1. With an increase in the severity of LT stress, under 3 °C and 0 °C treatments, the chla content in wheat leaves slightly increased during the middle stage of stress (4d) and significantly decreased during the later stage (6d). Under the T8 treatment, the chla content reached the maximum value of 3.22 mg·g^{-1}, showing a growth of 7.04% compared with the T2 treatment. Under the −3 °C treatment, the chla content in wheat leaves showed a continuous decreasing trend and reached the minimum value of 2.14 mg·g^{-1} under the T12 treatment, representing a decrease of 33.72% compared with the T3 treatment.

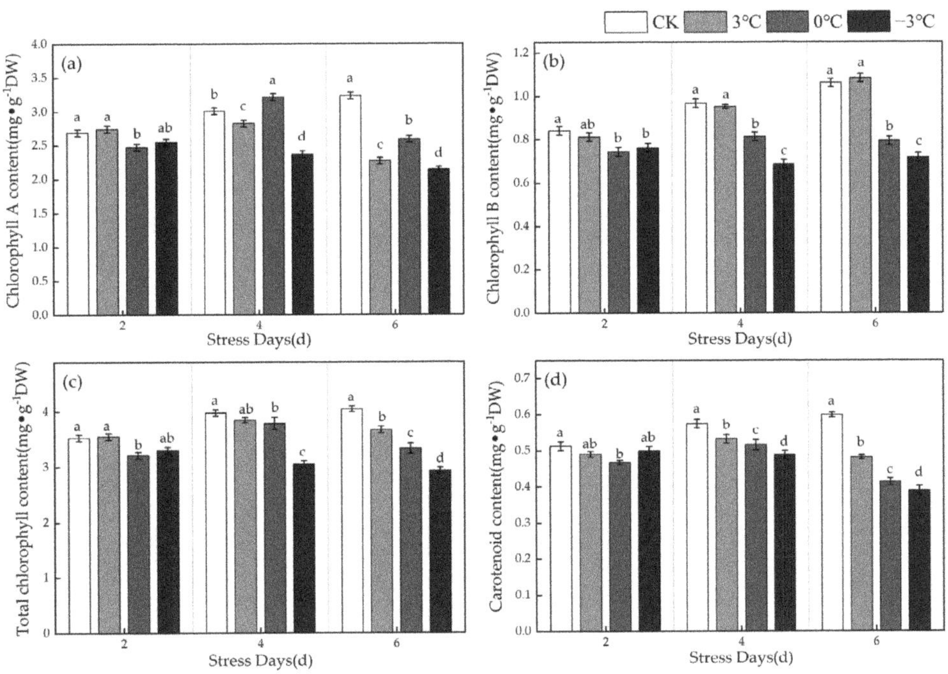

Figure 5. Changes in the leaf pigment content of winter wheat leaves under different LT (low temperature) stresses. (**a**) Chlorophyll a content. (**b**) Chlorophyll b content. (**c**) Total chlorophyll content. (**d**) Carotenoid content. Three replicates were used for determination, and each value is presented as "mean ± standard deviation (SD)". **Note:** Lowercase letters represent Duncan's test with $p < 0.05$.

Figure 5b shows changes in chlb content of winter wheat under different LT stress levels. From the figure, it can be observed that similar to chla, chlb in wheat leaves also exhibited an increasing trend under T1, T2, and T3 treatments. Under 3 °C treatment, there was not much difference in chlb content compared with T1, T2, and T3, as the severity of LT stress increased. Under 0 °C treatment, the chlb content in wheat leaves showed little change during the early to middle stages of stress (2–4d) but significantly decreased during the later stage (6d), compared with T1 and T2 treatments. Under the −3 °C treatment, the chlb content in wheat leaves continuously decreased with the increasing severity of LT stress. It reached the minimum value of 0.64 $mg \cdot g^{-1}$, representing a decrease of 40.72% under the T12 treatment, compared with the T3 treatment.

Figure 5c shows changes in the total chlorophyll content of winter wheat under different LT stresses. From the figure, it can be seen that under 3 °C and 0 °C treatments, the total chlorophyll content in wheat showed a slight increase during the early to middle stages of stress (2–4d) and a decrease during the later stage (6d), with a more pronounced decrease under 0 °C treatment compared with 3 °C treatment. Under the −3 °C treatment, the total chlorophyll content in wheat continuously decreased with the increasing severity of LT stress.

Figure 5d shows changes in car content of winter wheat under different LT stress levels. From the figure, it can be seen that during the early stage of stress (2d), there was not much difference in car content among the different LT treatments compared with T1. During the middle stage of stress (4d), the carotenoid content in wheat showed little change under different LT treatments compared with the early stage, but there was a significant difference compared with T2. As the severity of LT stress increased, the car content in wheat under all treatments decreased significantly. Under the T12 treatment, it reached the minimum value of 0.36, representing a decrease of 41.90% compared with the T3 treatment.

3.3. *Effect of LT Stress on the Activity of Protective Enzymes in Winter Wheat Leaves*

The temperature, stress duration, and the interaction between temperature and stress duration have a highly significant impact ($p < 0.05$) on the activity of catalase (CAT), superoxide dismutase (SOD), and peroxidase (POD) in winter wheat leaves (shown in Table S2).

The CAT activity in winter wheat leaves under different LT stress conditions is shown in Figure 6a. From the figure, it can be seen that CAT activity in winter wheat leaves significantly increases with the severity of LT stress under 3 °C and 0 °C treatments. The largest increase in CAT activity is observed under the 0 °C treatment, reaching a maximum value of 8.02 $U \cdot g^{-1} \cdot min^{-1}$ at T9, which is a 40.13% increase compared with the T3 treatment. Under −3 °C treatment, CAT activity in winter wheat leaves decreases significantly with the severity of stress. At T12, CAT activity is the lowest, which is 4.26 $U \cdot g^{-1} \cdot min^{-1}$, representing a 30.61% reduction compared with the T3 treatment.

The activity of SOD in winter wheat leaves under different levels of LT stress is shown in Figure 6b. From the graph, it can be observed that the LT stress at 3 °C has a minimal impact on the SOD activity in winter wheat leaves. However, as the severity of LT stress increases, there is a clear increasing trend in SOD activity. Under −3 °C treatment, the SOD activity in the leaves shows the most significant enhancement with the increasing severity of LT stress. In the T12 treatment, it reaches a maximum value of 296.5 2 $U \cdot g^{-1} \cdot min^{-1}$, which is a 66.14% increase compared with the T3 treatment. This indicates that winter wheat leaves are able to maintain effective scavenging of reactive oxygen species even under −3 °C LT stress.

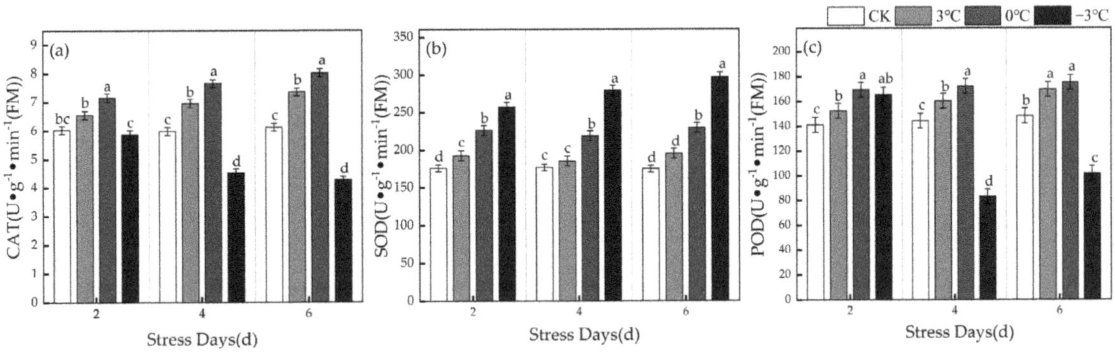

Figure 6. Changes in leaf protective enzyme activity of winter wheat leaves under different LT (low temperature) stresses. (**a**) The activity of CAT (catalase). (**b**) The activity of SOD (superoxide dismutase). (**c**) The activity of POD (peroxidase). Three replicates were used for determination, and each value is presented as "mean ± standard deviation (SD)". **Note:** Lowercase letters represent Duncan's test with $p < 0.05$.

The activity of POD in winter wheat leaves under different levels of LT stress is shown in Figure 6c. From the graph, it is evident that similar to CAT activity, there is a noticeable trend of increasing POD activity with the severity of LT stress, reaching a maximum value of 175.03 $U{\cdot}g^{-1}{\cdot}min^{-1}$ in the T9 treatment, which is an 18.21% increase compared with the T3 treatment. However, under −3 °C treatment, the POD activity in winter wheat leaves exhibits a significant increase in the early stages of stress (2d), followed by a notable decrease in the later stages of stress (4–6d), reaching its lowest value of 101.13 $U{\cdot}g^{-1}{\cdot}min^{-1}$ in the T12 treatment, representing a 31.85% reduction compared with the T3 treatment.

3.4. Effect of LT Stress on Endogenous Hormones in Winter Wheat Leaves

The temperature, duration of stress, and the interaction between temperature and stress duration have a significant impact on the content of zeatin (ZT), gibberellin (GA), indole-3-acetic acid (IAA), and abscisic acid (ABA) in winter wheat leaves ($p < 0.05$) (shown in Table S3).

Figure 7a shows changes in ZT content in winter wheat leaves under different LT stresses. From the figure, it can be observed that during the early to middle stages (2–4d) before cold stress, the ZT content in winter wheat leaves increased with the severity of LT stress and was significantly higher than in the control (CK) treatment. Under the T8 treatment, the ZT content in the leaves reached a maximum of 28.17 $\mu g{\cdot}g^{-1}$, showing a 27.16% increase compared with the T2 treatment. In the later stage of stress (6d), the ZT content in the leaves under all LT treatments sharply decreased. Under the T12 treatment, it reached the lowest value of 16.24 $\mu g{\cdot}g^{-1}$, showing a 35.58% decrease compared with the T3 treatment. This indicates that continuous 6d LT stress significantly affects the ZT content in winter wheat leaves.

Figure 7b shows changes in GA content in winter wheat leaves under different LT stresses. From the figure, it can be observed that during the early stage of LT stress (2d), the GA content in the leaves under various LT treatments did not differ significantly from the control (CK). However, as the severity of LT stress increased, there was a noticeable downward trend in the leaf GA content. Among them, the leaf GA content was most significantly affected by LT stress under the 3 °C treatment.

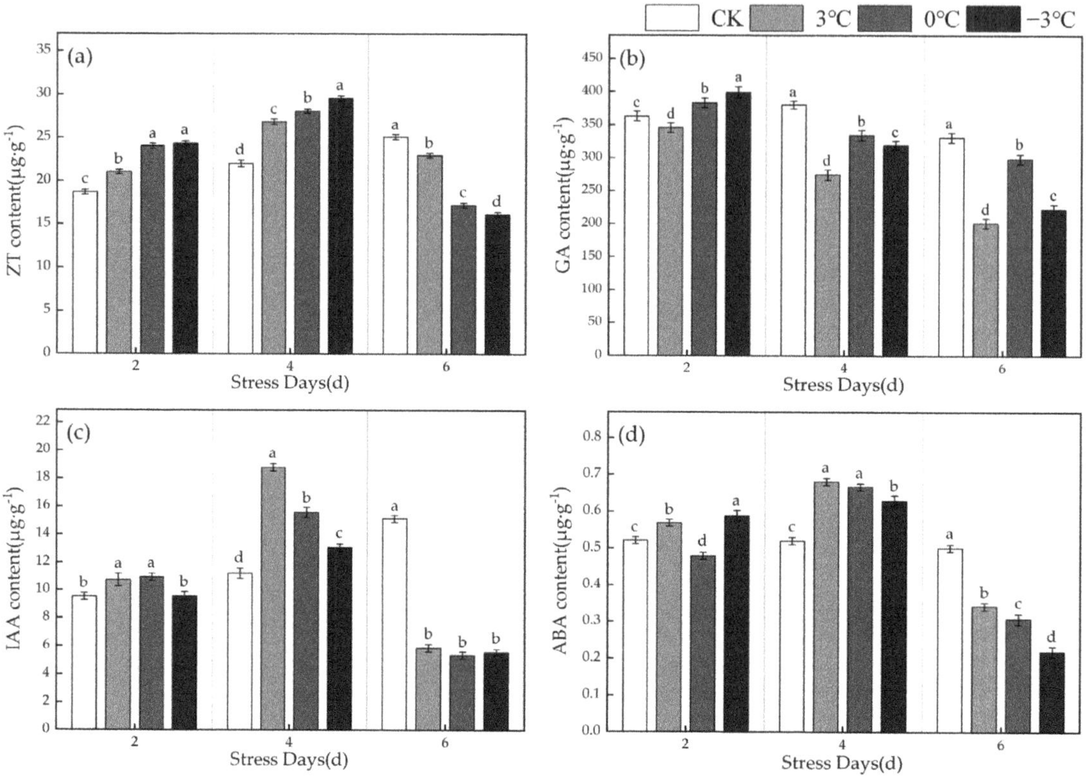

Figure 7. Changes in leaf endogenous hormone content of winter wheat leaves under different LT (low temperature) stresses. (**a**) The ZT (zeatin content) content. (**b**) The GA (gibberellin) content. (**c**) The IAA (indole-3-acetic acid) content. (**d**) The ABA (abscisic acid) content. Three replicates were used for determination, and each value is presented as "mean ± standard deviation (SD)". **Note:** Lowercase letters represent Duncan's test with $p < 0.05$.

Figure 7c illustrates changes in IAA content in winter wheat leaves under different LT stresses. From the figure, it can be observed that the IAA content in winter wheat leaves under T1, T2, and T3 treatments increased with time, indicating that the wheat leaves were in the growth stage. The IAA content in the leaves under various LT treatments showed an initial increase followed by a decreasing trend. Among them, the IAA content in the leaves under the T5 treatment reached a maximum value of 18.82 $\mu g \cdot g^{-1}$, an increase of 65.50% compared with T2. However, in the later stages of stress (6d), the IAA content in the leaves under all treatments remained at a lower level.

The effect of LT stress on the content of ABA in winter wheat leaves is shown in Figure 7d. It can be observed that with the increase in the LT stress, the leaf ABA content under each LT treatment shows a trend of increasing and then decreasing. Among them, the ABA content of winter wheat leaves under T5 treatment reached the maximum value of 0.68 $\mu g \cdot g^{-1}$, which increased by 30.83% compared with that of T2 treatment, while the ABA content of leaves under T12 treatment reached the minimum value of 0.22 $\mu g \cdot g^{-1}$, which decreased by 56.02% compared with that of T3.

3.5. Correlation Analysis of Various Physiological Indexes of Winter Wheat Leaves with Temperature and Duration Days

It can be seen that SOD activity and ZT content were negatively correlated with temperature, and all of them showed an increasing trend with the decrease in temperature. CAT activity, SOD activity, and chlb content were positively correlated with the number of days of duration, and all of them showed an increasing trend with the increase in the number of days of duration. Pn, Gs, and Tr showed highly significant correlations, and all of them decreased with the increase in the degree of LT stress. Also, chla content, chlb content, and chl(a + b) content showed highly significant correlations, all of which decreased with the increase in LT stress. CAT activity and POD activity showed highly significant correlations, all of which decreased with the increase in LT stress. ZT content, IAA content, and ABA content showed highly significant correlations, all of which decreased significantly with the increase in LT stress (Table 2).

Table 2. Correlation analysis.

Variants	Temperature	Stress Days	Pn	Gs	Tr	chla	chlb	chl (a + b)	car	POD	CAT	POD	ZT	GA	IAA	ABA
Temperature	1															
Stress days	0	1														
Pn	0.925 **	−0.134	1													
Gs	0.927 **	−0.265	0.943 **	1												
Tr	0.910 **	−0.247	0.929 **	0.901 **	1											
chla	0.242	−0.400	0.274	0.325	0.426	1										
chlb	0.054	0.291	0	−0.048	0.084	0.705 *	1									
chl(a + b)	0.199	−0.273	0.207	0.23	0.348	0.975 **	0.845 **	1								
Car	0.125	−0.326	0.088	0.167	0.292	0.848 **	0.710 **	0.862 **	1							
POD	0.224	−0.124	0.409	0.185	0.376	0.405	0.393	0.429	0.178	1						
CAT	0.199	0.214	0.397	0.135	0.352	0.447	0.501	0.494	0.221	0.910 **	1					
SOD	−0.898 **	0.116	−0.964 **	−0.888 **	−0.911 **	−0.422	−0.134	−0.360	−0.190	−0.543	−0.538	1				
ZT	−0.417	−0.469	−0.061	−0.044	0.071	0.243	−0.287	0.093	0.149	−0.021	0.050	−0.052	1			
GA	0.183	−0.749 **	0.157	0.262	0.382	0.446	0.125	0.376	0.570	0.180	0.029	−0.187	0.202	1		
IAA	0.236	−0.228	0.290	0.348	0.417	0.456	−0.098	0.314	0.480	0.030	0.068	−0.353	0.770 **	0.294	1	
ABA	0.122	−0.579 *	0.188	0.300	0.332	0.610 *	−0.042	0.447	0.535	0.079	0.082	−0.298	0.786 **	0.540	0.851 **	1

(1) ** denotes a highly significant correlation at $p < 0.01$; * denotes a significant correlation at $p < 0.05$. (2) Pn stands for net photosynthetic rate, Gs stands for stomatal conductance to water vapor, Tr stands for transpiration rate, chla stands for chlorophyll a content, chlb stands for chlorophyll b content, chl(a + b) stands for total chlorophyll content, car stands for carotenoid content, POD stands for peroxidase activity, CAT stands for catalase activity, SOD stands for superoxide dismutase activity, ZT stands for zeatin, GA stands for gibberellins, IAA stands for indole-3-acetic acid, and ABA stands for abscisic acid.

4. Discussion

Numerous studies have shown that during the jointing stages of winter wheat, its cold resistance is significantly reduced, making it unable to adapt to the adverse conditions of late spring frost [44,45]. The jointing stage is a critical period that affects the yield of winter wheat, and any LT damage occurring during this stage can cause irreparable losses to the yield [46,47]. In this study, we conducted experiments involving different LT stresses using the commonly cultivated winter wheat variety "Ji Mai 22" in China. We investigated the effects of different LT stresses on photosynthetic characteristics, senescence characteristics, and endogenous hormones of winter wheat leaves, and explored the relationship of each physiological index with temperature and days of stress through correlation analysis. This provides a scientific basis for the prevention of LT damage and the quantitative management of winter wheat growth and development.

Photosynthesis is an essential physiological process in the growth and development of winter wheat, which is affected by genetic and environmental factors, and its intensity can be reflected by various factors, such as stomatal conductance (Gs) and transpiration rate (Tr), which are closely related to the plant's status and metabolic level [48]. In this study, it was found that there was a highly significant positive correlation between Pn, Gs, and Tr in winter wheat leaves under LT stress, and they all decreased with the increase in the degree of LT stress. This is consistent with Fu's finding that LT stress limits the photosynthetic properties of winter wheat leaves [49]. This may be due to the fact that as the severity of LT stress increases, winter wheat decreases stomatal opening and transpiration rate to maintain temperature and leaf function, which leads to a decrease in net photosynthetic rate.

The leaf senescence characteristics of winter wheat can be observed from aspects such as chlorophyll content and protective enzyme activity [12]. In this study, it was found that

LT stress led to a decrease in chlorophyll content in winter wheat leaves, which is consistent with the results of recent studies that found a decreasing trend of chlorophyll with the increase in the degree of LT stresses [13,50]. On the one hand, LT stress reduces the activity of chlorophyll synthesis enzymes, thus inhibiting chlorophyll synthesis. On the other hand, LT slows down the metabolism of winter wheat, leading to insufficient raw materials for chlorophyll synthesis, resulting in a decrease in chlorophyll content. The contents of chla, chlb, and chl(a + b) were significantly correlated, and all of them decreased with the increase in the degree of LT stress, whereas the content of car was not significantly affected by LT stress, which may be attributed to the fact that LT stress is generally accompanied by drought, and drought will obviously cause car to rise [51].

Under LT stress, plants typically increase the activity of one or more antioxidant enzymes to adapt to the stress, and the increase in enzyme activity is generally associated with enhanced resistance [52]. Jin's research found that LT stress leads to an increase in SOD, POD, and CAT activity [53]. In this experiment, it was observed that with increasing severity of LT stress, SOD activity showed a continuous upward trend, while POD and CAT activities decreased in T11 and T12 treatments. In other words, POD and CAT exhibited an initial increase followed by a decrease with increasing LT stress severity. This may be attributed to the partial loss of leaf physiological functions when the severity of LT stress exceeds its tolerance limit, rendering the protective enzyme system ineffective. CAT activity and POD activity were highly correlated, while SOD activity was not significantly affected by LT stress, indicating that the SOD system in winter wheat leaves is more resistant to LT stress than CAT and POD.

The jointing stage is a crucial period for endogenous hormone regulation in winter wheat, and it is closely related to its yield [54]. This study found that as the severity of LT stress increased, the content of ZT in the leaves showed an initial increase followed by a decrease, while the GA content consistently decreased. This is consistent with the findings of Yang, who reported a significant decrease in ZT and GA content in winter wheat leaves under LT stresses [55]. The increase in ZT content in the early stages of leaf development was found to enhance the activity of SOD, thereby increasing the cold resistance of the leaves. However, as time progressed and temperature decreased, the growth of the winter wheat leaves slowed down, leading to a significant decrease in ZT synthesis. The decreasing trend of GA content indicated a reduction in the antiaging properties of the leaves under LT stress. In addition, this experiment also observed the content of IAA and ABA. The IAA content showed an initial increase (2–4d) followed by a decrease (4–6d) under LT stress. The increase in the early stage may be caused by the onset of LT stress when the leaves produce a certain degree of resistance, while the later decrease may be due to the fact that when the stress exceeds the leaf's ability to withstand stress, the leaf stops its metabolism. On the other hand, the enhancement of POD activity is due to the reduction of free radicals caused by LT stress at the same time and, also, to the degradation of the IAA. The ABA content also showed a trend of increasing and then decreasing, as well as the elevation of ABA content. On the one hand, LT promoted stomatal closure, which reduced water runoff and, at the same time, reduced the damage caused by LT stresses. On the other hand, the LT induced the production of some new proteins related to resistance, while the decrease might be due to the fact that the leaves did not metabolize when the temperature was low enough.

Although the study provides a detailed analysis of the changes in photosynthetic and senescence characteristics of winter wheat leaves under different LT stresses, there are still some limitations. In future experimental plans, we will continue to focus on selecting a more representative range of spring wheat, winter wheat, and winter wheat varieties. Additionally, we will attempt to conduct drought stress experiments. This will help overcome the limitations of using a single variety and not considering drought stress in our research. Through these experiments, we hope to gain a more comprehensive understanding and make comparisons of the physiological responses of winter wheat at the jointing stage to both low-temperature (LT) stress and drought stress.

5. Conclusions

With the decrease in temperature and the prolongation of stress, the changes in various winter wheat leaves were not linear. Among them, the photosynthetic parameters and carotenoids of wheat leaves showed a decreasing trend, while the chlorophyll b content showed an increasing trend. On the other hand, the chlorophyll a content, protective enzyme activity, zeaxanthin, auxin, and abscisic acid showed a trend of first increasing and then decreasing. The experiment discovered that, under 3 °C LT stress, the impact on the physiological parameters of the leaves was relatively minor, with a significant decrease observed only in pigment and endogenous hormone levels after 6 days of stress. Under 0 °C LT stress, the physiological parameters of wheat leaves significantly decreased after 4 days of stress, indicating that an LT environment of 0 °C or below for more than 4 days is not suitable for the growth of winter wheat. Under −3 °C LT stress, winter wheat leaves almost ceased photosynthesis, the protective enzyme system failed, and chlorophyll, carotenoids, and endogenous hormones remained at the lowest levels. This indicates that the temperature threshold for the growth of "Ji Mai 22" during the jointing stage is −3 °C. Considering the feasibility of future cold weather disasters, the research conducted in this paper focuses on the changes in photosynthetic parameters, aging characteristics, and endogenous hormone levels of wheat leaves under different levels of cold stress during the jointing stage. It provides an important theoretical basis for the classification of low-temperature disasters in the winter wheat variety "Ji Mai 22" during the jointing stage.

Supplementary Materials: The following supporting information can be downloaded at: https://www.mdpi.com/article/10.3390/agronomy13102650/s1. We provided additional information on the two-factor analysis of variance regarding the effects of temperature and duration on leaf chlorophyll content (Table S1), protective enzyme activity (Table S2), and endogenous hormone levels (Table S3) in winter wheat.

Author Contributions: Conceptualization, F.Z. and Z.Y.; methodology, Z.H.; software, F.Z.; validation, N.J., Z.Y. and Z.H.; formal analysis, F.Z.; investigation, F.Z.; resources, H.Z.; data curation, F.Z., H.Z. and N.J.; writing—original draft preparation, F.Z.; writing—review and editing, F.Z.; visualization, N.J.; supervision, H.Z.; project administration, Z.Y.; funding acquisition, Z.H. and Z.Y. All authors have read and agreed to the published version of the manuscript.

Funding: This research was funded by the National Key Research and Development Program Project 02, Topic: "Monitoring, Assessment, and Early Warning Forecast Research on Drought and Spring Frost Damage of wheat in the Huang-Huai-Hai Region" (2022YFD2300202).

Data Availability Statement: Data not readily available for public consumption due to privacy and other issues.

Acknowledgments: I would like to thank Supervisor Yang Zaiqiang and Huo Zhiguo for his guidance when I wrote this article, as well as Jiang Nan, Zhang Hanqi, Wu Xiangyi, and other colleagues for their help in revising the paper.

Conflicts of Interest: The authors declare no conflict of interest.

References

1. Seraya, T.; Bogatyreva, E.; Kirdun, T.; Machok, T.; Biryukova, O.; Belyavskaya, Y.A. Effect of fertilizer systems on winter wheat yield and grain quality in tillage systems involving traditional and surface treatment of sod–podzolic sandy loam soil. *Russ. Agricult. Sci.* **2022**, *48*, S44–S51. [CrossRef]
2. Ali, M.; Zhang, Y.; Rasheed, A.; Wang, J.; Zhang, L. Genomic prediction for grain yield and yield-related traits in chinese winter wheat. *Int. J. Mol. Sci.* **2020**, *21*, 1342. [CrossRef] [PubMed]
3. Kambona, C.M.; Koua, P.A.; Léon, J.; Ballvora, A. Intergenerational and Transgenerational Effects of Drought Stress on Winter Wheat (*Triticum aestivum* L.). *Physiol. Plant.* **2023**, *175*, e13951. [CrossRef] [PubMed]
4. Kim, M.; Park, J.; Kim, K.-M.; Kim, Y.; Kang, C.-S.; Son, J.; Ko, J.; Kim, K.-H. Low-Temperature effects on the growth and phytochemical properties of wheat sprouts. *Agriculture* **2022**, *12*, 745. [CrossRef]
5. Zhang, Y.F.; Liu, B.J. Detection of the Spatio-Temporal Differentiation Patterns and Influencing Factors of Wheat Production in Huang-Huai-Hai Region. *Foods* **2022**, *11*, 1617. [CrossRef] [PubMed]

6. Hu, X.; Ma, J.; Qian, W.; Cao, Y.; Zhang, Y.; Liu, B.; Tang, L.; Cao, W.; Zhu, Y.; Liu, L. Effects of low temperature on the amino acid composition of wheat grains. *Agronomy* **2022**, *12*, 1171. [CrossRef]
7. Wang, Z.F.; Song, L.Z.; Fan, Z.X.; Zhang, F.Y.; Liu, Y.T.; Wang, S.G.; Zhu, L.X.; Zhu, B.L.; Xu, M.Z.; Gao, H.Q.; et al. Photosynthesis and its effects on growth of winter wheat during winter. *Acta Agric. Nucl. Sin.* **2001**, *16*, 243–246. [CrossRef]
8. Zhai, J.Q.; Xun, J.B.; Xi, X.S. Characteristics and high-yield cultivation points of Jimai 22. Agric. *Eng. Technol.* **2020**, *40*, 56. [CrossRef]
9. Li, X.; Cai, J.; Liu, F.; Dai, T.; Cao, W.; Jiang, D. Cold priming drives the sub-cellular antioxidant systems to protect photosynthetic electron transport against subsequent low temperature stress in winter wheat. *Plant Physiol. Biochem.* **2014**, *82*, 34–43. [CrossRef]
10. Ge, J.; Liu, Z. Effects of low temperature stress on photosynthetic pigments, photosynthetic parameters and chlorophyll fluorescence characteristics of wheat at jointing stage. *J. Shanxi Agric. Sci.* **2021**, *49*, 1253–1256. [CrossRef]
11. Liu, N.; Yang, W.-X. Photosynthetic rate and water utilization of rainfed wheat with plastic mulching on the semiarid loess plateau, China. *Proc. Natl. Acad. Sci. India Sect. B Biol. Sci.* **2019**, *89*, 1047–1056. [CrossRef]
12. Fu, L.; Li, H.; Zhang, W.S.; Sheng, Y.; Han, S.Y. The effect of low temperature stress on the photosynthetic characteristics of winter wheat during different jointing processes. *New Agric.* **2021**, *13*, 69–71.
13. Li, W.Y.; Sun, M.X.; Ceng, F.L.; Wang, F.W. Hyperspectral estimation chlorophyll content in winter wheat leaves under low temperature stress. *Chin. J. Agrometeorol.* **2022**, *43*, 137–147. [CrossRef]
14. Sun, M.X. Hyperspectral Model Study on Physiological and Biochemical Parameters of Winter Wheat at Jointing Stage Under the Influence of Frost. Master's Thesis, Anhui Agricultural University, Anhui, China, 2022.
15. Li, Y.; Kang, J.H.; Wu, H.L. The effects of after-biooming high temperature stress on spring wheat yield and leaf protection enzyme activity. *Ningxia J. Agric. For. Sci. Technol.* **2017**, *58*, 1–4, 16.
16. Sheng, J. A Study on the Relationship between Endogenous Hormones and Grain Development and Plumpness in Wheat. Master's Thesis, Yangzhou University, Yangzhou, China, 2001.
17. Zhou, Y.; Zhu, X.; Guo, W.; Feng, C. Effects of climate change on wheat yield and nitrogen losses per unit of yield in the middle and lower reaches of the Yangtze River in China. *Atmosphere* **2023**, *14*, 824. [CrossRef]
18. Borovik, O.A.; Grabelnych, O.I.; Koroleva, N.A.; Pobezhimova, T.P.; Voinikov, V.K. The Influence of Carbohydrate Status and Low Temperature on the Respiratory Metabolism of Mitochondria from Etiolated Leaves of Winter Wheat. *J. Stress Physiol. Biochem.* **2014**, *10*, 118–130.
19. Liu, H.; Li, M.; Zhang, J.; Sun, H.; Long, Y.; Wu, L.; Zhang, Q. PCA based model on chlorophyll content diagnosis of winter wheat. *IFAC Pap. Online* **2018**, *51*, 643–647. [CrossRef]
20. Venzhik, Y.V.; Moshkov, I.E. The role of ultrastructural organization of cells in adaptation of winter wheat to low temperature. *Russ. J. Plant Physiol.* **2023**, *70*, 100. [CrossRef]
21. Mu, Q. Physiological Response Mechanism to Drought-Re-Watering and Water Regulation Strategy of Winter Wheat. Ph.D. Thesis, Northwest A&F University, Yangling, China, 2022. [CrossRef]
22. Zhang, J.; Sun, S.G.; Wang, L.M.; Wang, X.J.; Yang, Q.H.i; Chen, X.H. Physiological and biochemical characteristics and grain yield of winter wheat under low temperature at booting stage. *Acta Bot. Boreali Occident. Sin.* **2013**, *33*, 2249–2256.
23. Zhang, W.; Wang, J.; Huang, Z.; Mi, L.; Xu, K.; Wu, J.; Fan, Y.; Ma, S.; Jiang, D. Effects of low temperature at booting stage on sucrose metabolism and endogenous hormone contents in winter wheat spikelet. *Front. Plant Sci.* **2019**, *10*, 498. [CrossRef]
24. Li, Y.; Cui, D.; Huang, C.; Sui, X.; Fan, Q.; Chu, X. Dynamic changes of cell morphology and endogenous hormones during grain development of new wheat variety jimai 70. *Shandong Agric. Sci.* **2021**, *53*, 117–121. [CrossRef]
25. Wang, X.; Yu, J.; Yang, Y.; Cang, J.; Li, Z.F. Changes of endogenous hormones of winter wheat varieties with different cold-resistances under low temperature. *J. Triticeae Crops* **2009**, *29*, 827–831.
26. Wang, F.; Li, Y.X.; Jiang, L.P.; Zhao, M.H.; Lu, Z.M.; Yang, Y.C.; Zhang, J.Q.; Zhao, X.Y. Research progress on plant molecular response mechanisms under low temperature stress. *World For. Res.* **2020**, *33*, 15–21. [CrossRef]
27. Wang, G.Q.; Yang, C.W.; Tian, D.W.; Pan, G.Z. Study on High-Yielding Cultivation Techniques of Green High Quality Wheat in Huang-Huai-Hai Area. *Grain Sci. Technol. Econ.* **2018**, *43*, 96–98. [CrossRef]
28. Wang, L.Y.; Zhang, Y.S.; Dai, L.L. Artificial Selection Trend of Wheat Varieties Released in Huang-Huai-Hai Region in China Evaluated Using DUS Testing Characteristics. *Front. Plant Sci.* **2022**, *13*, 56–58. [CrossRef]
29. Pecina, M.; Ikic, I.; Tomasovic, S.; Mlinar, R.; Janjecic, Z. Targeting genotypes onto winter wheat growing environments using ammi model analysis. *Cereal Res. Commun.* **2008**, *36*, 1603–1606.
30. Wei, T.T.; Yang, Z.Q.; Wang, L.; Zhao, H.L.; Li, J.S. Simulation model of hourly air temperature inside glass greenhouse and plastic greenhouse. *Chin. J. Agrometeorol.* **2018**, *39*, 644–655. [CrossRef]
31. Xie, W.Q.; Wang, S.S.; Yan, X.D. Diurnal temperature range in winter wheat–growing regions of China: CMIP6 model evaluation and comparison. *Theor. Appl. Climatol.* **2023**, *152*, 123–134. [CrossRef]
32. Deng, X.; Xu, X.X.; Sun, Q.; Zhang, Y.L.; Zhu, Z.X.; Hao, T.J.; Gao, G.L.; He, X.Y.; Wang, X.L.; Zhao, C.X. Photosynthetic characteristics and transcriptome analysis of winter wheat seedlings under different salt concentration stress. *J. Plant Physiol.* **2023**, *25*, 1–11. [CrossRef]
33. Ye, Z.P. Photosynthesis affects light and CO_ Research progress on 2-response models. *J. Plant Ecol.* **2010**, *34*, 727–740. [CrossRef]
34. Li, H.S. *Experimental Principles and Techniques of Plant Physiology and Biochemistry*; Higher Education Press: Beijing, China, 2000.

35. Bai, Y.H.; Zhao, Z.Y.; Wang, Y.J.; Li, J.Y.; Zhao, F.; Chen, G.L. Changes rule of protective enzyme activity during fruit growth of Goutou jujube. *J. Shanxi Agric. Sci.* **2020**, *48*, 345–348.
36. Wang, M.; Zhang, X.; Jia, W.; Zhang, C.; Boczek, T.; Harding, M.; Liu, Y.; Li, M.; Zhang, S.; Lei, S. Circulating glutathione peroxidase and superoxide dismutase levels in patients with epilepsy: A meta-analysis. *Seizure* **2021**, *91*, 278–286. [CrossRef] [PubMed]
37. Zhao, Y.; Li, Y.S.; Gao, X.F. A new method for accurate determination of peroxidase activity based on fluorescence decrease of guaiacol. *Chin. J. Anal. Chem.* **2015**, *43*, 1040–1046. [CrossRef]
38. Wang, Q.; Liu, C.W.; Xu, W.J. Ultraviolet spectrophotometry measurement of catalase activity in maize. *Chin. Agricl. Sci. Bull.* **2016**, *32*, 159–165.
39. Yang, T.X.; Wei, A.Z.; Zheng, Y.; Yang, H.; Yang, X.N.; Zhang, R. Simultaneous determination of 8 endogenous hormones in apricot floral bud by high performance liquid chromatography. *Anal. Chem.* **2007**, *35*, 1359–1361. [CrossRef]
40. Zhang, Z.C.; Sun, Z.H.; Zheng, X.N.; Liu, H.F. Determination of four endogenous hormones in the young fruit of korla fragrant pear by high performance liquid chromatography (HPLC). *Xinjiang Agric. Sci.* **2017**, *54*, 886–892. [CrossRef]
41. López-Carbonell, M.; Moret, A.; Nadal, M. Changes in cell ultrastructure and zeatin riboside concentrations in Hedera helix, Pelargonium zonale, Prunus avium, and Rubus ulmifolius leaves infected by fungi. *Plant Dis.* **1998**, *82*, 914–918. [CrossRef]
42. Yan, J.; Zhao, L.; Zhang, Y.; Liu, M.; Yang, Y.; Liu, Z.; Zhang, L. Effects of microclimatic factors on stomatal conductance of plants in vertical greenery systems in humid subtropical areas. *Sustain. Cities Soc.* **2022**, *85*, 104056. [CrossRef]
43. Wang, Q.; Jin, J. Leaf transpiration of drought tolerant plant can be captured by hyperspectral reflectance using PLSR analysis. *iForest Biogeosci. For.* **2015**, *9*, 30. [CrossRef]
44. Lu, Q.; Guo, F.; Xu, Q.; Cang, J. LncRNA improves cold resistance of winter wheat by interacting with miR398. *Funct. Plant Biol.* **2020**, *47*, 544–557. [CrossRef]
45. Cao, Y.Y.; Ge, C.B.; Qi, S.L.; Liao, P.A.; Huang, J.; Wang, J.; Li, L.L.; Qiao, J.L.; Zhang, Z.Y. Physiological response to low temperature stress and cold resistance evaluation of different winter wheat varieties (lines) at jointing stage. *Jiangsu Agric. Sci.* **2022**, *50*, 59–66. [CrossRef]
46. Xie, Z.; Jiang, D.; Cao, W.; Dai, T.; Jing, Q. Relationships of endogenous plant hormones to accumulation of grain protein and starch in winter wheat under different post-anthesis soil water statusses. *Plant Growth Regul.* **2003**, *41*, 117–127. [CrossRef]
47. Wu, Y.; Gong, Z.; Ji, L.; Ma, J. Estimating daily minimum grass temperature to quantify frost damage to winter wheat during stem elongation in the central area of Huang-Huai plain in China. *Environ. Sci. Pollut. Res.* **2023**, *30*, 61072–61088. [CrossRef]
48. Fu, Y.S.; Tan, Z.; Li, W.Y. Effects of low temperature at jointing stage on grain yield and fluorescence parameters of flag leaves in wheat grain filling. *Ecol. Sci.* **2022**, *41*, 33–40. [CrossRef]
49. Xu, Y.S.; Gao, Y.; Chen, S.; Li, W.B.; Li, Y.J.; Shi, Y. Effects of compound water retaining agent on photosynthesis, senescence and yield of flag leaf of winter wheat in saline-alkali land. *Agric. Res. Arid Areas* **2023**, *41*, 38–49. [CrossRef]
50. Wang, T.; Gao, M.; Cao, C.; You, J.; Zhang, X.; Shen, L. Winter wheat chlorophyll content retrieval based on machine learning using in situ hyperspectral data. *Comput. Electron. Agric.* **2022**, *193*, 106728. [CrossRef]
51. Zhang, S.H. Study on the Synthesis Mechanism of Carbohydrate, Carotenoid and Oil During the Development of Maize Kernel and Its Related Application. Ph.D. Thesis, Tianjin University, Tianjin, China, 2021.
52. Chen, C. Relationship Between Sugar Metabolism, Antioxidant Activity and Cold Tolerance in Winter Wheat Under Low Temperature Conditions. Master's Thesis, Northeast Agricultural University, Harbin, China, 2014. [CrossRef]
53. Jin, L.H.; Feng, Q.; Pan, S.Q.; Miao, S.J.; Qiao, Y.F. Effects of warming on protective enzyme activities of winter wheat in different soils. *J. Agric. Env. Sci.* **2023**, *42*, 1228–1237. [CrossRef]
54. Dong, H.; Zhong, Y.L.; Qi, L.C.; Wang, H.T.; Zhang, Y.Q.; Song, H.; Dong, Z.R. Effect of tillage methods on endogenous hormone content and yield of winter wheat. *J. Triticeae Crops* **2015**, *35*, 542–547. [CrossRef]
55. Chen, S.Y.; Niu, J.F.; Zhang, X.Y. Effects of straw mulching temperature effect on agronomic and physiological traits of winter wheat. *Chin. J. Ecol. Agric.* **2022**, *30*, 820–830. [CrossRef]

agronomy

Article

A Quantitative Reconstruction of Nutrient Changes of Quaternary Red Soils (Luvisols) Affected by Land-Use Patterns

Ying-Ying Jiang [1], Zhong-Xiu Sun [2,*], Ruo-Meng Wang [2], Hong-Ling Wang [1] and Jia-Qing Wang [1,*]

[1] College of Life Engineering, Shenyang Institute of Technology, Shenfu Reform and Innovation Demonstration Zone, Shenyang 113122, China; yingying9111@126.com (Y.-Y.J.); wanghongling@situ.edu.cn (H.-L.W.)
[2] College of Land and Environment, Shenyang Agricultural University, Shenyang 110866, China
* Correspondence: zhongxiusun@syau.edu.cn (Z.-X.S.); jiaqing5212@163.com (J.-Q.W.); Tel.: +86-15734005989 (Z.-X.S.)

Abstract: The Quaternary red soil widely distributed in China is an important arable land resource. A quantitative understanding of nutrient changes of Quaternary red soils under different land-use patterns is the necessary premise for effective regulation, management, and sustainable utilization. In this study, five typical Quaternary red soil profiles under different land-use patterns were taken as the research object in Chaoyang City, Liaoning Province, China. The results showed that: (1) Buried Quaternary red soils were minimally affected by external disturbances. The contents of nitrogen (around 0.02%), phosphorus (ranging from 0.06% to 0.07%), and potassium (ranging from 3.12% to 3.50%) were at relatively low levels and homogeneously distributed with depth. (2) The total nitrogen content of red soils under each land-use pattern showed an increasing trend in the upper part of the profile (A and B horizons), and a sequence of woodland (CL-04) > grassland (CL-03) > arable land (CL-05) = sparse forest–grassland (CL-02). The nitrogen content in the lower part of different land-use patterns was about 0.02%. The phosphorus content of the topsoil layer remained unchanged (ranging from 0.05% to 0.06%), while the subsoil phosphorus decreased to varying extents. The potassium experienced leaching in both topsoil and subsoil layers, with the topsoil losses being lower than that in the subsoil. The range of total potassium content in the grassland (CL-03) ranged from 2.64% to 4.21%, from 3.91% to 4.44% for sparse forest–grassland (CL-02), from 2.41% to 2.63% for woodland (CL-04), and 2.85% to 2.92% for arable land (CL-05), respectively. The variation in nutrient content was related to the vegetation type, coverage rate, artificial fertilization method and species, etc. The accumulative mass change in the sparse forest–grassland increased by 384.16 g/100 cm^2, and the other land-use patterns showed a decreasing trend of arable land (83.71 g/100 cm^2) > woodland (83.71 g/100 cm^2) > grassland (83.71 g/100 cm^2), with the topsoil leaching losses being smaller than those in the subsoil layer. The characteristics of windbreak, sand fixation, and soil and water conservation of the sparse forest–grassland could well hold the nutrient-rich loess sediments, resulting in increased nutrients in the Quaternary red soil, which is a reasonable land-use pattern for the Chaoyang area.

Keywords: soil nutrient; nutrient fluxes; horizon evolution; arable land; management

1. Introduction

Quaternary red soils have experienced intense weathering, leaching, desilication, and iron and aluminum enrichment under warm and humid climate conditions during the Early and Middle Pleistocene of the Quaternary or even earlier periods [1,2]. Quaternary red soils retain their original characteristics formed during their formation periods, as well as features derived from the combined effects of various factors such as biological and human influences [3]. Quaternary red soils are widely distributed in China, mainly in areas south of 30° N [4], and are important arable land resources. In northern regions, Quaternary red soils are mostly buried underground and are exposed to varying degrees at the surface due to erosion and other processes. The quality of red soils varied with the influence of different land-use patterns [3]. Unscientific land use and management can decrease the Quaternary red soil quality, such as soil erosion, tillage pan formation, structural damage, and fertility decline, threatening regional agricultural and economic sustainable development [3,5]. Therefore, understanding the changes and evolution of Quaternary red soil under different land-use patterns is a prerequisite for effectively regulating, managing, and sustainably utilizing red soil resources [6,7].

As an essential component of soil quality, the soil nutrient composition affects the soil ecosystem and soil quality [8]. Its distribution, form, content, transformation, and migration are influenced by natural factors such as parent material and climatic conditions, and it is also affected by human factors such as different land-use patterns [9]. Land-use patterns of human intervention directly influence the soil nutrient composition [10]. Appropriate and rational land-use patterns can improve soil nutrient composition and enhance soil resistance to external environmental factors. Conversely, irrational land-use and management can alter the earth's biogeochemical cycles [11], cause soil property and quality degradation, result in incalculable damage, and threaten the land resources on which humans depend. Therefore, a quantitative reconstruction of the changes in the nutrient composition of Quaternary red soils under different land-use patterns is of great significance for understanding the evolution of the red soil quality and its regulation and management.

Quaternary red soils are widely distributed in China. In southern China, the area of Quaternary red soils is more than 101.87×10^4 km^2, which has been studied deeply. The total area of Quaternary red soils in northern China is 1013 km^2, which has been poorly studied. Chaoyang City in Liaoning Province, a low-lying hilly area with a wide distribution of Quaternary red soils, was selected as the typical research area for this study. The soil reconstruction model was used to study the evolution of Quaternary red soils under different land-use patterns. The research results will provide a theoretical basis for agricultural production practice in the Quaternary red soil distribution area.

2. Materials and Methods

2.1. The Study Area

The Quaternary red soil widely distributed area, Beipiao, Chaoyang City, Liaoning Province in northeast China (Figure 1), was selected by observing field investigation results and consulting the China Soil Series, Liaoning Volume, and Liaoning Soil. It is a mountainous and hilly area and is located in the transition between the Inner Mongolia Plateau and the eastern coastal plain [12]. The area has been influenced by a northern temperate continental monsoon, with a current average temperature of 5.4–8.7 °C and a current annual precipitation of 450–580 mm [13,14]. The Quaternary red soil can be classified as Argosols, Alfisols, and Luvisols in the Chinese Soil Taxonomy [15], Soil Taxonomy [16], and the World Reference Base for Soil Resources [17], respectively. The buried Quaternary red soil (MC-02), 210 cm at depth, without the influence of the modern climate and human activities, was investigated as the reference base. It had 2.5 YR Hue, silty loam texture, blocky structure, hard consistence (dry), and showed few clay coatings and iron-manganese nodules.

2.2. Sample Collection and Pretreatment

Typical Quaternary red soils of sparse forest–grassland (CL-02), grassland (CL-03), woodland (CL-04), and arable land (CL-05) were collected (Figure 2). The main plants of each land-use pattern are in Table 1. They are in a relatively flat and stable part derived from the same stratum and thus share the same topography, parent material, climate, biology, and formation time with the buried Quaternary red soil (MC-02). The upper part (0–68 cm) of the MC-02 is loess, and the underlying lower part (68–210 cm) is the buried Quaternary red soil. Therefore, the underlying lower part (68–210 cm) was used as the reference base.

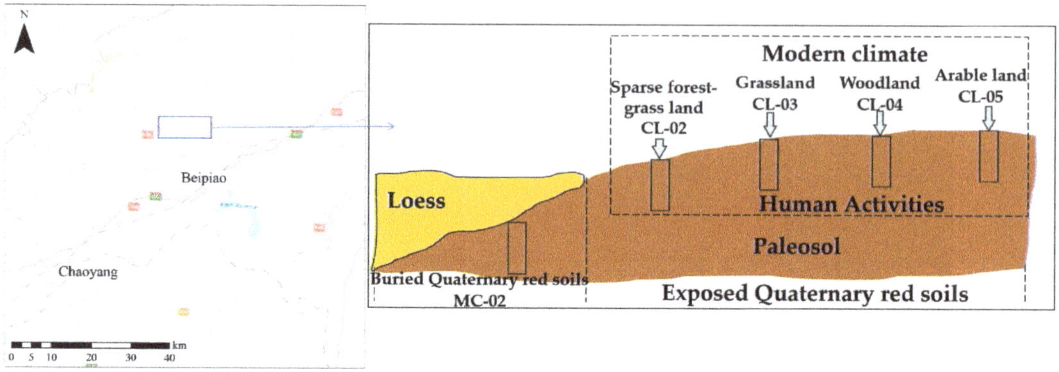

(**a**) (**b**)

Figure 1. (**a**) The location map of Chaoyang City, Liaoning Province, China. The schematic map was plotted based on the World Topographic Map (2016) using Arc GIS 10.2.2. (**b**) The sampling points schematic distribution map.

Figure 2. The Quaternary red soil view under different land-use patterns.

Table 1. The main plants of different land-use patterns.

ProfileNo.	Land-Use Patterns	Vegetation Coverage	The Main Plants
CL-02	Sparse forest–grassland	30%	The herbaceous plants: *Themeda triandra Forssk.* The woody plants: *Vitex negundo* var.; *heterophylla (Franch.) Rehd., Ulmus davidiana* var.; *japonica (Rehd.) Nakai, and Pinus tabuliformis Carrière.*
CL-03	Grassland	30%	The herbaceous plants: *Themeda triandra Forssk*
CL-04	Woodland	35%	The woody plants: *Vitex negundo* var.; *heterophylla (Franch.) Rehd.; Ulmus davidiana* var.; *japonica (Rehd.) Nakai, and Pinus tabuliformis Carrière.*
CL-05	Arable land	-	*Zea mays* L.

A typical soil profile was investigated and samples from pedogenic horizons were collected under different land-use patterns. The morphologies of Quaternary red soil profiles were described according to the Manual of Soil Description and Sampling [18]. The soil samples were transported to the laboratory and airdried. The soils were also sieved through 100 mesh for upcoming analysis. The investigated profiles were derived from loess and determined to have a parent material uniformity [3].

2.3. Laboratory Methods

2.3.1. Soil Total Nitrogen

The 0.04–0.05 g 100 mesh soil sample was first packed and compacted, and measured by an elemental analyzer produced by the Elementar Analysensysteme (Vario EL III, Elementar Company, Langenselbold, Germany) for the soil total nitrogen content (%) [19].

2.3.2. Soil Total Phosphorus and Total Potassium

The 0.25 g 100 mesh soil sample was thoroughly mixed with 2 g sodium hydroxide in the nickel crucible. Then, it was put into a high-temperature electric furnace and melted at 720 °C for 15 min. The melt material was finally dissolved and made up to 100 mL volumetric bottle for the determination of the total phosphorus [20] and total potassium [21].

2.4. Soil Quantitative Reconstruction Model

The soil reconstruction model (SR) was used to quantitatively calculate gains or losses of a specific component in a volume of 100 cm^3 of a weathered soil layer during pedogenesis [22]. The reconstruction results of a specific component in the unit area can be quantitatively compared with each other. The unit volume factor (UVF) can describe the changes in soil volume.

The equations are as follows.

$$UVF = \frac{BD_w \times C_{iw}}{BD_{pm} \times C_{ipm}} \quad (1)$$

$$D_{jw} = BD_w \times C_{jw} - UVF \times \left(BD_{pm} \times C_{jpm}\right) \quad (2)$$

Note: BD_w represents the bulk density of a weathered soil layer; BD_{pm} represents the bulk density of the parent material of the weathered soil layer; C_{iw} represents the content of a stable, referenceable component in the weathered soil layer; Ci_{pm} represents the content of a stable, referenceable component i in the parent material; C_{jw} represents the content of component j in the weathered soil layer; C_{jpm} represents the content of component j in the parent material of the weathered soil layer; D_{jw} represents the change in the content of component j in the weathered soil layer during pedogenesis.

2.5. Data Processing and Analysis

The Descriptive Statistics and Calculated Variable in SPSS Statistics 17.0 were used for data statistics. The SigmaPlot 12.5 was used for plotting graphs.

3. Results

The buried red soil has a Hue of 2.5 YR, a value of 6 to 7, and a Chroma of 6 to 8. It has a silty loam texture, blocky structure, hard consistency in dry conditions, a few clay coatings, and iron–manganese nodules. These morphologies distribute uniformly with depth. Compared to the buried red soil, exposed red soils under different land-use patterns have similar structure and consistency, while their color and texture slightly vary with depth. Their topsoil (A horizon) and subsoil (B horizon) increase to 5 YR in Hue, with a darkening value between 4 and 6. The texture of red soils under the sparse forest–grassland, grassland, and forestland varies from silty loam to silt clay loam/clay loam with depth, while arable land remains unchanged and distributes uniformly for silty loam with depth. The morphology in terms of color and texture can relate to changes in total nitrogen content, phosphorus content, and potassium content with soil depth. The darkening color showed by value and finer texture relates to increased total nitrogen, unchanged phosphorus, and decreased potassium content, and also relates to their fluxes showing complex nonlinearity relationships.

3.1. Distribution Characteristics of the Total Nitrogen in Quaternary Red Soils under Different Land-Use Patterns

The nitrogen content of the buried Quaternary red soil (MC-02) is around 0.02% and homogeneously distributed with depth. The buried Quaternary red soils were exposed at the surface, and affected by different land-use patterns, resulting in significant changes in the soil's total nitrogen content within the profile. The soil total nitrogen in woodland (CL-04) and grassland (CL-03) varied greatly, ranging from 0.027% to 0.158% and 0.017% to 0.078%, respectively. The arable land (CL-05) and sparse forest–grassland (CL-02) have relatively smaller ranges of soil total nitrogen, from 0.019% to 0.057% and 0.018% to 0.055% (Figure 3), respectively. The total nitrogen content of Quaternary red soils under different land-use patterns showed a decreasing trend with depth. However, the total nitrogen content of red soils under each land-use pattern showed an increasing trend in the upper part of the profile (A and B horizons), and a sequence of woodland (CL-04) > grassland (CL-03) > arable land (CL-05) = sparse forest–grassland (CL-02). The nitrogen content in the lower part of different land-use patterns was about 0.02%, which was basically consistent with the buried Quaternary red soil (Figure 3).

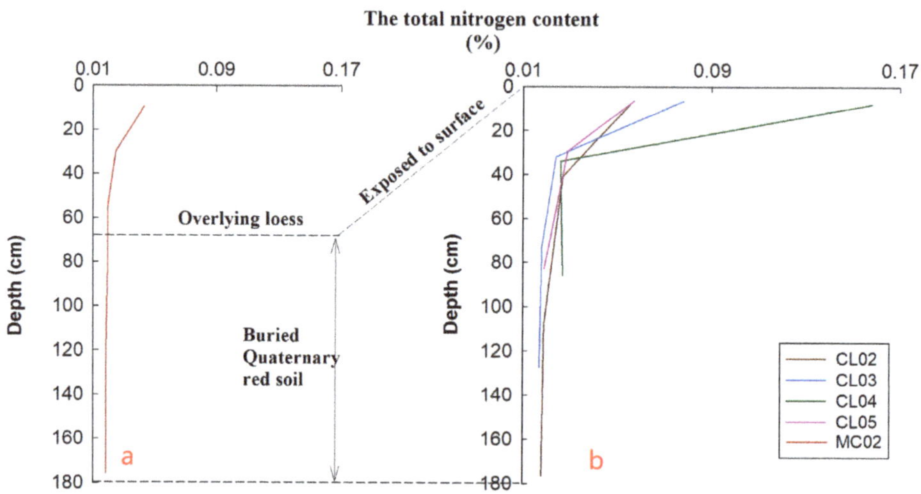

Figure 3. Distribution characteristics of the soil total nitrogen in Quaternary red soils. (**a**) Buried Quaternary red soil; (**b**) exposed Quaternary red soils under different land-use patterns.

3.2. Distribution Characteristics of the Total Phosphorus in Quaternary Red Soils under Different Land-Use Patterns

The total phosphorus content in buried Quaternary red soils (MC-02) did not change significantly with the depth, ranging from 0.06% to 0.07%. The total phosphorus content under different land-use patterns changed slightly, ranging from 0.05% to 0.06%, which was greater than that of the buried Quaternary red soil. The total phosphorus content of red soils under different land-use patterns showed that the upper part was greater than the lower part (Figure 4).

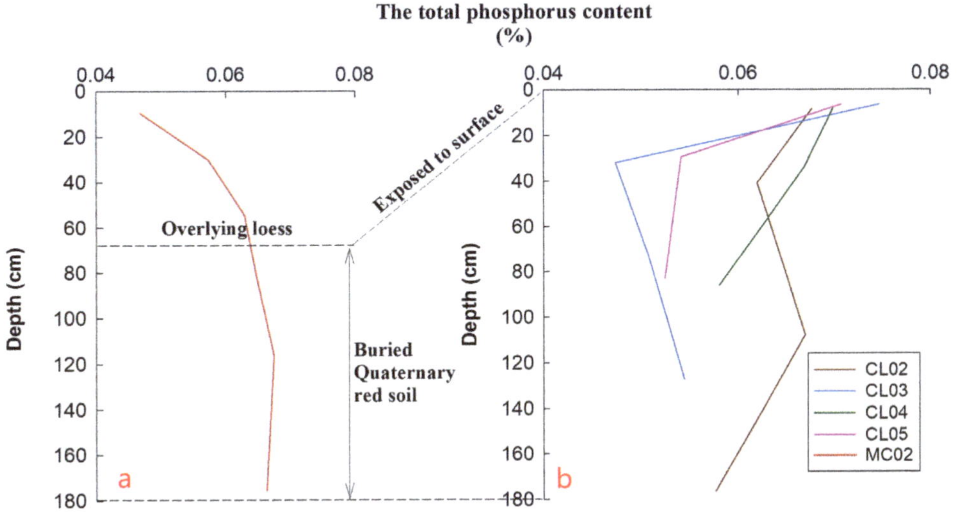

Figure 4. Distribution characteristics of the total phosphorus in Quaternary red soils. (**a**) Buried Quaternary red soil; (**b**) exposed Quaternary red soils under different land-use patterns.

3.3. Distribution Characteristics of the Total Potassium in Quaternary Red Soils under Different Land-Use Patterns

The potassium content in the buried Quaternary red soils (MC-02) first increased and then decreased with soil depth, ranging from 3.12% to 3.50% (Figure 5). The potassium content in the soil below 120 cm was uniformly distributed with a depth of 3.12%. The range of total potassium content in the grassland (CL-03) was the largest, ranging from 2.64% to 4.21%. The content of total potassium in sparse forest–grassland (CL-02) varied from 3.91% to 4.44%. The total potassium content of the other two land-use patterns was smaller than that of the buried Quaternary red soil, which ranged from 2.41% to 2.63% for woodland (CL-04) and 2.85% to 2.92% for arable land (CL-05), respectively (Figure 5).

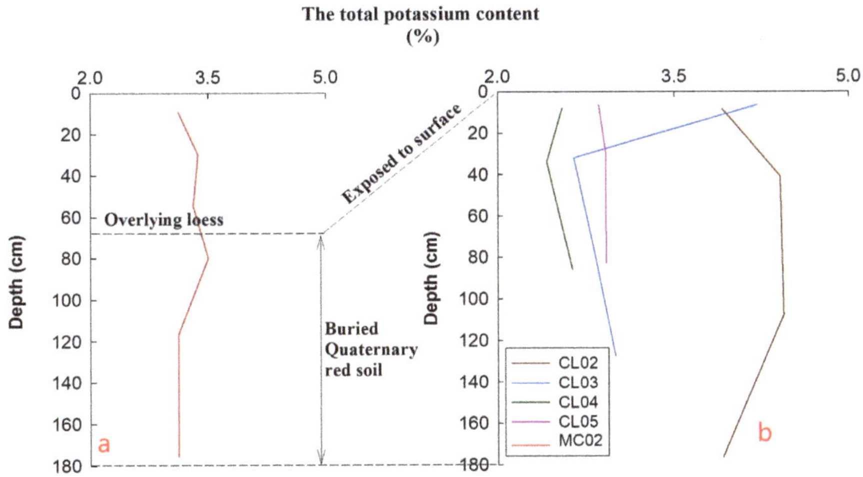

Figure 5. Distribution characteristics of the total potassium in Quaternary red soils. (**a**) Buried Quaternary red soil; (**b**) exposed Quaternary red soils under different land-use patterns.

3.4. Results of the SR Model

Compared with the buried Quaternary red soils (MC-02), the soil nitrogen content in profiles under different land-use patterns had increased in the surface layer (layer A). Among them, the woodland (CL-04) had the largest increase, with 0.19 g/100 cm^3. The increase in sparse forest–grassland (CL-02), grassland (CL-03), and arable land (CL-05) was smaller, which were 0.05 g/100 cm^3, 0.07 g/100 cm^3, and 0.06 g/100 cm^3, respectively. There was almost no change in the soil total nitrogen of subsoil layers (B layer). The soil phosphorus content in subsoil layers (layer B) under different land-use patterns reduceded by 0.01 g/100 cm^3–0.03 g/100 cm^3. The soil potassium content varied greatly under different land-use patterns, among which the overall potassium content in the sparse forest–grassland (CL-02) increased by 1.05 g/100 cm^3 in the surface layer (A) and 1.98 g/100 cm^3 in the subsoil layer (B). The potassium content in the surface soil of the grassland (CL-03) increased by 1.29 g/100 cm^3, while that in the subsoil decreased by 0.56 g/100 cm^3. The changes in woodland (CL-04) were similar to the arable land (CL-05). And the potassium content in all layers showed a decreasing trend, with the surface layer decreasing by 0.55 g/100 cm^3 and 0.53 g/100 cm^3, and the potassium content in the subsoil layer was smaller than that in the surface layer, decreasing by 0.74 g/100 cm^3 and 0.70 g/100 cm^3, respectively.

4. Discussion

The soil nutrient content is an important indicator for evaluating soil fertility levels [9]. The plant growth and land productivity are limited by the soil nutrient availability [11]. The soil nutrient content can usually reflect the conditions of parent material, climate, and vegetation, and its vertical distribution can be used to explore the process of nutrient input, output, and circulation [23]. At the same time, soil nutrients can also reflect the ecosystem health [23]. Therefore, the characteristics of soil nutrients have become an important research field in pedology, agronomy, ecology, and environmental science [11,24].

4.1. Changes in the Total Nitrogen in Quaternary Red Soils under Different Land-Use Patterns

The soil nitrogen accumulated through the biological nitrogen fixation process and non-biological nitrogen fixation process [25]. Biological nitrogen fixation, the main nitrogen fixation, is the fixation of atmospheric molecular nitrogen into ammonia by nitrogen-fixing microorganisms in plants under normal temperature and pressure and the catalysis of corresponding enzymes, which is further transformed into amino acids and protein compounds [26]. Subsequently, with the death of plants, as residues such as fallen leaves and dead branches, nitrogen is transformed into the soil component. Different land-use types have different land cover vegetation. The types and contents of soil organic matter transformed by plant root exudates, dead leaves, and other residues lead to changes in soil microbial communities, resulting in the differentiation of soil nitrogen mineralization and transformation, thus affecting the soil nitrogen content distribution [27].

The buried Quaternary red soils (MC-02) are minimally affected by external factors (such as human activities) due to being buried, with less external nitrogen input and weaker soil microbial activity. Therefore, the nitrogen content in the buried Quaternary red soils was only about 0.02% and distributed homogeneously with depth (Figure 3). Compared with the buried Quaternary red soils, the surface of sparse forest–grassland (CL-02), grassland (CL-03), and woodland (CL-04) have plant invasion (herbs and shrubs). The external nitrogen fixed in plants returned to the soil with the death and decay of plants, thus increasing the nitrogen content in the surface soil to 0.05 g/100 cm^3, 0.07 g/100 cm^3, and 0.19 g/100 cm^3, respectively (Table 2). Although the arable land (CL-05) was covered by crops during the growing season, most of the soil nitrogen was taken away following the crop harvest. The soil surface nitrogen content increased by 0.06 g/100 cm^3, which was caused by nitrogen input into the soil through artificial fertilization and other human activities. These land-use patterns directly affected the nitrogen content in the topsoil [27], while the subsoil layer was minimally affected. So, the subsoil nitrogen content under different land-use patterns remained almost unchanged when compared to the buried Quaternary red soil (MC-02). The total accumulative change of soil nitrogen under different land-use patterns showed a trend of woodland (CL-04) (4.50 g/100 cm^3) > sparse forest–grassland (CL-02) (1.48 g/100 cm^3) > arable land (CL-05) (0.97 g/100 cm^3) > grassland (CL-03) (0.72 g/100 cm^3). The increase in woodland nitrogen was the largest, mainly due to its high plant coverage and a potentially large return rate. The large amounts of residues were accumulated and mineralized, which returned nutrients to the soil and improved the soil texture and moisture penetration to increase the activity of microorganisms and enhance the effect of soil nitrogen fixation [27,28]. At the same time, the root decomposition products in the soil produced acidic substances, promoting the transformation of insoluble minerals and other substances into effective nitrogen components, thus increasing the nitrogen content in the soil [28]. While the arable land and grassland had lower vegetation coverage and return rates, resulting in smaller increases of nitrogen content when compared to the buried Quaternary red soil.

Table 2. The quantitative reconstruction results of nutrient changes using the soil reconstruction model (SR).

Profile No.	Horizon	Depth (cm)	TN (g/100 cm^3)	TP (g/100 cm^3)	TK (g/100 cm^3)	MC (g/100 cm^3)
MC-02	2Btr2	90–100	0.00	0.00	0.00	0.00
CL-02	Ah	0–17	0.05	0.00	1.05	1.10
	Btr1	17–65	0.02	0.00	2.45	2.47
	Btr2	65–151	0.00	0.00	2.15	2.15
	Btr3	151–202	0.00	−0.02	1.24	1.22
	Variation of the layer B		0.00	−0.01	1.98	1.97
Variation of cumulative mass (g/100 cm^2)			1.48	−1.02	383.70	384.16
CL-03	Ah	0–13	0.07	0.01	1.29	1.37
	BA	13–51	0.00	−0.02	−0.59	−0.61
	Btr1	51–95	0.00	−0.03	−0.77	−0.80
	Btr2	95–160	0.00	−0.02	−0.41	−0.43
	Variation of the layer B		0.00	−0.02	−0.56	−0.59
Variation of cumulative mass (g/100 cm^2)			0.72	−3.49	−65.84	−68.61
CL-04	Ah	0–16	0.19	0.01	−0.55	−0.35
	Btr1	16–52	0.01	0.00	−1.14	−1.13
	Btr2	52–120	0.02	−0.01	−0.52	−0.51
	Variation of the layer B		0.01	−0.01	−0.74	−0.73
Variation of cumulative mass (g/100 cm^2)			4.50	−0.56	−85.54	−81.60
CL-05	Ap	0–13	0.06	0.00	−0.53	−0.47
	Bt	13–46	0.01	−0.02	−0.38	−0.39
	Btr	46–120	0.00	−0.03	−0.84	−0.87
	Variation of the layer B		0.00	−0.03	−0.70	−0.73
Variation of cumulative mass (g/100 cm^2)			0.97	−3.16	−81.52	−83.71

Note: MC indicates the mass change of total selected nutrients; TN indicates the total nitrogen; TP indicates the total phosphorus; TK indicates the total potassium.

4.2. Changes in the Total Phosphorus of Quaternary Red Soils under Different Land-Use Patterns

In the primary ecosystem, the form and circulation of phosphorus in soils are affected by factors such as the parent material, climate, environment, time, and topography. During the transition from native ecosystems to systems disturbed by human factors, the form, content, and distribution of soil phosphorus were affected by the input–output of soil and by human activities [29]. There is no gaseous form of phosphorus in the soil. The main ways of soil phosphorus input are plant return, artificial application of chemical fertilizers, and organic fertilizers. Soil phosphorus output is mainly through two ways:

plant absorption and runoff infiltration. Soil structure, soil properties, vegetation coverage, slope, fertilization method, fertilizer amount, and precipitation are factors affecting soil phosphorus output [30–32]. It is difficult for phosphorus to leach down in the soil, surface runoff is the main way of soil phosphorus loss [30].

The phosphorus content of buried Quaternary red soils varied between 0.05% and 0.07% with a homogeneous change with depth due to the fact that it was barely affected by external factors such as runoff (Figure 4). After being exposed to the land surface, the phosphorus of Quaternary red soils had been changed to varying degrees by different land-use patterns. Plants absorb the necessary phosphorus for their growth from deep soil layers. Although a small portion of phosphorus was returned to the surface soil through crop residue mineralization, most of the phosphorus remained in plants and was taken away through crop harvest. Additionally, surface runoff caused a part of phosphorus leaching [33], the topsoil phosphorus remained essentially balanced under different land-use patterns. The phosphorus in the subsoil decreased due to limited leaching. The losses of subsoil showed a trend of grassland (3.49 g/100 cm^3) > arable land (3.16 g/100 cm^3) > sparse forest–grassland (1.02 g/100 cm^3) > woodland (0.56 g/100 cm^3). Due to the low vegetation coverage rate of grassland and basically no coverage of arable land after the autumn harvest, the Quaternary red soils were vulnerable to erosion caused by wind and water, resulting in relatively high phosphorus loss in the profile. The woodland with a large surface coverage and stable soil structure can hold more phosphorus in soil when compared to others.

4.3. *Changes in the Total Potassium in Quaternary Red Soils under Different Land-Use Patterns*

Different land-use patterns lead to differences in soil physical and chemical properties and then affect the migration and transformation of potassium [34]. The main input modes of soil potassium are atmospheric dust deposition and external application of potassium fertilizers, the main output modes are leaching with the soil solution [35].

The potassium content of the buried Quaternary red soil varied between 3.12% and 3.51% and showed a homogeneous change with depth due to being hardly affected by external factors. Under different land-use patterns, potassium leaching occurred in topsoil and subsoil layers to varying degrees of Quaternary red soil profiles except for sparse forest–grassland, which had a greater loss in the subsoil than that in the topsoil. This could be explained that the potassium is highly soluble and easily migrates with soil solution, causing a large amount of potassium to leach downward out of the soil. In addition, plants absorb potassium from the subsoil layer to meet their growth and development [36]. Compared to buried Quaternary red soils, the potassium content of sparse forest–grassland increased by 383.70 g/100 cm^3, while others showed a decreasing trend of woodland (85.5 g/100 cm^3) > arable land (81.52 g/100 cm^3) > grassland (65.84 g/100 cm^3). The woodland had the largest loss of potassium. This may be due to the surface vegetation with a high coverage had absorbed a considerable amount of potassium from the soil, and the absorbed amount of potassium was greater than the returned amount [37]. The decrease in arable land was relatively larger, which may be due to that the crops absorbed potassium and were taken away for harvest, resulting in a very small return of soil potassium, leading to the decrease in potassium in soil [38]. At the same time, the low vegetation coverage of arable land was vulnerable to soil erosion, which was also a reason for the decrease in soil potassium [38]. As a combination of woodland and grassland, sparse forest–grassland had the characteristics of windbreak, sand fixation, and soil and water conservation, which could well hold a large amount of potassium-rich loess sediments, resulting in an increasing trend of potassium in the profile [39].

4.4. *The nutrient Evolution of Quaternary Red Soils under Different Land-Use Patterns*

Different land-use patterns have a significant impact on the soil nutrient composition and distribution [40]. The total nutrient mass change (MC) of Quaternary red soils has decreased with a wide range, which was due to the serious soil erosion in the Chaoyang area.

Compared to the buried Quaternary red soil, total nutrients in the topsoil and subsoil layers under different land-use patterns (except for the sparse forest–grassland) have experienced different degrees of losses. The loss of the topsoil layer was lower than that of the subsoil layer. This was because some nutrients could migrate downward with the soil solution to leach out of the soil, and the lost nutrients were not replenished in time [41]. In addition, plants absorbed nutrients from deep soil, and the returned amount was less than the absorbed amount, which was also a reason for the decrease in soil nutrients. Although fertilization would supply some nutrients in arable land, the applied fertilizer generally compounds fertilizer and organic fertilizer and does not provide a specific supplement for a specific element in soil. At the same time, soil erosion would also aggravate the nutrient leaching. The surface vegetation of the sparse forest–grassland consisted of woody and herbaceous plants, which could effectively prevent wind erosion, stabilize the sand, conserve water and soil, and efficiently retain nutrient-rich loess deposits in the soil. As a result, the accumulative mass change in sparse forest–grassland increased by 384.16 g/100 cm^2. However, the cumulative mass change of other land-use patterns showed a decreasing trend of arable land (83.71 g/100 cm^2) > woodland (83.71 g/100 cm^2) > grassland (83.71 g/100 cm^2).

5. Conclusions

Different land-use patterns can cause the redistribution of soil nutrients by affecting surface vegetation status and soil microbial species, etc., thus affecting land productivity and soil quality. Therefore, the quantitative research on the evolution characteristics of soil nutrients under different land-use patterns is of great significance for selecting and optimizing land-use patterns and improving soil productivity and land management levels.

(1) The buried Quaternary red soil was minimally affected by external disturbances. The contents of nitrogen, phosphorus, and potassium were at relatively low levels and homogeneously distributed with depth.

(2) Compared to the buried Quaternary red soil, the topsoil nitrogen content of Quaternary red soils increased under different land-use patterns, while the subsoil nitrogen content remained relatively stable. The phosphorus content of the topsoil layer remained unchanged, while the subsoil phosphorus decreased to varying extents. The potassium experienced leaching in both topsoil and subsoil layers, with the topsoil losses being lower than that in the subsoil.

(3) The variation in nutrient content was related to the vegetation type, coverage rate, artificial fertilization method, species, etc.

(4) Compared to the buried Quaternary red soil, their daytime counterparts showed different degrees of nutrient leaching loss in both topsoil and subsoil layers except for the spare forest–grassland, with the topsoil leaching losses being smaller than those in the subsoil layer.

(5) The characteristics of windbreak, sand fixation, and soil and water conservation of sparse forest–grassland could well hold the nutrient-rich loess sediments, resulting in increased nutrients, which is a reasonable land-use pattern in the Chaoyang area.

Author Contributions: Conceptualization, Y.-Y.J., Z.-X.S. and J.-Q.W.; Data Curation, Y.-Y.J. and Z.-X.S.; Formal Analysis, Y.-Y.J., Z.-X.S. and R.-M.W.; Funding Acquisition, Y.-Y.J., Z.-X.S., H.-L.W. and J.-Q.W.; Investigation, Y.-Y.J. and Z.-X.S.; Methodology, Y.-Y.J., Z.-X.S. and R.-M.W.; Project Administration, Y.-Y.J. and Z.-X.S.; Resources, Y.-Y.J. and Z.-X.S.; Software, Y.-Y.J., Z.-X.S. and J.-Q.W.; Supervision, Y.-Y.J. and Z.-X.S.; Validation, Y.-Y.J., Z.-X.S., H.-L.W. and J.-Q.W.; Visualization, Y.-Y.J. and Z.-X.S.; Writing—Original Draft, Y.-Y.J., Z.-X.S., R.-M.W. and J.-Q.W.; Writing—Review and Editing, Y.-Y.J., Z.-X.S. and J.-Q.W. All authors have read and agreed to the published version of the manuscript.

Funding: This research was funded by the Applied Basic Research Program of Liaoning Province (No. 2022JH2/101300167), the National Natural Science Foundation of China (No. 42277285), the Basic Scientific Research Project of Liaoning Provincial Department of Education (No. LJKZ1341), and the Applied Basic Research Program of Liaoning Province (No. 2023020244-JH2/1016).

Data Availability Statement: All data are true, valid, and can be made available.

Acknowledgments: The authors sincerely thank all the students and staff who provided input to this study. Our acknowledgments also extend to the anonymous reviewers for their constructive reviews of the manuscript.

Conflicts of Interest: The authors declare no conflict of interest. The funders had no role in the design of the study; in the collection, analyses, or interpretation of data; in the writing of the manuscript; or in the decision to publish the results.

References

1. Wang, D.P.; Wang, Q.B.; Han, C.L.; Wang, H.Q.; Liu, S.H. Study on fractal features of paleosol in Liaoning Province. *J. Anhui Agric. Sci.* **2007**, *35*, 6186–6189.
2. Han, C.L.; Wang, Q.B.; Sun, F.J.; Liu, S.H.; Chen, H. Properties and taxonomy of Quaternary paleo-latosol-like soils in Chaoyang area of Liaoning province. *Acta Pedol. Sin.* **2010**, *47*, 836–846.
3. Duan, S.Y.; Sun, Z.X.; Wang, Q.B.; Jiang, Y.Y.; Sun, F.J. Comparison of aggregate composition in quaternary red soil under different land use pattens. *Chin. J. Soil Sci.* **2020**, *51*, 587–596.
4. Sun, Z.X.; Owens, R.P.; Han, C.L.; Chen, H.; Wang, X.L.; Wang, Q.B. A quantitative reconstruction of a loess–paleosol sequence focused on paleosol genesis: An example from a section at Chaoyang, China. *Geoderma* **2016**, *266*, 25–39. [CrossRef]
5. Duan, S.Y.; Sun, Z.X.; Wang, Q.B.; Jiang, Y.Y.; Han, C.L.; Zhang, Y.W.; Lv, M.F.; Sun, F.J.; Chen, L.M. Characteristics of soil organic carbon distribution of quaternary red soils under different land use patterns. *Chin. J. Soil Sci.* **2021**, *52*, 1078–1084.
6. Wang, Q.B.; Han, C.L. Discussion on the Status of Plaeo-latosol-like Soils in the Chinese Soil Taxonomy. *J. Shenyang Agric. Uni.* **2008**, *39*, 3–6.
7. Sun, Z.X.; Jiang, Y.Y.; Wang, Q.B.; Owens, R.P. A fractal evaluation of particle size distributions in an eolian loess-paleosol sequence and the linkage with pedogenesis. *Catena* **2018**, *165*, 80–91. [CrossRef]
8. Zhou, J.M.; Shen, R.F. *Dictionary of Soil Science*; Science Press: Beijing, China, 2013.
9. Anne, K.; Tambet, K.; Priit, P.; Ain, K. Modeling topsoil phosphorus-From observation-based statistical approach to land-use and soil based high-resolution mapping. *Agronomy* **2023**, *13*, 1183.
10. Alatengxihuri; Zeng, X.B.; Bai, L.Y.; Li, L.F.; Sun, N.; Gao, J. Effect of different land use on farmland soil nutrients. *Res. Agric. Mod.* **2010**, *31*, 492–495.
11. Zhang, J.; Li, Z.; Luo, Y.; Wang, X.; Yang, D.; Zhang, X. Thickness of a Compost Layer on the Distribution of water and nutrients in a Surface-Drip-Irrigated Sandy Soil Column. *Agronomy* **2023**, *13*, 1181. [CrossRef]
12. Han, C.L.; Liu, S.H.; Wang, Q.B.; Wang, H.Q. Basic properties and genetic characteristic research of Quaternary paleosol in Chaoyang city of Liaoning. *Chin. J. Soil Sci.* **2009**, *6*, 1233–1239.
13. Sui, J.Y. Analysis on precipitation features during the flood season of Chaoyang in Liaoning Province in recent 58 Years. *J. Anhui Agric. Sci.* **2011**, *39*, 14931–14932.
14. Zhang, F.M.; Zong, Y.F. The characteristics of temperature change in Chaoyang area of Liaoning Province. *Anhui Agric. Sci. Bull.* **2015**, *21*, 150–160.
15. Chinese Soil Taxonomy Research Group, Institute of Soil Science Chinese Academy of Sciences; Cooperative Research Group on Chinese Soil Taxonomy. *Keys to Chinese Soil Taxonomy*, 3rd ed.; University of Science and Technology of China Press: Hefei, China, 2001. (In Chinese)
16. Soil Survey Staff. *Keys to Soil Taxonomy*, 12th ed.; United States Department of Agriculture (USDA), Natural Resources Conservation Service (NRCS), U.S. Government Printing Office: Washington, DC, USA, 2014.
17. IUSS Working Group WRB. *World Reference Base for Soil Resources 2014, Update 2015 International Soil Classification System for Naming Soils and Creating Legends for Soil Maps*; World Soil Resources Reports No. 106; Food and Agriculture Organization (FAO): Rome, Italy, 2015.
18. Zhang, G.L.; Li, D.C. *Manual of Soil Description and Sampling*; Science Press: Beijing, China, 2017. (In Chinese)
19. Li, J.W.; Jiao, X.G.; Sui, Y.Y.; Cheng, S.J. The Belonging of Dark Brown Soil in the East of Jilin Province in Chinese Soil Taxonomy. *Chin. Agric. Sci. Bull.* **2011**, *27*, 74–79.
20. *GB/T 9837-88*; Determination of Soil Total Phosphorus. Standards Press of China: Beijing, China, 1988.
21. *GB/T 9836-88*; Determination of Soil Total Potassium. Standards Press of China: Beijing, China, 1988.
22. Stolt, M.H.; Baker, J.C. Quantitative comparison of soil and saprolite genesis: Examples from the Virginia Blue Ridge and Piedmont. *Southeast. Geol.* **2000**, *39*, 129–150.
23. Manuel, D.B.; Fernando, T.M.; Gallardo, A.; Bowker, M.A. Decoupling of soil nutrient cycles as a function of aridity in global drylands. *Nature* **2013**, *502*, 672–676.
24. Brewer, R. *Fabric and Mineral Analysis*; Robert, E., Ed.; Krieger Publishing Company: New York, NY, USA, 1976.
25. Kahmen, A.; Buchmann, W.N. Foliar δ15N values characterize soil N cycling and reflect nitrate or ammonium preference of plants along a temperate grassland gradient. *Oecologia* **2008**, *156*, 861–870. [CrossRef]
26. Dorji, T.; Field, D.J.; Odeh, I.O. Soil Aggregate Stability and Aggregate associated Organic Carbon under different Land Use/Land Cover Types. *Soil Use Manag.* **2020**, *36*, 308–319. [CrossRef]

27. Richter, D.D.; Markewitz, D.; Wells, C.G.; Allen, H.L.; April, R.; Heine, P.R.; Urrego, B. Soil chemical change during three decades in an old-field loblolly pine (*Pinus taeda* L.) ecosystem. *Ecology* **1994**, *75*, 1463–1473. [CrossRef]
28. Maithani, K.; Arunachalam, A.; Tripathib, R.S.; Pandey, H.N. Nitrogen mineralization as influenced by climate, soil and vegetation in a subtropical humid forest in northeast India. *For. Ecol. Manag.* **1998**, *109*, 91–101. [CrossRef]
29. Debicka, M.; Jamroz, E.; Bekier, J.; Ćwieląg-Piasecka, I.; Kocowicz, A. The Influence of Municipal Solid Waste Compost on the Tranformations of Phosphorus Forms in Soil. *Agronomy* **2023**, *13*, 1234. [CrossRef]
30. Correll, D.L. The role of phosphorus in the eutrophication of receiving water: A review. *J. Environ. Qual.* **1998**, *27*, 261–266. [CrossRef]
31. Daniel, T.C.; Sharpley, A.N.; Lemunyon, J.L. Agricultural phosphorus and eutrophication: A symposium overview. *J. Environ. Qual.* **1998**, *27*, 251–257. [CrossRef]
32. Gburek, W.J.; Sharpley, A.N. Hydrologic controls on phosphorus loss from upland agricultural watersheds. *J. Environ. Qual.* **1998**, *27*, 267–277. [CrossRef]
33. Jiang, Y.Y.; Sun, Z.X.; Owens, P.R.; Adhikari, K.; Wang, Q.B.; Dorantes, M.J.; Libohova, Z. Spatial distribution of soil phosphorus, calcium, and pH after long-term broiler litter application. *J. Environ. Qual.* **2019**, *48*, 594–602. [CrossRef]
34. Tesfahunegn, G.; Gebru, T. Variation in soil properties under different cropping and other land-use systems in Dura catchment, Northern Ethiopia. *PLoS ONE* **2020**, *15*, e0222476. [CrossRef] [PubMed]
35. Negasa, T.; Ketema, H.; Legesse, A.; Sisay, M.; Temesgen, H. Variation in soil properties under different land use types managed by smallholder farmers along the toposequence in Southern Ethiopia. *Geoderma* **2017**, *290*, 40–50. [CrossRef]
36. Burst, M.; Chauchard, S.; Dambrine, E.; Dupouey, J.L.; Amiaud, B. Distribution of soil properties and forest-grassland interfaces: Influence of permanent environmental factors or land-use after-effects. *Agric. Ecosyst. Environ.* **2020**, *289*, 106739. [CrossRef]
37. Reitemeier, R.F. Soil Potassium. *Adv. Agron.* **1951**, *3*, 113–164.
38. Li, W.; Liu, X.-M.; Hu, Y.; Teng, F.Z.; Chadwick, O.A. Potassium isotope fractionation during chemical weathering in humid and arid Hawaiian regoliths. *Geochim. Cosmochim. Acta* **2022**, *333*, 39–55. [CrossRef]
39. Jiang, T. Study on the Effect of Sand Fixation and Soil Improvement by Different Land Use Methods in the Sand-Blown Area of Northwest Liaoning Province. Master's Thesis, Chinese Academy of Agricultural Sciences (CAAS), Beijing, China, 2012; pp. 1–29.
40. Zhao, W.; Huang, L. Stoichiometric characteristics and influencing factors of soil nutrients under different land use types in an alpine mountain region. *Acta Ecol. Sin.* **2022**, *42*, 4415–4427.
41. Fentie, S.F.; Jembere, K.; Fekadu, E.; Wasie, D. Land Use and Land Cover Dynamics and Properties of Soils under Different Land Uses in the Tejibara Watershed, Ethiopia. *Sci. World J.* **2020**, *2020*, 1479460. [CrossRef] [PubMed]

agronomy

MDPI

Article

Spatial Distribution Characteristics of Soil C:N:P:K Eco-Stoichiometry of Farmland and Grassland in the Agro-Pastoral Ecotone in Inner Mongolia, China

Yanli Zhang [1], Miao Liu [2], Li Han [1], Jinhu Yang [1], Xinyao Zhao [1], Jiahui Qu [3], Lijun Li [1,*], Yunlong Bai [4], Dong Yan [4] and Guannan Hou [5]

[1] College of Agronomy, Inner Mongolia Agricultural University, Hohhot 010019, China; zyanli2010@126.com (Y.Z.); imauhanli@163.com (L.H.); yjh0019@emails.imau.edu.cn (J.Y.); zxy0520@emails.imau.edu.cn (X.Z.)
[2] College of Grassland Science and Technology, China Agricultural University, Beijing 100193, China; liumiaocau@cau.edu.cn
[3] Inner Mongolia Academy of Agriculture & Animal Husbandry Sciences, Hohhot 010031, China; nmgqujiahui@163.com
[4] Inner Mongolia Agriculture and Animal Husbandry Technology Popularization Center, Hohhot 010019, China; baiyunlong2007@163.com (Y.B.); nmgtfzjsk@163.com (D.Y.)
[5] College of Humanities and Social Sciences, Beihang University, Beijing 100191, China; houguannan666@126.com
* Correspondence: lijun-li@imau.edu.cn; Tel.: +86-15848154170

Abstract: Ecological stoichiometry (ES) is an important index that reflects the balance of various elements in ecological processes. Therefore, it is of great significance to understand the soil nutrient cycle to clarify the environmental control of soil carbon (C), nitrogen (N), phosphorus (P), and potassium (K). In this study, we analyzed the spatial distribution of soil C, N, P, and K contents and the C:N:P:K stoichiometric characteristics of 0–20 cm and 20–40 cm of farmland and grassland in four agro-pastoral areas in Inner Mongolia. Spearman correlation was used to analyze the effects of environmental factors on the soil C:N:P:K stoichiometric relationship. The results showed that there was no fixed Redfield ratio for the soil stoichiometric relationship of farmland and grassland in Inner Mongolia, and the values were 15:2:1:9 to 145:10:1:26 and 25:1:1:29 to 228:15:1:65, respectively. The stoichiometric relationships between farmland and grassland were consistent with the law of geographical and spatial heterogeneity. The ratios of C:N, C:P, C:K, N:P, and N:K showed an *N* distribution from west to east, while the ratio of P:K showed a V distribution. The stoichiometric relationships in grassland soil were mainly affected by soil organic carbon and total nitrogen content, while those in farmland were mainly affected by total nitrogen and total phosphorus content. The annual mean precipitation has a significant effect on stoichiometric relationships in farmland, while the annual mean temperature has a more significant effect on grassland. In conclusion, the spatial distribution difference in the soil stoichiometric relationship in the agro-pastoral ecotone of Inner Mongolia was more significant than the difference in the land use pattern. The influences of annual mean temperature and annual mean precipitation on soil ecological stoichiometry were in accordance with the geographical spatial similarity law. Compared with grassland, the stoichiometric relationship of farmland soil was greatly affected by fertilization, and farmland in this region was mainly limited by carbon and nitrogen. Thus, field management should be carried out according to local conditions. This study is of great significance as it promotes the rational utilization of land resources and the sustainable development of agriculture.

Keywords: C:N:P:K stoichiometry; spatial distribution; land use pattern; agro-pastoral ecotone; environmental factors

1. Introduction

Ecological stoichiometry (ES) is a hot topic in current biogeochemical cycle research, which combines the basic principles of chemistry, physics, and biology to study the proportional relationship and the fluxes of various chemical elements in ecological processes [1,2]. In ES research, the Redfield ratio postulates a consistent C:N:P molar ratio of 100:16:1 in marine phytoplankton and open oceanic waters [3]. The C:N:P ratios for all soil layers and organic-rich soil (0–10 cm) in China were 60:5:1 and 134:9:1, respectively [4], and the C:N:P of forests, grasslands, and deserts (0–10 cm) in China was 55:4:1 [5]. However, the law of spatial heterogeneity means that things have spatial differences [6]. Studies in black soil areas in China showed that there were spatial differences in C:N:P stoichiometric characteristics, which are affected by environmental factors [7]. However, the spatial and temporal variation pattern of soil ecological stoichiometry is not completely clear [8]. Do the C:N:P:K stoichiometric relationships of soil follow a certain proportion or the law of spatial heterogeneity at the regional scale in the agro-pastoral ecotone of Inner Mongolia? What is the spatial distribution?

ES is an important predictive index to reflect the composition of organic matter, biogeochemical cycle, and soil quality, and it is also an important index to judge the mineralization and fixation of C, N, P, and K elements [9], which play an active role in ecosystem interactions and represent an important means of understanding the nutrient regulation factors of plant–litter–soil interaction [10–12]. The soil C:N ratio (R_{CN}) is a sensitive index reflecting changes in the soil environment or soil quality. It can be used as an index to measure the mineralization ability of soil C and N, the decomposition rate of organic matter, and the status of nutrient balance. Generally speaking, soil R_{CN} is inversely proportional to the decomposition rate of organic matter. Soil C:P ratio (R_{CP}) is a characterization parameter of phosphorus availability. The smaller the soil R_{CP}, the higher the phosphorus availability in the soil. The N:P (R_{NP}) in soil is used as an indicator of nutrient-constrained productivity and general biogeochemical status. As a necessary element for plant growth, potassium (k) plays a very important role in the material chemical cycle of the earth, and it has gradually attracted the attention of ecological researchers. In addition, the balance of N, P, and K is beneficial to carbon sequestration in surface soil [1,13]. The study on the spatial distribution of nutrient content and the stoichiometric relationship of C, N, P, and K and their environmental control in the agro-pastoral ecotone of Inner Mongolia has important guiding significance for understanding soil nutrient limitation in each region and adopting reasonable production management measures.

Land use change is a local environmental issue of global importance [14]. Human beings have changed biogeochemical cycles at different scales through farmland reclamation, and the differences in the stoichiometric ratios of C, N, and P in space and time have different impacts on biota [15–17]. Located between semi-humid and semi-dry areas, the agro-pastoral ecotone in Internal Mongolia is a transitional belt for farmland and grassland ecosystems and is the most sensitive and unstable area associated with the surrounding environment. From 1947 to 2021, the cultivated land area in the Inner Mongolia Autonomous Region increased from 3,967,000 hm^2 to 11,567,000 hm^2 (Statistics Bureau of Inner Mongolia Autonomous Region, 2023), and the change in cultivated land is mainly transformed from woodland and grassland [18]. At present, research on the eco-stoichiometry of soil in the agro-pastoral ecotone of Inner Mongolia is mainly focused on the county level (such as Dalate Banner [19], Yijinhuoluo Banner [20], Duolun County [21], Wengniute Banner [22], etc.) or land types (such as wetlands [23], sandy land [24,25], and grassland [26]), but there are few studies on a large scale in space. However, due to the large longitude span of the agro-pastoral ecotone in Inner Mongolia, there are differences in the topography, soil characteristics, and agricultural planting structures among the agro-pastoral areas.

Previous studies have mainly focused on the C:N:P stoichiometric relationship, while our study innovatively compared the stoichiometry of soil C:N:P:K between farmland and grassland at different scales in the agro-pastoral ecotone of Inner Mongolia. It aimed to clarify the stoichiometric ratio of soil C:N:P:K of different land use patterns and its spatial distribution, analyze soil nutrient deficiency in different regions, and determine the influence of environmental factors on stoichiometry. This provides a theoretical basis that reveals the interaction and balance between C, N, P, and K elements and has important practical significance for understanding the impact of human activities on ecosystem processes and services and exploring agricultural production strategies according to local conditions.

2. Materials and Methods

2.1. Study Site

This study area is the agro-pastoral ecotone from Hulunbuir to Bayannur in Inner Mongolia, China (105°53′–115°31′ E, 40°51′–53°20′ N). According to the Land and Spatial Planning of Inner Mongolia Autonomous Region (2021–2035), the agro-pastoral areas along the Yellow River mainstream plain (I), the foothills of Yinshan Mountain (II), the West Liaohe River Basin (III), and the foothills of Daxing'an Mountains (IV) are, respectively, located from west to east.

The mean annual precipitation in area I is 150–380 mm, and the mean annual temperature is 3.7–7.6 °C. The topography of this area is dominated by plateaus, mountains, hills, and plains, with an elevation of 986–2280 m. Irrigation and silt soil, saline-alkali soil, chestnut soil, and meadow soil are the main soil types, and the main crops are wheat, sunflower, and corn [27]. The mean annual precipitation in area II is 150–500 mm, and the mean annual temperature is 0–6.7 °C. The topography of this area is dominated by plateaus, mountains, and hills, with an elevation of 1150–2350 m. Cinnamon soil and chestnut soil are the main soil types, and the main crops are wheat, rapeseed, sunflower, potato, and miscellaneous grain [28]. The mean annual precipitation in area III is 305–485 mm, and the mean annual temperature is 0–7 °C. The topography of this area is dominated by plateaus, mountains, hills, and plains, with an elevation of 300–2000 m. Soil types include brown soil, black soil, irrigation and silt soil, meadow soil, aeolian sandy soil, and alkaline soil, and the main crops are corn, sunflower, potato, miscellaneous grains, and beans [29]. The mean annual precipitation in area IV is 270–467 mm, and the mean annual temperature is −2–6 °C. The topography of this area is dominated by plateaus, hills, and plains, with an elevation of 150–1800 m. Black soil is the main soil type, and the main crops are corn, soybeans, wheat, and rapeseed [30].

2.2. Data Collection and Sample Analysis

Soil samples were taken from typical farmland and nearby grassland in the agro-pastoral areas of Inner Mongolia during the flourishing period of plant growth. The sampling period of wheat samples in area I was in June, and that of other samples was from August to September. The samples of Hangjinhou Banner, Linhe District, Wulatezhong Banner, Wuchuan County, Wengniute Banner, Keerqin District, Zhalantun City, Yakeshi City, and Labudalin Farm Ranch were sampled in 2019; the samples of Wuyuan County, Dalate Banner, Tumotezuo Banner, Kalaqin Banner, Naiman Banner, Zhalaite Banner, and Sanhe Hui Township were sampled in 2020; and the samples of Taipusi Banner and Chahaeryouyiqian Banner were sampled in 2021. There were 36 farmland sampling sites and 14 grassland sampling sites (Figure 1). In each plot, 4 points with similar terrain and environmental conditions were selected using the diagonal sampling method, and soil samples were collected at 0~20 cm and 20~40 cm soil depths. A total of 288 farmland samples and 112 grassland samples were obtained. Soil samples were air-dried and carefully selected to remove organic matter and fine roots for soil property analysis. Each mixed soil sample was divided into four parts according to the quartering method. One part was selected to pass the 80-mesh sieve to determine the soil pH value and electric conductivity (EC), and the other part was selected to pass the 100-mesh sieve to determine the contents of soil organic carbon (SOC), total nitrogen (TN), total phosphorus (TP), and total potassium (TK). The determination of soil properties follows all standard schemes [31]. Meteorological factors were obtained from the records of 18 national weather stations near each sampling site, and annual mean temperature (MAT), annual mean precipitation (MAP), annual mean relative humidity (MARH), and annual mean sunshine duration (MASD) (1981–2021) were regarded as climate factors. In addition, the data on longitude and latitude for each sampling site were determined using the Global Positioning System (GPS).

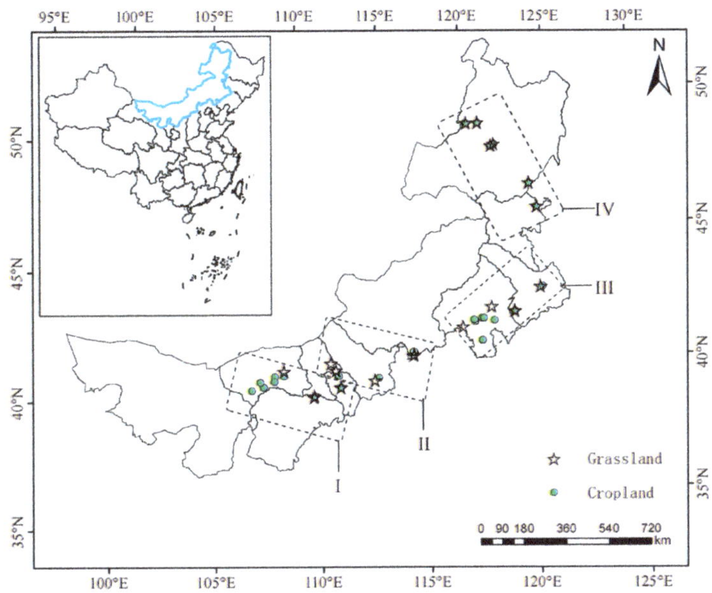

Figure 1. The location map of the samples.

2.3. Data Analysis

The ratios of C:N, C:P, C:K, N:P, N:K, and P:K were calculated on a molar mass ratio [32]. The data were tested using a Shapiro–Wilk normal test with SPSS (version 26, IBM SPSS, Somers, NY, USA), and the differences were analyzed via a nonparametric test. The box chart was drawn using Origin (version 2023, OriginLab, Northampton, MA, USA) to show the data on nutrient contents and stoichiometric ratios, and then all the indexes were plotted via Spearman correlation analysis.

3. Results

3.1. Comparison of C, N, P, and K Contents between Farmland and Grassland

Soil nutrients in farmland and grassland were significantly different among agro-pastoral areas ($p < 0.05$) (Figure 2). On the whole, the contents of SOC and TN in farmland were lower than those in grassland, while the contents of TP and TK were higher than those in grassland, and the difference in TP content in the 0–20 cm soil layer in area I was significant ($p < 0.01$). Compared with the grassland, SOC, TN, and TK contents in farmland were significantly lower in area I than in other areas, and TP contents were significantly higher than in other areas ($p < 0.05$). SOC, TN, and TP contents in area III were significantly lower than those in other areas, while TN and TK contents in area IV were significantly higher than those in other areas ($p < 0.05$).

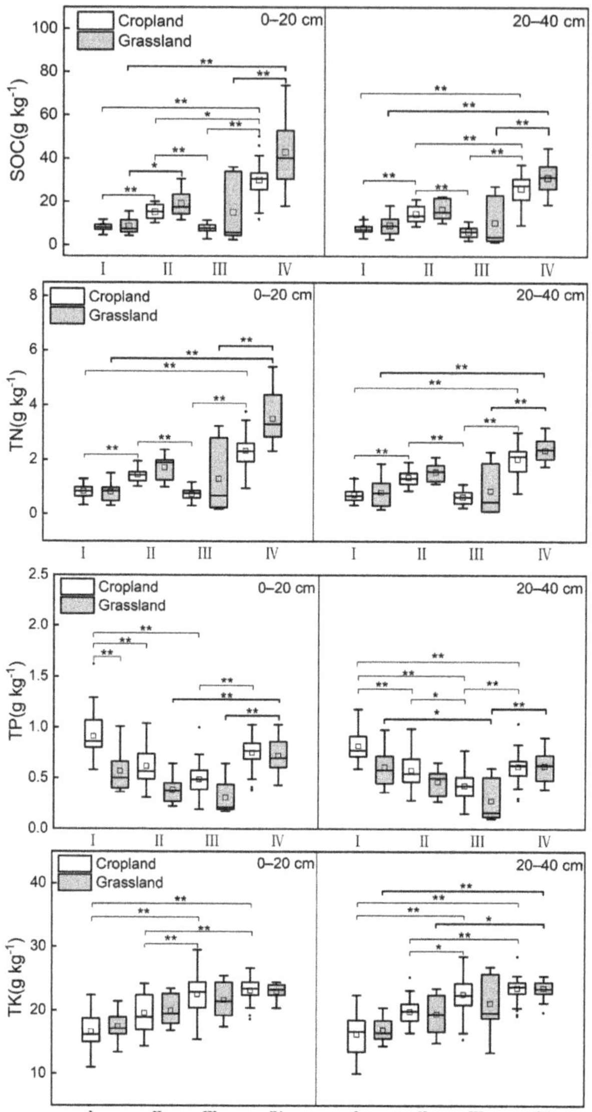

Figure 2. Box plots of soil organic carbon (SOC), total nitrogen (TN), total phosphorus (TP), and total potassium (TK) contents in the 0–20 cm and 20–40 cm soil layers in four agro-pastoral areas. Thin lines represent the comparison between farmlands in different regions, while thick lines indicate the comparisons between grasslands in different regions, and red lines represent the comparison between farmland and grassland in the same region. The dots represent outliers and the boxes represent the mean value. The difference between treatments was analyzed via a Kruskal–Wallis test; * and ** denote significant differences at 0.05 and 0.01 probability levels.

3.2. Eco-Stoichiometric Characteristics of Soil C:N:P:K in Farmland and Grassland

The differences in the eco-stoichiometric ratios of farmland and grassland (except R_{CN}) among agro-pastoral areas were significant ($p < 0.05$) (Figure 3). The change trends of farmland and grassland were the same among areas other than R_{PK}, while other indexes showed *N* changes from west to east. On the whole, the R_{CN}, R_{CP}, R_{NP}, and R_{NK} of farmland were lower than those of grassland, and the R_{PK} of farmland was higher than that of grassland. The R_{CK} values of area I and area III were higher than those of grassland, and the opposite was true for area II and area IV. Compared with grassland, the R_{CP}, R_{CK}, and R_{NK} values of farmland in area I were significantly lower than those in other areas; R_{PK} was significantly higher than that in other areas ($p < 0.05$); R_{CP}, R_{CK}, and R_{PK} of farmland in area II were significantly higher than those in other areas; R_{NP} and R_{NK} of farmland in area III were significantly lower than those in other areas; and R_{CN} in area IV was significantly higher than that in other areas ($p < 0.05$).

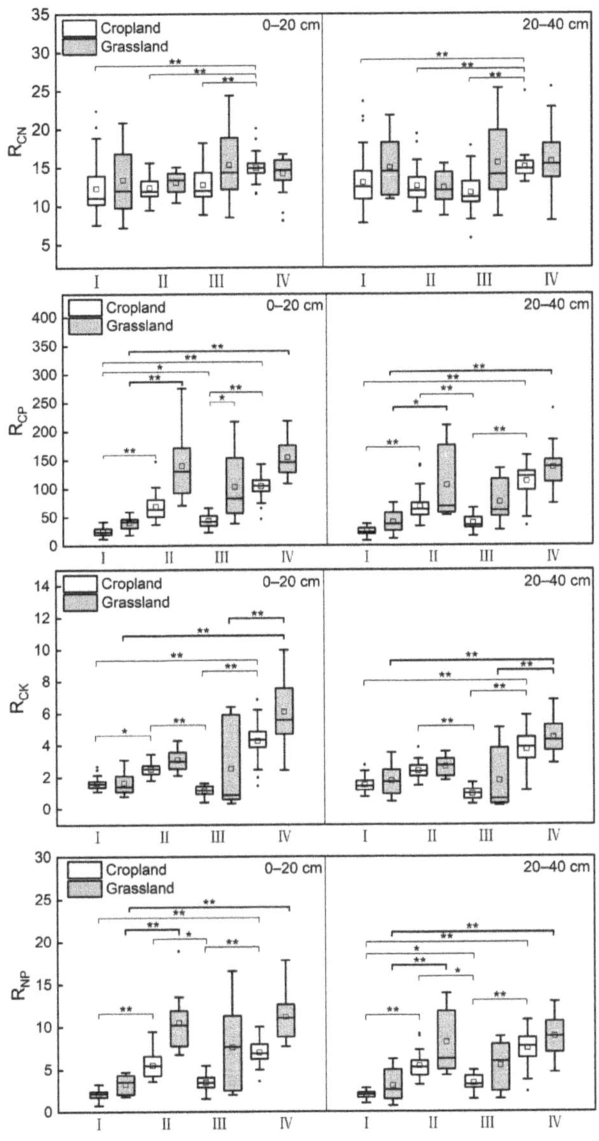

Figure 3. *Cont.*

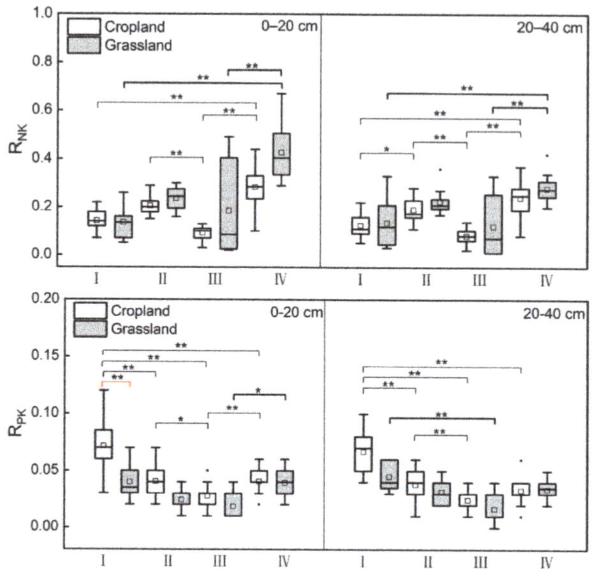

Figure 3. Box map of the eco-stoichiometric ratio of soil C, N, P, and K in 0–20 cm and 20–40 cm soil layers in four agro-pastoral areas. R_{CN}, R_{CP}, R_{CK}, R_{NP}, R_{NK}, and R_{PK} stand for ratios of C:N, C:P, C:K, N:P, N:K, and P:K, respectively. Thin lines represent the comparison of farmlands in different regions; thick lines indicate the comparison between grasslands in different regions; and red lines represent the comparison between farmland and grassland in the same region. The dots represent outliers and the boxes represent the mean value. The difference between treatments was analyzed using a Kruskal–Wallis test; * and ** denote significant differences at 0.05 and 0.01 probability levels.

3.3. Comparison of Nutrient Content and Stoichiometry in the Study Area with Other Scales

Regarding the overall scale of agro-pastoral ecotone in Inner Mongolia, the SOC and TN levels of farmland were lower than those of grassland, but the level of TP was higher, the farmland nutrients were at a middle level, and the level of grassland was at an upper-middle level. The level of TP in farmland in areas I, II, and III was higher than that in grassland, while the levels of SOC and TN in areas II and III were lower than those in grassland. The nutrient level of area III was at the middle and lower levels, and the nutrient level in area IV was the highest (Figure 4).

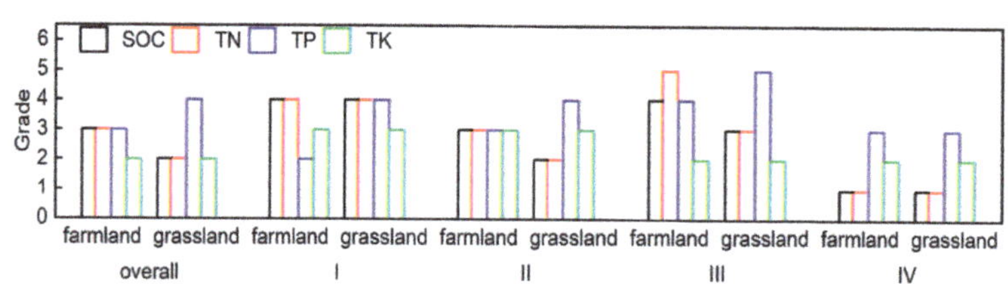

Figure 4. The nutrients of each area are classified according to the grading system of the second nationwide condition census soil survey [33] (Table A1).

Compared with the global soil C:N:P, the R_{CN} of the overall area and four regions in this study was higher, except for the farmland in area III, and the R_{CP} and R_{NP} of the farmland in area I and area III and the grassland in all regions were lower. Across the whole study area, compared with the northwest agro-pastoral ecotone, the C:N:P of farmland was higher and the grassland was lower. Compared with the yellow soil area of western Shanxi, the R_{CN} of farmland was lower, and the other values were higher. Compared with Hebei Province, the R_{CN} of grassland was lower, and the other values were higher. Compared with Qinghai Province, the R_{CP} and R_{NP} of farmland were higher, and the other values were lower. Compared with the study in the same region, the R_{CN} in area II was lower, and the R_{CP} and R_{NP} were higher. Compared with the research in the same region, R_{CN} in area III was higher and R_{CP} and R_{NP} were lower. R_{CN} and R_{NP} values in farmland were lower than those in the northern wind and sandy area, and conversely, R_{CP} in farmland and grassland was higher (Figure 5).

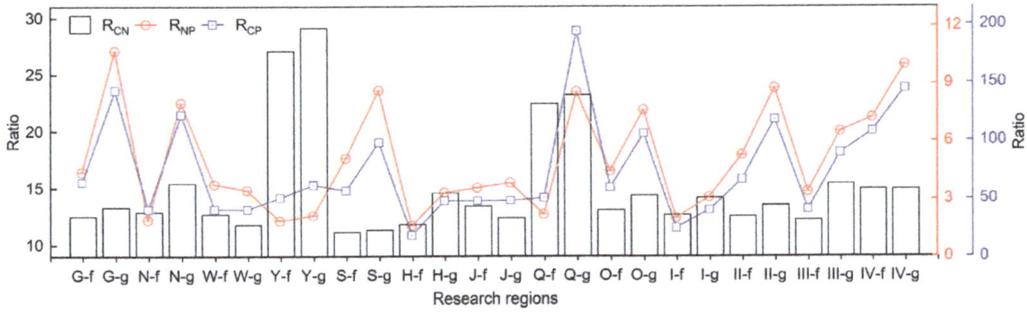

Figure 5. Comparison of soil C:N:P in this research and other study regions. R_{CN}, R_{CP}, and R_{NP} stand for ratios of C:N, C:P, and N:P, respectively; f and g mean farmland and grassland soil, respectively. G, N, W, Y, S, H, J, Q, and O refer to global level [34], northwest agro-pastoral ecotone [35], windy and sandy areas of northern China [25], agro-pastoral area at the northern foot of Yinshan Mountain [36], Horqin sandy land [24], Hebei province [37], yellow soil area of western Shanxi [38], eastern Qinghai Province [39], and the overall scale of this study, respectively. I, II, III, and IV refer to the four areas in this study.

3.4. Relationships between Soil C:N:P:K Stoichiometric and Environmental Factors

Regarding the Spearman correlation between soil stoichiometry and environmental factors (Figure 6), the soil stoichiometry of farmland in area I was significantly correlated with MAP, MARH, and MASD ($p < 0.05$), while the soil C:N:P of grassland was significantly correlated with MAT and MARH ($p < 0.05$). The soil stoichiometry of farmland (except R_{CK}) in area II was significantly correlated with climate factors ($p < 0.05$), and the soil stoichiometry of grassland was significantly correlated with MAT, MAP, and MARH ($p < 0.05$). The R_{CP} and R_{NP} of farmland in area III were significantly correlated with climate factors ($p < 0.01$), and the soil stoichiometry of grassland (except R_{CN}) was significantly correlated with MAT and MASD ($p < 0.01$), while MASD was opposite compared to other climate factors in this region. In area IV, the R_{CN} of farmland was significantly correlated with MARH and MASD ($p < 0.01$); the R_{CK} was significantly correlated with MASD ($p < 0.05$); the R_{NK} was significantly correlated with MAT and MAP ($p < 0.05$); and the R_{CN}, R_{CK}, R_{NK}, and R_{PK} of grassland were significantly correlated with climate factors ($p < 0.05$). In addition, MARH had a different correlation compared to other climate factors in this region. In areas I, II, and III, the correlation between stoichiometry and soil electric conductivity was more significant than pH. The correlations between the contents of C, N, P, and K and the soil stoichiometry of farmland in areas I and II and grassland in area III were significant ($p < 0.05$).

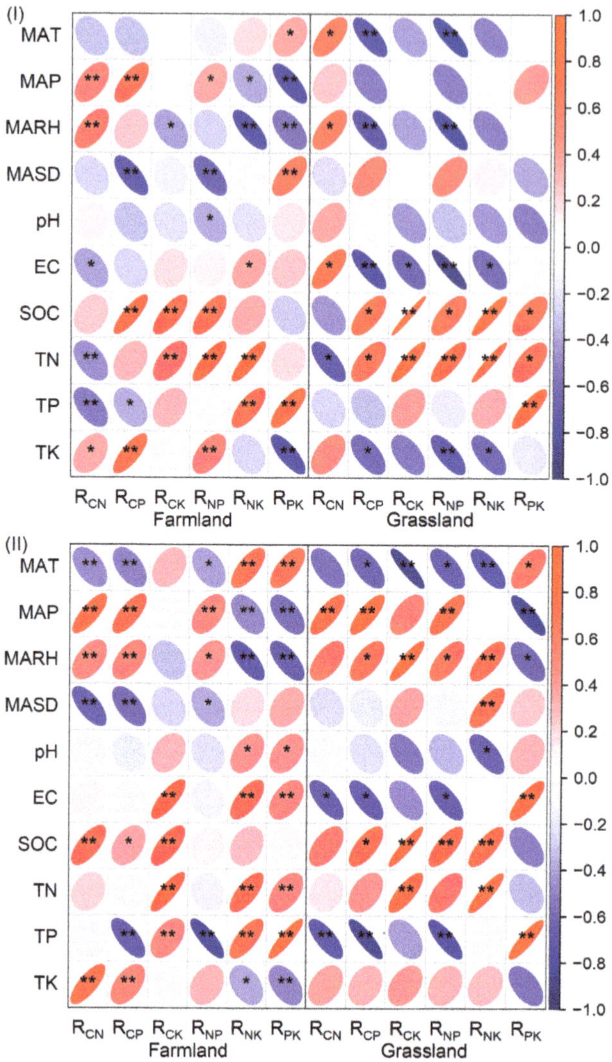

Figure 6. *Cont.*

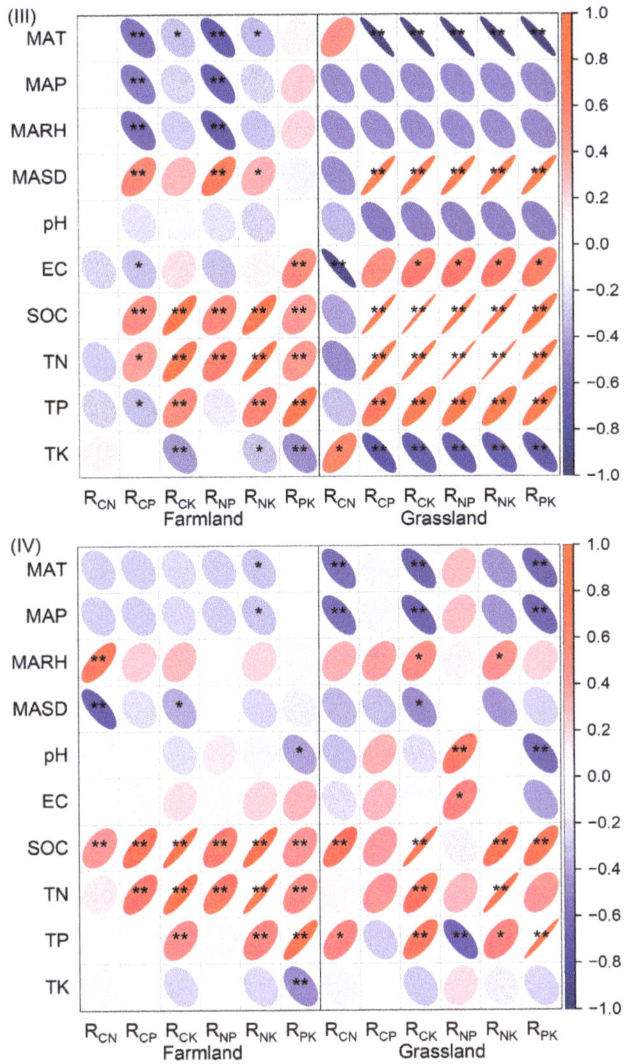

Figure 6. Spearman correlation of soil stoichiometry with nutrient content and environmental factors of farmland and grassland in four agro-pastoral areas. The red oval indicates a positive correlation, blue indicates a negative correlation, a darker color and narrower shape refers to a greater correlation coefficient. MAT, MAP, MARH, and MASD stand for annual mean temperature, annual mean precipitation, annual mean relative humidity, and annual mean sunshine duration from 1981 to 2021, respectively. pH and EC refer to the pH and electric conductivity values of soil. SOC, TN, TP, and TK stand for the organic carbon, total nitrogen, total phosphorus, and total potassium contents of soil. * and ** denote significant differences at 0.05 and 0.01 probability levels.

4. Discussion

4.1. The Spatial Difference in C:N:P:K Stoichiometry Was More Significant than That of Land Use Patterns

On the whole-region scale of the agro-pastoral ecotone, the SOC, TN, R_{CN}, R_{CP}, and R_{NP} of grassland were higher than those of farmland, while TP, TK, and R_{PK} were lower, indicating that grassland had a stronger nutrient accumulation capacity than farmland. This was similar to the research results of other scholars in agro-pastoral ecozones [21,24,39–41]. However, it was inconsistent with previous results in the windy and sandy areas of northern China and the yellow soil area of western Shanxi (Figures 4 and 5), which may be due to the single soil type and low nutrient contents in the study areas. Soil microorganisms in farmland need nutrients to supply their own propagation, and soil available nitrogen content was lower, so R_{CN} was higher than that in grassland.

Li et al. showed that soil C and N were more sensitive to grassland transformation than P through a meta-analysis of 92 studies [42]. It was believed that crop harvest leads to a reduction in soil C in the farmland ecosystem, that tillage destroys soil structure and accelerates N loss, and that fertilization leads to higher P and K content in farmland. Historically, soils have lost 40–90 Pg C through tillage and disturbance globally [43]. Studies conducted in the agro-pastoral ecotones of northern China [44], northeast China [45,46], and Castelluccio di Norcia (central Italy) [47] areas have also shown that the conversion of grassland to farmland will result in the loss of SOC and a decrease in soil chemical characteristics and basic soil fertility. However, in our study, land use types only had significant effects on TP content, R_{CP}, and R_{PK} ($p < 0.05$), while spatial environmental heterogeneity in the agro-pastoral ecotone of Inner Mongolia had more significant effects on soil nutrient contents and eco-stoichiometric ratios (Figures 2 and 3). This is because soil eco-stoichiometry is jointly regulated by land use patterns, soil properties, human disturbance, climate, and topography factors [48]. The spatial variation coefficient of SOC and TN content in farmland and grassland soil was higher (Table A2), while the variation coefficients of TP and TK were lower because the accumulation of C and N elements was related to the decomposition of organic matter and the self-reproduction ability of soil microorganisms. Therefore, it was greatly influenced by environmental factors. P and K elements were mainly affected by fertilization and soil parent materials, so the variability was small [49].

4.2. Stoichiometric Characteristics Demonstrate Constraints on Agricultural Production in Each Region

The 0–40 cm soil C:N:P:K values of farmland and grassland on the whole scale in the agro-pastoral ecotone of Inner Mongolia were 44:3:1:25 and 82:6:1:36, respectively. The C:N:P of farmland was lower than that of Chinese soil (60:5:1), while that of grassland was higher [4]. The R_{CN} ratio between 12 and 16 indicated that organic matter was well decomposed. At the same time, a study of forest soil showed that $R_{CN} < 25$ indicated a high risk of nitrate leaching. $R_{CP} < 200$ indicated net mineralization, and $R_{NP} < 10$ represented N limitation [50,51]. The ranges of R_{CN}, R_{CP}, and R_{NP} in this study area were 12.2–15.44, 24.49–145.09, and 2–9.99, respectively, which indicated that soil organic matter in farmland and grassland in this study area could be decomposed well and that the phosphate mineralization rate was high, but nitrogen was limited.

The C:N:P:K values of farmland and grassland in area I were 23:2:1:15 and 37:3:1:24, respectively, being lower than the C:N:P of the national soils. However, the R_{CN} was higher than the global level, and the R_{CP} and R_{NP} were lower, which was similar to the results of Dalat Banner [19]. This indicated that, compared with other research areas, the decomposition rate of soil organic matter in the farmland of area I was slower, and the availability of nitrogen was lower, but the availability of phosphorus was higher [52]. In addition, compared with grassland, the TP content of farmland was two grades higher, and the contents of C, N, and K were at a lower-than-medium level. This may be due to the loss of nitrogen caused by flood irrigation and soil leaching in the Yellow River and the increase in phosphorus caused by over-fertilization, resulting in soil N limitation and P saturation. Therefore, the rational application of phosphorus fertilizer and the management measures to improve the efficient use of soil nitrogen are more conducive to the effective use of agricultural resources in this area, aiding in the achievement of soil element balance.

The soil C:N:P:K values for farmland and grassland in area II were 62:5:1:26 and 100:8:1:36, respectively. The C:N:P values were higher than the national soils, the R_{CN} was higher than on the global level, and the R_{CP} and R_{NP} of grassland were lower. The soil nutrient contents of farmland were at a medium level; C and N were one grade lower than those of grassland, and P was one grade higher. The results indicated that the soil mineralization rate in this area was higher, but it was still limited by a smaller degree of nitrogen. The study in 2019 at the northern foot of Yinshan showed that grassland reclamation changed soil physical structure and increased microbial activity, thereby increasing soil respiration, accelerating the mineralization and decomposition of organic matter, and accelerating the nitrogen loss rate [36]. However, compared with them, the R_{CN} was lower and the R_{NP} was higher in our study. This may be because our study included the southern and northern foothills of Yinshan Mountain. The wind erosion intensity of soil in the southern foothills was weaker than that in the northern foothills, and the soil available nitrogen content was generally higher than that in the northern foothills. Therefore, cultivation measures to improve soil carbon sequestration capacity and available nitrogen are more beneficial to the stoichiometric balance of the region.

The soil C:N:P:K values of farmland and grassland in area III were 39:3:1:40 and 87:6:1:81, respectively. The C:N:P of farmland was lower than that of the whole country, while that of grassland was higher. The R_{CN} of grassland was higher than the global level, the R_{CP} and R_{NP} were lower, and the R_{CN}, R_{CP}, and R_{NP} of farmland were lower than the global level. The nutrient content of farmland was lower, and the TP content was one grade higher than that of grassland. The results showed that the decomposition rate of soil organic matter in this area was fast, which was not conducive to the accumulation of organic carbon, and the net mineralization rate of phosphorus was fast, but there was still a strong nitrogen limitation, and the degree of nutrient deficiency was N > C > P. However, the results were at odds with other studies on sandy land, which may be due to the study scale being different [53]. Therefore, reasonable fertilization combined with cultivation measures to improve the ability of soil water and fertilizer retention is beneficial to the material circulation of farmland ecosystems in this region.

The soil C:N:P:K values of farmland and grassland in area IV were 102:7:1:28 and 141:10:1:29, respectively. The C:N:P was higher than the national soil, and the grassland C:N:P (134:9:1) was higher than the national 0–10 cm soil layer. The R_{CN}, R_{CP}, and R_{NP} values of the farmland soil layer were higher than the global level, and the R_{CN} and R_{CP} of grassland soil were higher, but the R_{NP} was lower. The contents of C, N, and K in farmland and grassland soil were at an upper level, and the contents of P were at a medium level. The results indicated that there was no nutrient deficiency in this area, but the net phosphorus mineralization rate was significantly lower. The R_{CN} of the 0–20 cm soil layer in farmland was higher than that in grassland, which may be due to the low average temperature in the region and the slow decomposition of organic carbon [4,45]. Combined with measures such as crop rotation, fallow, and residue, the surface soil has a strong ability to retain organic carbon.

Early research in Brigelo, Queensland, found that continuous tillage and planting can maintain the availability of soil nitrogen better than continuous grazing [54]. Recent studies in Hokkaido, Japan [55], and Bavaria, Germany, showed that the short-term conversion of grassland to cultivated land increased the diversity of soil bacterial community structure, and combined with the nitrogen fixation of leguminous plants, the organic carbon in the soil increased. Therefore, the dry farming areas in the agro-pastoral ecotone of Inner Mongolia should promote grain–grass rotation, increase the application of organic fertilizer and straw returning to the field, and rationally allocate fertilizer according to the vegetation types and the actual situation in the growth stage, which will help to balance the soil eco-stoichiometry.

4.3. Effects of Environmental Factors on the Eco-Stoichiometry of Farmland and Grassland

Through Spearman correlation analysis, we found that the effects of C, N, P, and K contents on soil eco-stoichiometry in this study area followed the law of geographical correlation [56] (Figure 6). The C content in areas III and IV significantly affected the N,

P, and K cycles. K content significantly affected the C, N, and P cycles only in areas I and II. However, N and P content significantly affected the element ecological cycle in the whole region, and R_{CN} decreased with an increase in P content; therefore, it was not conducive to the accumulation of organic matter. This may be because area I was adjacent to area II, area III was adjacent to area IV, and area I and area III were river basins, so the soil properties between regions were similar. Therefore, the ecological chemical cycle of elements in grassland soil was mainly controlled by C and N, while farmland was mainly controlled by N and P, and R_{CN}, R_{CP}, and R_{NP} were the limiting indexes of soil nutrient content. Similar results had been obtained for the Yellow River Wetland in Baotou [19], Horqin Sandy Land [24], and Chongqing Mountain [57]. In this study, soil pH was only significantly correlated with R_{NP}, R_{NK}, and R_{PK}, which was inconsistent with the research conducted in paddy fields [58] and in the alpine region of the Loess Plateau [59]; however, fertilization caused soil stoichiometry in farmland to be greatly affected by soil pH [60]. This may be related to the land type, geographical environment, and research scale in the study area, so that the conclusions drawn have regional limitations.

The effects of MAT and MAP on soil eco-stoichiometry also had geographical gradients. Soil eco-stoichiometry in areas I and II was affected by MAT and MAP, while MAT and MAP in areas III and IV had the same effects on soil eco-stoichiometry. This may be because areas I and II were located along the Yinshan Mountains in the east–west direction, while areas III and IV were along the Daxing'anling Mountains in the north–south direction. Therefore, the climate and environment of areas I and II and areas III and IV were similar. At the same time, temperature and precipitation were significantly correlated with soil stoichiometry in farmland and grassland, and this was inconsistent with the findings of the northeast black land study [7]. Compared with the mean value of a single year, the average temperature and precipitation in the long-term series of this study can better reflect the long-term climate patterns. Moreover, the influences of temperature and precipitation change on soil stoichiometry in different ecosystems are inconsistent, which contradicts the results of studies on forest and grassland [61]. This may be because farmland is a semi-natural ecosystem and thus different from the natural ecosystem due to human factors.

5. Conclusions

There is no single Redfield-like ratio [3] in farmland and grassland soils across the Inner Mongolia agro-pastoral ecotone, and the R_{CN}, R_{CP}, and R_{NP} of farmland are lower than those of natural grassland. Land use patterns have a significant impact on the cycle of P elements. The stoichiometric relationship between farmland and grassland follows the geographical gradient in the agro-pastoral ecotone, and the trend is consistent. The difference in stoichiometric spatial distribution is more significant than that of the land use pattern.

The deficiency degree of soil nutrients in farmland is C > N, but P is saturated in the agro-pastoral area along the Yellow River mainstream plain, and N > C > P in the agro-pastoral area along the foot of Yinshan Mountain and the West Liaohe River Basin. The effect of agricultural production on the accumulation of C, N, P, and K in soil was not obvious in the agro-pastoral area along the foothills of the Daxing'anling Mountains.

The effect of grassland nutrient content on stoichiometry is related to the similarity of soil properties, while the man-made input of chemical elements destroys the stoichiometric balance of farmland soil, so the stoichiometric relationship of farmland soil is greatly affected by element contents. The effect of MAT and MAP on soil stoichiometry is related to the environmental heterogeneity between regions. Farmland is more significantly affected by MAP, while grassland is more significantly affected by MAT.

In the future, based on this study's results, the C:N:P:K stoichiometric relationships between soil and plants in farmland can be combined to explore the stoichiometric relationship mechanism in biomass allocation under different agricultural management measures, which can be very useful in improving agricultural production efficiency.

Author Contributions: Conceptualization, Y.Z. and L.L.; methodology, Y.Z. and M.L.; software, Y.Z.; validation, Y.Z., L.L. and M.L.; formal analysis, Y.Z., L.H. and J.Q.; investigation, Y.Z., L.H., J.Y., X.Z., Y.B., D.Y. and G.H.; resources, L.L.; data curation, Y.Z.; writing—original draft preparation, Y.Z. and J.Q.; writing—review and editing, Y.Z., M.L. and L.L.; visualization, Y.Z. and M.L.; supervision, L.L.; project administration, L.H. and J.Y.; funding acquisition, L.L. All authors have read and agreed to the published version of the manuscript.

Funding: This This research was funded by the National Key Research and Development Program of China "Research and Demonstration of Water–Heat Matching and Productivity Improvement Technology" (2022YFD1500904-3), the Inner Mongolia Autonomous Region Science and Technology major special project "Research and Demonstration on Breeding of New Oat Varieties, Green Cultivation Technology, and Nutritive Function Products" (2021ZD0002), and the Ordos Major Science and Technology Project "Research and Demonstration of Saline-Alkali Land Biological Improvement Technology and Efficient Utilization Model" (2022EEDSKJZDZX011).

Data Availability Statement: Data will be made available on request. All relevant data are within the paper.

Acknowledgments: We would like to thank the Oat Research Team for providing experimental equipment and are grateful to anonymous reviewers who all gave very helpful editorial comments.

Conflicts of Interest: The authors declare no conflicts of interest.

Appendix A

Table A1. Nutrient classification system of the second nationwide condition census soil survey.

Index	Unit	Nutrient Classifications					
		1	2	3	4	5	6
SOC	g/kg	>23.2	17.4–23.2	11.6–17.4	5.8–11.6	3.48–5.8	<3.48
TN	g/kg	>2	1.5–2	1–1.5	0.75–1.00	0.5–0.75	<0.5
TP	g/kg	>1.0	0.8–10	0.6–0.8	0.4–0.6	0.2–0.4	<0.2
TK	g/kg	>25	20–25	15–20	10–15	5–10	<5

Table A2. Descriptive statistics.

Variables	Farmland (N = 144)					Grassland (N = 56)				
	Min	Max	Mean	SD	CV(%)	Min	Max	Mean	SD	CV(%)
MAT (°C)	−1.85	8.69	4.61	3.36	72.98	−1.85	7.71	3.96	3.48	87.77
MAP (mm)	135.97	532.33	340.15	97.04	28.53	210.79	532.33	376.39	71.72	19.06
MARH (%)	46.39	68.46	53.46	6.58	12.31	46.39	68.46	55.41	6.35	11.45
MASD (h)	2491.63	3222.56	2902.34	185.65	6.40	2491.63	3098.61	2840.74	168.35	5.93
pH	5.97	8.65	7.70	0.59	7.62	6.15	8.55	7.42	0.73	9.79
EC (μS cm^{-1})	45.15	989.00	156.53	144.10	92.06	34.35	217.50	90.09	41.94	46.55
SOC (g kg^{-1})	2.49	43.77	14.24	9.60	67.45	2.01	58.47	21.66	14.79	68.25
TN (g kg^{-1})	0.27	3.39	1.24	0.70	56.45	0.14	4.10	1.80	1.10	61.15
TP (g kg^{-1})	0.18	1.18	0.65	0.21	31.82	0.14	1.00	0.52	0.22	41.91
TK (g kg^{-1})	12.05	27.48	20.56	3.65	17.73	14.05	25.86	20.82	3.14	15.09
R_{CN}	8.06	21.93	13.07	2.42	18.52	8.76	24.85	14.41	3.48	24.17
R_{CP}	12.37	144.82	59.77	35.57	59.51	17.73	231.23	105.90	55.01	51.95
R_{CK}	0.35	5.96	2.22	1.31	58.91	0.26	7.88	3.32	2.06	62.04
R_{NP}	0.86	10.17	4.46	2.22	49.70	1.30	15.49	7.63	3.80	49.81
R_{NK}	0.03	0.39	0.17	0.08	48.81	0.01	0.48	0.24	0.13	54.69
R_{PK}	0.01	0.11	0.04	0.02	46.70	0.01	0.06	0.03	0.01	42.33

References

1. Sardans, J.; Janssens, I.A.; Ciais, P.; Obersteiner, M.; Peñuelas, J. Recent advances and future research in ecological stoichiometry. *Perspect. Plant. Ecol. Evol. Syst.* **2021**, *50*, 125611–125634. [CrossRef]
2. Elser, J.; Sterner, R.; Gorokhova, E.A.; Fagan, W.; Markow, T.; Cotner, J.; Harrison, J.; Hobbie, S.; Odell, G.; Weider, L. Biological stoichiometry from genes to ecosystems. *Ecol. Lett.* **2000**, *3*, 540–550. [CrossRef]
3. Redfield, A.C. The biological control of chemical factors in the environment. *Am. Sci.* **1958**, *46*, 205–221.
4. Tian, H.Q.; Chen, G.S.; Zhang, C.; Melillo, J.M.; Hall, C.A. Pattern and variation of C: N: P ratios in China's soils: A synthesis of observational data. *Biogeochemistry* **2010**, *98*, 139–151. [CrossRef]
5. Zhang, J.; Li, M.; Xu, L.; Zhu, J.; Dai, G.; He, N. C: N: P stoichiometry in terrestrial ecosystems in China. *Sci. Total Environ.* **2021**, *795*, 148849–148857. [CrossRef] [PubMed]
6. Goodchild, M.F. The validity and usefulness of laws in geographic information science and geography. *Ann. Assoc. Am. Geogr.* **2004**, *94*, 300–303. [CrossRef]
7. Chen, Q.Q.; Shi, Z.; Chen, S.C.; Gou, Y.X.; Zhuo, Z.Q. Role of Environment Variables in Spatial Distribution of Soil C, N, P Ecological Stoichiometry in the Typical Black Soil Region of Northeast China. *Sustainability* **2022**, *14*, 2636. [CrossRef]
8. Feng, D.F.; Bao, W.K. Review of the temporal and spatial patterns of soil C:N:P stoichiometry and its driving factors. *Chin. J. Appplied Environ. Biol.* **2017**, *23*, 400–408.
9. Graham, H. Ecological Stoichiometry: Biology of elements from molecules to the biosphere. *J. Plankton Res.* **2003**, *25*, 1183. [CrossRef]
10. Downing, J.A. Marine nitrogen: Phosphorus stoichiometry and the global N: P cycle. *Biogeochemistry* **1997**, *37*, 237–252. [CrossRef]
11. Zhang, L.X.; Bai, Y.F.; Han, X.G. Application of N: P stoichiometry to ecology studies. *J. Integr. Plant Biol.* **2003**, *45*, 1009–1018.
12. Aitkenhead, J.A.; Mcdowell, W.H. Soil C: N ratio as a predictor of annual riverine DOC flux at local and global scales. *Glob. Biogeochem. Cycles* **2000**, *14*, 127–138. [CrossRef]
13. Lou, Y.; Xu, M.; Wang, W.; Sun, X.; Zhao, K. Return rate of straw residue affects soil organic C sequestration by chemical fertilization. *Soil Tillage Res.* **2011**, *113*, 70–73. [CrossRef]
14. Liu, C.; Xu, Y.Q.; Sun, P.L.; Huang, A.; Zheng, W.R. Land use change and its driving forces toward mutual conversion in Zhangjiakou City, a farming-pastoral ecotone in Northern China. *Environ. Monit. Assess.* **2017**, *189*, 505. [CrossRef]
15. Ptacnik, R.; Jenerette, G.D.; Verschoor, A.M.; Huberty, A.F.; Solimini, A.G.; Brookes, J.D. Applications of ecological stoichiometry for sustainable acquisition of ecosystem services. *Oikos* **2005**, *109*, 52–62. [CrossRef]
16. Li, C.; Zhao, L.; Sun, P.; Zhao, F.; Kang, D.; Yang, G.; Han, X.; Feng, Y.; Ren, G. Deep soil C, N, and P Stocks and stoichiometry in response to land use patterns in the loess hilly region of China. *PLoS ONE* **2016**, *11*, e0159075. [CrossRef] [PubMed]
17. Kim, D.-G.; Kirschbaum, M.U.F.; Eichler Löbermann, B.; Gifford, R.M.; Liáng, L.L. The effect of land-use change on soil C, N, P, and their stoichiometries: A global synthesis. *Agric. Ecosyst. Environ.* **2023**, *348*, 108402–108416. [CrossRef]
18. Tao, Z.P.; Wang, S.Q.; Sun, P.L.; Li, K.D.; Tian, W.; Han, X.X. Spatio-temporal differentiation and driving factors of cropland in the agro-pastoral ecotone of Northern China. *Arid Land Geogr.* **2022**, *45*, 153–163.
19. Wang, H.; Guo, Y.F.; Yao, Y.F.; Qi, W.; Qin, F.C.; Wang, J.K. Ecological stoichiometry of soil carbon, nitrogen and phosphorus under different land use patterns. *Southwest China J. Agric. Sci.* **2020**, *33*, 995–1000. [CrossRef]
20. Ma, Y.F.; Wang, W.M.; Jia, B.Q. Spatial variability analysis on soil nutrients in semi-arid agro-pastoral transition area—A case study in EjinHolo Banner, Inner Mongolia. *J. Arid Land Resour. Environ.* **2007**, *21*, 123–130.
21. Liu, Q.Y.; Tong, Y.P. Effects of land use type on soil nutrient distribution in northern agro-pastoral ecotone. *Chin. J. Appl. Ecol.* **2005**, *16*, 1849–1852. [CrossRef]
22. Xie, H.L.; Li, B.; Liu, L.M.; Zhang, X.S. Study on spatial feature of soil nutrients based on integration of spatial statistical analysis and GIS in farming-pastoral zone—A case study in Wengniute County, Inner Mongolia. *J. Soil Water Conserv.* **2006**, *20*, 73–76. [CrossRef]
23. Yang, R.; Sai, N.; Su, L.; Shang, H.J.; Liu, Y.H.; Guo, Y.S. Soil C, N and P contents and ecological stoichiometric characteristics in Baotou Yellow River wetland, Inner Mongolia. *Acta Ecol. Sin.* **2020**, *40*, 2205–2214.
24. Cao, W.J. Stoichiometric Characteristics and Their Spatial Distribution Patterns in Surface Soils Across the Horqin Sandy Land of China. Master's Thesis, Lanzhou Jiaotong University, Lanzhou, China, 2021. [CrossRef]
25. Sun, X.D.; Ning, Z.Y.; Yang, H.L.; Zhang, Z.Q.; Li, Y.L. The Stoichiometry of carbon, nitrogen and phosphorus in soil in typical desertified regions, North China. *J. Desert Res.* **2018**, *38*, 1209–1218.
26. Chen, L.L.; Wang, K.X.; Baoyin, T. Effects of grazing and mowing on vertical distribution of soil nutrients and their stoichiometry (C:N:P) in a semi-arid grassland of North China. *Catena* **2021**, *206*, 105507–105514. [CrossRef]
27. Qu, J.H.; Li, L.J.; Zhao, P.Y.; Han, D.Y.; Zhao, X.Y.; Zhang, Y.L.; Han, L.; Wang, Y. Impact of phosphorous fertilization on rape and common vetch intercropped fodder and soil phosphorus dynamics in North China. *Agric. Week* **2022**, *12*, 1949. [CrossRef]
28. Han, D.Y.; Li, L.J.; Zhao, X.Y.; Qu, J.H.; Yang, J.H.; Wang, Q.J.; Luo, S.J.; Han, L. Effects of nitrogen application on yield, quality, water and nitrogen use efficiency of intercropping oat and common vetch. *J. Northwest AF Univ. (Nat. Sci. Ed.)* **2023**, *51*, 40–51. [CrossRef]
29. Yang, J.H.; Li, L.J.; Zhang, Y.L.; Han, D.Y.; Han, L.; Zhao, X.Y.; Luo, S.J.; Zhang, H.J. Effects of oat and common vetch intercropping and fertilization on forage yield and quality in Horqin Sandy Land. *Agric. Res. Arid Areas* **2023**, *41*, 179–189.

30. Wu, P.B.; Li, L.J.; Zhang, Y.L.; Li, X.T.; Yang, F. Effects of rotation and fertilization on soil organic carbon and its fractions and soil nutrients. *Chin. J. Soil Sci.* **2020**, *51*, 416–422. [CrossRef]
31. Bao, S.D. *Agrochemical Analysis of Soil*, 3rd ed.; China Agriculture Press: Beijing, China, 2000; pp. 30–33, 42–48, 76–78, 101–102.
32. Cleveland, C.C.; Liptzin, D. C:N:P stoichiometry in soil: Is there a "Redfield ratio" for the microbial biomass? *Biogeochemistry* **2007**, *85*, 235–252. [CrossRef]
33. National Soil Census Office. *Chinese Soil*, 1st ed.; China Agriculture Press: Beijing, China, 1998; pp. 878–904, 922.
34. Xu, X.F.; Thornton, P.E.; Post, W.M. A global analysis of soil microbial biomass carbon, nitrogen and phosphorus in terrestrial ecosystems. *Glob. Ecol. Biogeogr.* **2013**, *22*, 737–749. [CrossRef]
35. Liu, X.; Ma, J.; Ma, Z.W.; Li, L.H. Soil nutrient contents and stoichiometry as affected by land-use in an agro-pastoral region of northwest China. *Catena* **2017**, *150*, 146–153. [CrossRef]
36. Gao, J.L.; Luo, F.M.; Gao, Y.; Dang, X.H.; Meng, Z.J.; Chen, X.N.; Duan, N. Ecological soil C, N, and P stoichiometry of different land use patterns in the agriculture-pasture ecotone of Northern China. *Acta Ecol. Sin.* **2019**, *39*, 5594–5602. [CrossRef]
37. Cao, X.H.; Long, H.Y.; Zhou, J.G.; Zhu, A.X.; Liu, H.B.; Lei, Q.L.; Qiu, W.W. Spatial variation of ecological stoichiometry characteristics of topsoil carbon, nitrogen and phosphorus in Hebei Province, China. *Acta Ecol. Sin.* **2017**, *37*, 6053–6063. [CrossRef]
38. Hu, Y.W.; Sun, R.X.; Shen, M.S.; Shi, Z.L.; Liu, C.; Xu, Q.T.; Liu, J.T.; Zhang, J.J. Effects of land use types on the stoichiometric characteristics of soil C:N:P and the physical and chemical properties of soil in western Shanxi loess region. *Arid Zone Res.* **2021**, *38*, 990–999. [CrossRef]
39. Zhao, W.; Huang, L.M. Stoichiometric characteristics and influencing factors of soil nutrients under different land use types in an alpine mountain region. *Acta Ecol. Sin.* **2022**, *42*, 4415–4427. [CrossRef]
40. Shi, C.Y.; Ma, L. Effect of different land use on soil nutrient in Northern agriculture-pasturage eco-zone—A case in saibei management area of Zhangjiakou City. *J. Hebei Norm. Univ. (Nat. Sci. Ed.)* **2009**, *33*, 815–819.
41. Wang, F.-P.; Wang, X.-C.; Yao, B.-Q.; Zhang, Z.-H.; Shi, G.-X.; Ma, Z.; Chen, Z.; Zhou, H.-K. Effects of land-use types on soil organic carbon stocks: A case study across an altitudinal gradient within a farm-pastoral area on the eastern Qinghai-Tibetan Plateau, China. *J. Mt. Sci.* **2018**, *15*, 2693–2702. [CrossRef]
42. Li, S.C.; Xu, J.H.; Tang, S.M.; Zhan, Q.W.; Gao, Q.H.; Ren, L.T.; Shao, Q.Q.; Chen, L.; Du, J.; Hao, B. A meta-analysis of carbon, nitrogen and phosphorus change in response to conversion of grassland to agricultural land. *Geoderma* **2020**, *363*, 114149–114156. [CrossRef]
43. Smith, P. Land use change and soil organic carbon dynamics. *Nutr. Cycl. Agroecosystems* **2007**, *81*, 169–178. [CrossRef]
44. Rong, Y.P.; Ma, L.; Johnson, D.A.; Yuan, F. Soil respiration patterns for four major land-use types of the agro-pastoral region of northern China. *Agric. Ecosyst. Environ.* **2015**, *213*, 142–150. [CrossRef]
45. Ding, F.; Hu, Y.L.; Li, L.J.; Li, A.; Shi, S.W.; Lian, P.Y.; Zeng, D.H. Changes in soil organic carbon and total nitrogen stocks after conversion of meadow to cropland in Northeast China. *Plant Soil* **2013**, *373*, 659–672. [CrossRef]
46. Qi, Y.C.; Dong, Y.S.; Peng, Q.; Xiao, S.S.; He, Y.T.; Liu, X.C.; Sun, L.G.; Jia, J.Q.; Yang, Z.J. Effects of a conversion from grassland to cropland on the different soil organic carbon fractions in Inner Mongolia, China. *J. Geogr. Sci.* **2012**, *22*, 315–328. [CrossRef]
47. Francioni, M.; D'ottavio, P.; Lai, R.; Trozzo, L.; Budimir, K.; Foresi, L.; Kishimoto-Mo, A.W.; Baldoni, N.; Allegrezza, M.; Tesei, G. Seasonal soil respiration dynamics and carbon-stock variations in mountain permanent grasslands compared to arable lands. *Agric. Week* **2019**, *9*, 165. [CrossRef]
48. Chapin, F.S., III; Matson, P.A.; Mooney, H.A. *Principles of Terrestrial Ecosystem Ecology*; Springer: New York, NY, USA, 2002; pp. 1–17. [CrossRef]
49. Wang, S.Q.; Yu, G.R. Ecological stoichiometry characteristics of ecosystem carbon, nitrogen and phosphorus elements. *Acta Ecol. Sin.* **2008**, *28*, 3937–3947.
50. Bui, E.N.; Henderson, B.L. C: N: P stoichiometry in Australian soils with respect to vegetation and environmental factors. *Plant Soil* **2013**, *373*, 553–568. [CrossRef]
51. Gundersen, P.; Callesen, I.; De Vries, W. Nitrate leaching in forest ecosystems is related to forest floor CN ratios. *Environ. Pollut.* **1998**, *102*, 403–407. [CrossRef]
52. Zhang, H.; Ouyang, Z.C.; Zhao, X.M. Effects of different land use types on ecological stoichiometry characteristics of carbon, nitrogen and phosphorus in farmland soils in Jiangxi Province, China. *Acta Sci. Circumstantiae* **2019**, *39*, 939–951. [CrossRef]
53. Tang, X.; Hu, J.; Lu, Y.; Qiu, J.; Dong, Y.; Li, B. Soil C, N, P stocks and stoichiometry as related to land use types and erosion conditions in lateritic red soil region, south China. *Catena* **2022**, *210*, 105888–105898. [CrossRef]
54. Robertson, F.A.; Myers, R.J.K.; Saffigna, P.G. Carbon and nitrogen mineralization in cultivated and grassland soils in subtropical Queensland. *Soil Res.* **1993**, *31*, 611–619. [CrossRef]
55. Mukumbuta, I.; Shimizu, M.; Hatano, R. Short-term land-use change from grassland to cornfield increases soil organic carbon and reduces total soil respiration. *Soil Tillage Res.* **2019**, *186*, 1–10. [CrossRef]
56. Miller, H.J. Tobler's first law and spatial analysis. *Ann. Assoc. Am. Geogr.* **2004**, *94*, 284–289. [CrossRef]
57. Zhang, L. Response of C, N, P Eco-Stoichiometry of Typical Mountain Soils to Changes in Land Use Patterns: A Case Study of Tianfu Town, Beibei District, Chongqing. Master's Thesis, Southwest University, Chongqing, China, 2021. [CrossRef]
58. Dai, Y.T.; Zhou, P.; Guo, X.B.; Luo, P.; Chen, X.B.; Wu, J.S. Role of environmental factors on concentrations and ratios of subsoil C–N–P in subtropical paddy fields. *J. Soils Sediments* **2023**, *23*, 1999–2010. [CrossRef]

59. Liu, R.S.; Wang, D.M. Soil C, N, P and K stoichiometry affected by vegetation restoration patterns in the alpine region of the Loess Plateau, Northwest China. *PLoS ONE* **2020**, *15*, 241859–241865. [CrossRef]
60. Sun, L.Q. The Study of Farmland Soil C:N:P Stoichiometry Characteristics in China. Master's Thesis, China University of Geosciences, Beijing, China, 2018.
61. Sun, Y.; Wang, C.T.; Chen, H.Y.H.; Luo, X.S.; Qiu, N.W.; Ruan, H.H. Asymmetric responses of terrestrial C:N:P stoichiometry to precipitation change. *Glob. Ecol. Biogeogr.* **2021**, *30*, 1724–1735. [CrossRef]

agronomy

MDPI

Article

Excavation of Genes Response to Heat Resistance by Transcriptome Analysis in Bottle Gourd (*Lagenaria siceraria* (Mol.) Standl.)

Min Wang [1,2,†], Wenrui Liu [1,2,†], Qingwu Peng [1], Shaoqi Shi [1,2], Ying Wang [3], Liqin Cao [2], Biao Jiang [1,2], Yu'e Lin [1], Tianyue Zhao [2], Xiaojuan Cui [2] and Songguang Yang [1,2,*]

[1] Vegetable Research Institute, Guangdong Academy of Agricultural Sciences, Guangzhou 510640, China; wangmin1989sun@163.com (M.W.); liuwr10@126.com (W.L.); pengqingwu@gdaas.cn (Q.P.); 13610200357@163.com (S.S.); jiangbiao@gdaas.cn (B.J.); linyue@gdaas.cn (Y.L.)

[2] Guangdong Key Laboratory for New Technology Research of Vegetables, Guangzhou 510640, China; 17835548789@163.com (L.C.); syxzty2020@163.com (T.Z.); 15630897219@163.com (X.C.)

[3] Institute of Vegetables, Zhejiang Academy of Agricultural Sciences, Hangzhou 310012, China; wangying@zaas.ac.cn

* Correspondence: yangsongguang@gdaas.cn

† These authors contributed equally to this work.

Abstract: Heat stress, as a negative factor, severely threatens the quality and production of bottle gourd, which prefers to grow in a warm environment. To understand which genes are involved in the resistance to heat stress in bottle gourd (*Lagenaria siceraria* (Mol.) Standl.), we analyzed the characteristics of two genetic bottle gourd varieties, "Mei feng"-MF (heat resistant) and "Lv long"-LL (heat sensitive). Under heat stress, MF plants exhibited a higher survival rate, lower relative electrolytic leakage, and decreased stomatal aperture compared with LL. In addition, RNA-Seq was carried out on the two varieties under normal conditions and heat stress. The results revealed a total of 1485 up-regulated and 946 down-regulated genes under normal conditions, while 602 genes were up-regulated and 1212 genes were down-regulated under heat stress. Among these genes, several differentially expressed genes (DEGs) involved in the MAPK (mitogen-activated protein kinase) signaling pathway and members of bHLH (basic helix-loop-helix) transcription factors showed significant up- or down-regulation after heat stress. Next, to validate these findings, we conducted quantitative real-time PCR (qRT-PCR) analysis, which confirmed the expression patterns of the genes detected through RNA-Seq. Collectively, the DEGs between the two contrasting cultivars identified in our study provide novel insight into excavating helpful candidate genes associated with heat tolerance in bottle gourd.

Keywords: bottle gourd; RNA-Seq; DEGs; MAPK; bHLH; heat stress

1. Introduction

Heat stress (HS) caused by high temperatures severely impacts plant growth, endangering ecosystem quality and agricultural production in many areas around the world. When plants are exposed to transitory or constantly high temperatures, there are multiple changes in physiological and biochemical facts, leading to drastic reductions in crop yield and quality [1]. In order to survive unfriendly conditions, plants have evolved several intrinsic strategies to adapt to HS [2]. Plant hormone changes, heat shock protein (HSP) protection, and reactive oxygen species (ROS) scavenging are involved in conferring tolerance to plants, especially at high temperatures [3–5]. The abscisic acid (ABA), ethylene, and jasmonic acid (JA) can affect plant adaptation to high temperatures [6], especially ABA. ABA is required for the accumulation of ascorbate peroxidase (APX1) during HS [7]. Previous studies have reported that ABA-deficient and -insensitive mutants are sensitive to high temperatures, while overexpressing ABRE (abscisic acid responsive elements) binding factors (ABF) can enhance plants' thermotolerance [8]. Hydrogen peroxide exerts roles in increasing the ABA-dependent expression of heat shock protein 70 (HSP70) and enhancing the tolerance of plants to heat stress [9]. In rice, the *HEAT TOLERANCE AT SEEDLING STAGE* (*OsHTAS*) gene, encoding a ring finger ubiquitin E3 ligase, can increase heat tolerance by regulating H_2O_2-induced stomatal closure [10].

Most vegetables prefer to grow under warm conditions; thus, they do not possess the ability to tolerate HS. For example, when the temperature is over 35 °C, cucumber plants bear severe thermal damage [11]. However, previous studies have shown that many genes are also involved in HS responses in vegetables. For instance, ectopic expression of spinach betaine aldehyde dehydrogenase (BADH) in tobacco improves plant thermotolerance [12]. Pepper *heat shock protein 24* (*CaHSP24*), which is significantly induced by high temperatures, plays a crucial role in heat tolerance [13]. *HSP70*, which is prominently induced by ABA, increases heat tolerance in cucumber [14]. Overexpression of *CsCaM3* (Calmodulin 3) in cucumber plants leads to improved heat tolerance by regulating the expression of heat-responsive genes [15].

African-native bottle gourd (*Lagenaria siceraria* (Mol.) Standl.), also known as calabash or opo squash, is grown worldwide for its medicinal, decorative, and grafting rootstock properties [16]. Previous studies have demonstrated that bottle gourd exhibits some level of drought and salinity tolerance compared to other cucurbits [17,18]; hence, it is widely used as a rootstock for watermelon and melon. Bottle gourd prefers to grow in a warm environment (25–35 °C), without higher temperature resistance. When exposed to high temperature conditions (normally > 40 °C), especially in the seedling stage, the development, quality, and yield are seriously affected. However, few genes have been explored related to HS resistance in bottle gourd, which restricts crucial gene application in heat resistance genetic breeding of bottle gourd. Our study aims to explore the responses of some important genes to heat stress by transcriptome sequencing, which will not only provide insights into the molecular mechanisms underlying HS responses in bottle gourd but also highlight the involvement of genes in conferring heat tolerance.

2. Materials and Methods

2.1. Plant Materials and Heat Stress Treatment

In 2018, a total of 120 bottle gourd varieties were exposed to high temperatures (45 °C/40 °C in day/night) for 7 days in a phytotron room. Combining plant phenotype and survival rate, we obtained several heat-resistant and -sensitive varieties, including MF and LL (data unpublished). MF with heat tolerance and LL with heat sensitivity were used in this study. MF and LL were homozygous inbred lines derived from the native variety "Fuzhou bottle gourd" and "Hubei bottle gourd", respectively. "Fuzhou bottle gourd" is a commonly cultivated inbred variety in Fujian province, while "Hubei bottle gourd" is an inbred cultivar grown in the central region of China. The seeds were soaked in warm water at 50 °C for 20 min, followed by soaking in room temperature water for 10 h. Subsequently, the seeds were germinated in an incubator at a temperature range of 30~32 °C under dark conditions for 2 days. Subsequently, the pre-germinated MF and LL seeds were grown in an artificial climate box maintained at 30 °C/24 °C day/night temperature (5500 lux, 60% RH). When plants were grown to one true leaf stage (14-day seedling stage), they were exposed to heat treatment for 7 days (45 °C/40 °C in day/night) and recovered at room temperature for 3 days. Each cultivar contained 24 plants, with three biological replicates.

2.2. Stomatal Aperture Analyses

To view the stomatal aperture using scanning electron microscopy (SEM), the abaxial epidermises of true leaves of HS (consecutively exposed to 45 °C for 24 h) or mock-treated plants were prepared following standard procedures [19]. Leaf samples were immersed in 5% glutaraldehyde overnight, washed in 0.1 M phosphate buffer (pH 7.0), and then treated in 1% osmium tetroxide for 2 h. The samples were dehydrated using serially diluted ethanol (70% to 100%) and isoamyl acetate. Specimens were dried in a Hitachi HCP-2 critical point dryer (Hitachi High-Tech, Tokyo, Japan) with liquid CO_2 and mounted onto double-coated carbon conductive tapes attached to specimen holders. Samples were sputter-coated with gold-platinum and examined using a Hitachi TM-1000 TSEM (Hitachi High-Tech, Tokyo, Japan). Twenty stomata were randomly chosen for measurement in each leaf. Five plants were analyzed per treatment, and two independent experiments were performed, showing similar results.

2.3. Transcriptome Sequencing

Total RNA was extracted from the true leaves of MF and LL according to the instruction manual of the TRIzol Reagent (Life Technologies, Carlsbad, CA, USA). RNA integrity and concentration were checked by gel electrophoresis and Nanodrop quantification. mRNA was isolated by the NEBNext Poly (A) mRNA Magnetic Isolation Module (NEB, MA, UK; E7490). The cDNA library was constructed following the manufacturer's instructions of the NEBNext Ultra RNA Library Prep Kit for Illumina (NEB, ipswitch, MA, UK; E7530) and NEBNext Multiplex Oligos for Illumina (NEB, ipswitch, MA, UK; E7500). In brief, the enriched mRNA was fragmented into approximately 200 nt RNA inserts, which were used to synthesize the first-strand cDNA and the second cDNA. End-repair/dA-tail and adaptor ligation were performed on the double-stranded cDNA. The suitable fragments were isolated using Agencourt AMPure XP beads (Beckman Coulter, Inc., Brea, CA, USA) and enriched by PCR amplification. Finally, the constructed cDNA libraries of the bottle gourd leaves were sequenced on a flow cell using an Illumina HiSeq™ sequencing platform.

2.4. Bioinformatic Analysis of RNA-Seq Data

The low-quality reads, like adaptors, unknown nucleotides > 5%, or Q20 < 20% (percentage of sequences with sequencing error rates < 1%) were removed by perl script. The clean reads were mapped to the Bottle Gourd reference genome (http://cucurbitgenomics.org/v2/organism/29, (accessed on 10 October 2023)) using TopHat2 software (http://ccb.jhu.edu/software/tophat, (accessed on 10 October 2023)). The transcript level of each gene was calculated using FPKM values (fragments per kilobase of exon per million fragments mapped) by Cufflinks 2.2.1 software. Genes with a false discovery rates (FDR) < 0.05 and fold change values > 1.5 were regarded as differentially expressed genes (DEGs). Only genes with an absolute value of log2 ratio $\geq$ 1.5 and FDR significance score < 0.05 were used for subsequent analysis.

For the gene annotation and classification assay, the BLASTX program was used in various protein databases, including the National Center for Biotechnology Information (NCBI) non-redundant protein (Nr) database and the Swiss-Prot database with a cut-off E-value of 10^{-5}. Genes were retrieved based on the best BLAST hit (highest score), along with their protein functional annotation. To annotate the genes with Gene Ontology (GO) terms, the Nr BLAST results were imported into the WEGO program [20]. The obtained annotation was enriched and refined using TopGo (R package). Pathway assignments were made by mapping the unigenes onto KEGG categories (http://www.genome.jp/kegg/ (accessed on 10 October 2023)) [21]. Raw sequencing data were archived under accession numbers SRR24414144-SRR24414143-SRR24414132 (MF-CK), SRR24414127-SRR24414126-SRR24414125 (LL-CK), SRR24414140-SRR24414139-SRR24414138 (MF-HS), and SRR24414137-SRR24414136-SRR24414135 (LL-HS) in the NCBI Sequence Read Archive (SRA, http://www.ncbi.nlm.nih.gov/Traces/sra, (accessed on 10 October 2023)).

2.5. Quantitative Real-Time PCR (qRT-PCR)

About 2 μg total RNA was used to generate cDNA using the TransScript® One-Step gDNA Removal and cDNA Synthesis SuperMix (TransGen, Beijing, China) according to the manufacturer's instructions. As a template, 150 ng of synthesized cDNA was used to perform RT–qPCR analysis. PCR reactions were performed in a total volume of 20 μL, with 1 μL of each primer (final concentration 200 nM) and 10 μL of SYBR Green Master

Mix (Takara Bio, Beijing, China). The PCR program included an initial denaturation step at 94 °C for 3 min, followed by 40 cycles of 30 s at 94 °C and 35 s at 60 °C. Each sample was quantified in triplicate and normalized using TuB-α gene [22] as internal controls. Three biological replicates were performed, and three technical repeats were performed for each biological replicate. The primer pairs for RT–qPCR are listed in Table S1. The relative expression levels were normalized to the TuB-α and calculated using the $2^{-\Delta\Delta CT}$ method.

3. Results

3.1. MF Is a Resistant Cultivar to Heat Stress

MM and LL (24 plants for three biological replicates, respectively) were grown to the one true leaf stage under normal conditions (Figure 1A). Then, they were transferred to high temperature conditions (45 °C) for 7 days (Figure 1B) and recovered for 3 days (Figure 1C). Before treatment, both MM and LL thrived exuberantly; however, the top growth point of LL showed wilting symptoms after heat stress, and its cotyledons and true leaves turned chlorotic and yellow (Figure 1A,B). After recovery, most LL plants were dead, while some MF plants exhibiting continued growth (Figure 1C). Approximately 82% of MF plants survived after heat stress treatment in comparison with about 21% of LL plants (Figure 1D). In addition, we found that the relative electrolytic leakage in the surviving seedlings was significantly higher in LL compared to MF (Figure 1E). Previous studies reported that the stomatal aperture exerts crucial roles in abiotic stress [22,23]. In order to determine the difference in stomatal aperture between MF and LL, we used scanning electron microscopy (SEM) to carry out the assay. The results showed that the stomatal opening was similar in MF and LL plants under normal conditions. However, when the plants were treated with HS, the stomatal opening was noticeably smaller in MF than in LL (Figure 2A,B). Taken together, these findings suggest that LL and MF are susceptible and resistant cultivars, respectively, in response to HS.

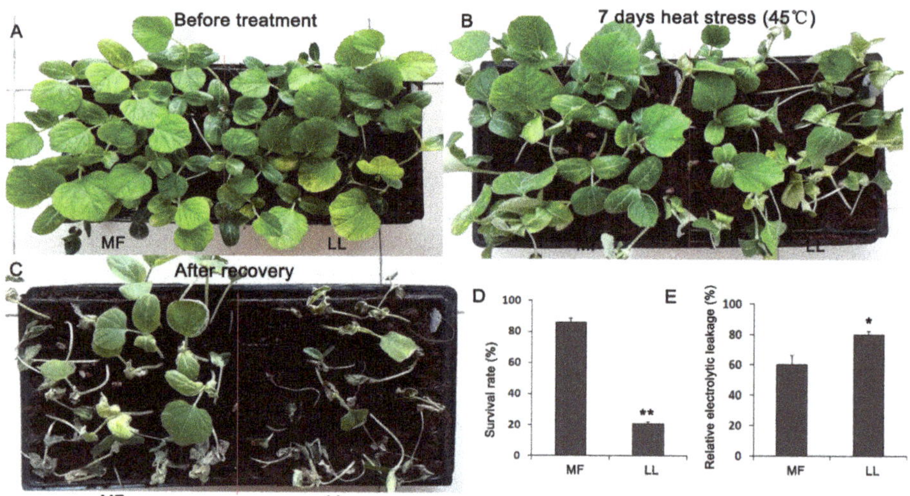

Figure 1. Phenotypes of MF and LL under HS treatment. (**A**) MF and LL plants before treatment. (**B**) MF and LL plants exposed to HS treatment for 7 days. (**C**) MF and LL plants after recovery. (**D**) Survival rates of MF and LL after heat stress treatment. (**E**) Relative electrolytic leakage under normal conditions. Data are presented as the mean ± SD (n = 9). ** $p < 0.01$, * $0.01 < p < 0.05$; Student's t-test.

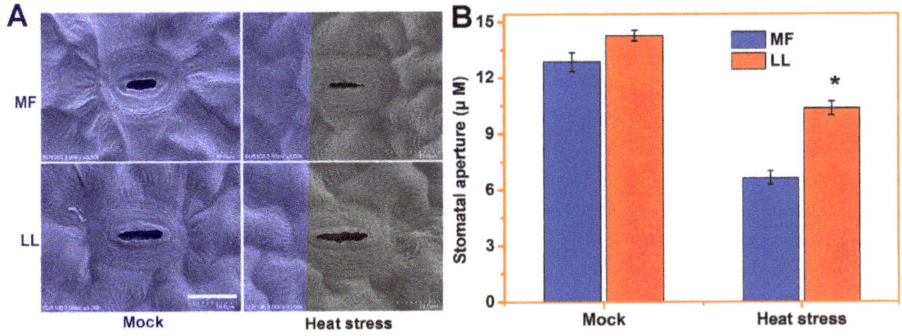

Figure 2. HS decreases the stomatal aperture in MF leaves. Representative SEM image of stomata of MF and LL true leaves in response to HS (**A**); the stomatal aperture data are shown in (**B**). Values are the means ± SD. Data from two independent experiments showed similar results. The results of one representative experiment are shown, $n = 20$ for each treatment group. The asterisk indicates a significant difference between the treatments (Student's t-test, $p < 0.05$).

3.2. Heat Stress-Induced Transcriptional Changes

In order to study the molecular response to heat stress between MM and LL, we carried out transcriptome sequencing of these two varieties under normal conditions (CK) and after 24 h heat stress treatment (H). Three biological replicates from each treatment were obtained. More than 43 million reads and over 92% of clean reads were obtained from each sample after removing the low-quality and adaptor-containing reads (Table 1). At least 6.9 Gb clean bases were acquired from each sample (Table 1). The number of base recognition accuracies over 99.9% was no less than 6.2 Gb (Table 1). Over 92.3% of the clean reads were totally mapped to the reference bottle gourd genome [23], with more than 97.6% being uniquely mapped (Table S2). After normalization using total reads, we finally identified 2431 (normal condition, CK) and 1814 (heat stress, H) differentially expressed genes (DEGs) (Figure 3). Among them, a total of 1485 genes were up-regulated and 946 genes were down-regulated before HS treatment (Table S3, Figure 3), while 602 genes were up-regulated and 1212 genes were down-regulated in response to HS (Table S4, Figure 3). Furthermore, to validate the RNA-seq results, a total of 21 genes, which showed significantly up- or down-regulation with the absolute value Log2 fold change > 1 after HS treatment (Table S1), were randomly selected and used for qRT-PCR analysis. The results showed that a strong positive correlation (two tailed, $R^2 = 0.938$) was detected between MF and LL (Figure 4), indicating the accuracy of the RNA-seq data.

Table 1. Analysis of raw data between MF and LL under normal conditions (CK) and 24 h after heat stress (H).

Sample	Reads No.	Bases (bp)	Q30 (bp)	Clean Reads No.	Clean Data (bp)	Q30 (%)	Clean Reads %	Clean Data %
MF_CK_1	48713312	7,355,710,112	7,019,875,701	45173544	6,821,205,144	95.43	92.73	92.73
MF_CK_2	45757598	6,909,397,298	6,587,698,893	42487732	6,415,647,532	95.34	92.85	92.85
MF_CK_3	43690728	6,597,299,928	6,275,922,552	40610484	6,132,183,084	95.12	92.94	92.94
LL_CK_1	43525386	6,572,333,286	6,259,514,252	40395314	6,099,692,414	95.24	92.8	92.8
LL_CK_2	49867998	7,530,067,698	7,180,347,054	46316818	6,993,839,518	95.35	92.87	92.87
LL_CK_3	46377310	7,002,973,810	6,575,537,624	43003502	6,493,528,802	93.89	92.72	92.72
MF_H_1	48711970	7,355,507,470	7,003,097,638	45215592	6,827,554,392	95.2	92.82	92.82
MF_H_2	50998488	7,700,771,688	7,335,822,954	47313752	7,144,376,552	95.26	92.77	92.77
MF_H_3	48670696	7,349,275,096	7,003,542,935	45160744	6,819,272,344	95.29	92.78	92.78
LL_H_1	46504370	7,022,159,870	6,697,944,713	43154332	6,516,304,132	95.38	92.79	92.79
LL_H_2	47952574	7,240,838,674	6,914,489,745	44463080	6,713,925,080	95.49	92.72	92.72
LL_H_3	50213476	7,582,234,876	7,255,490,965	46615128	7,038,884,328	95.69	92.83	92.83

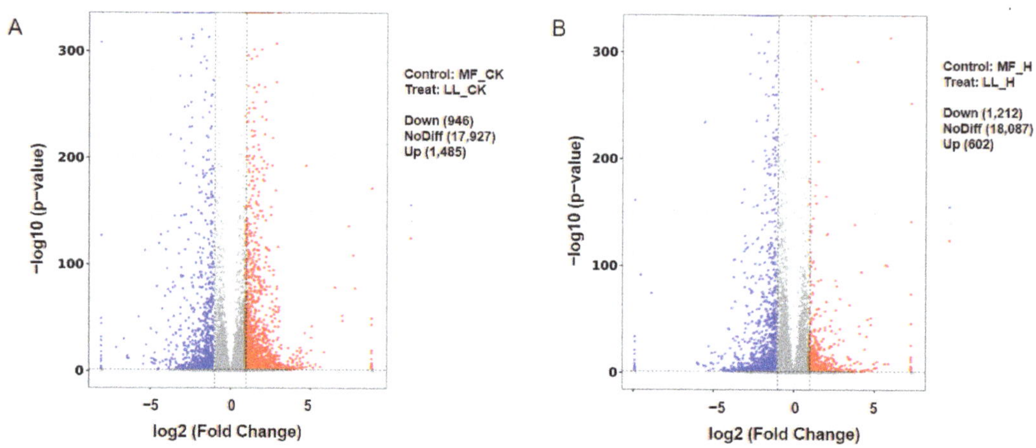

Figure 3. Differentially expressed gene (DEG) analysis in leaves under normal conditions (CK) and 24 h after heat stress (H). (**A**). DEGs under normal conditions. (**B**). DEGs after 24 h of heat stress. Red, brown, and blue represent up-regulated, no-difference, and down-regulated gene expression, respectively.

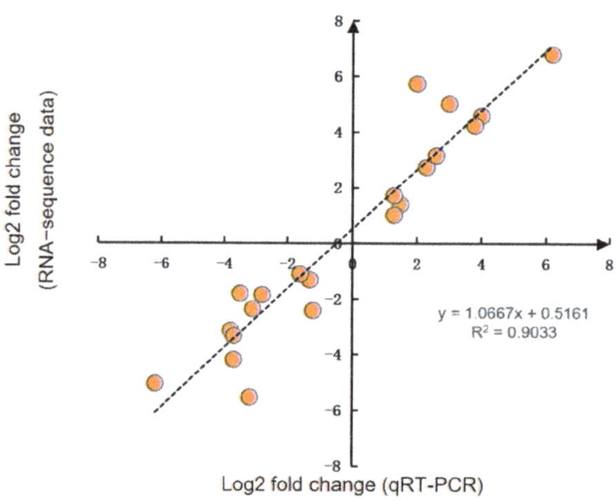

Figure 4. Validation of DEGs by qRT-PCR. The orange circles represent different gene expressions.

3.3. *Enriched Gene Categories Related to Heat Stress*

In order to understand which categories are represented in DEGs in comparison with the bottle gourd genome, all 2431 (CK) and 1814 (H) genes were further analyzed for Gene Ontology (GO) functional annotations. The results showed that three categories "cellular components (CC)", "molecular function (MF)", and "biological process (BP)" were classified under normal conditions (Table S5) and heat stress (Table S6) between MF and LL. The number of CC and BP were prominently decreased after heat treatment, while the number of genes in the MF category were significantly increased, especially in endopeptidase inhibitor activity, peptidase inhibitor activity, and so on (Figure S1).

Next, to examine the obtained DEGs associated with different pathways, the KEGG pathway database was used. The results showed that the main pathways under normal conditions were "phenylpropanoid biosynthesis", "plant hormones signal transduction", and "amino sugar and nucleotide sugar metabolism" (Figure 5A). When exposed to heat stress, genes related to the "MAPK signaling pathway" and "photosynthesis" were mostly enriched (Figure 5B), while the pathway of "phenylpropanoid biosynthesis" was also enriched in heat stress.

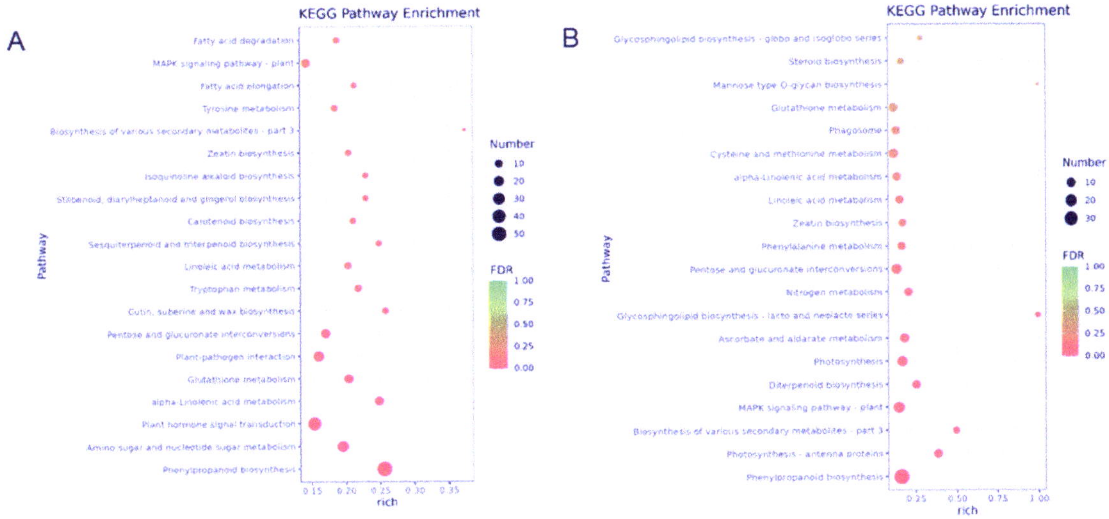

Figure 5. KEGG analysis of DEGs under normal conditions (CK) and 24 h after heat stress (H). (**A**). DEGs under normal conditions. (**B**). DEGs after 24 h of heat stress. High and low FDR (false discovery rate) are represented by green and red, respectively.

3.4. Expression of bHLH-Related Genes under Heat Stress

In order to study the number and types of transcription factors (TFs) that changed after heat stress treatment, transcriptome data were further analyzed. The results showed that the expression of various TFs, including bHLH, ERF (ETHYLENE RESPONSE FACTOR), and NAC (NAM, ATAF, and CUC), were significantly changed under CK (Figure S2A). After heat stress treatment, most TFs, such as bHLH and ERF, were down-regulated compared to normal conditions (Figure S2B). Indeed, the basic helix-loop-helix (bHLH) superfamily plays important roles in normal plant growth and development, as well as function in response to biotic or abiotic stress [24,25]. Thus, we analyzed the expression of members of the bHLH family using RNA-Seq data. As shown in Table S7, almost all bHLH family genes exhibited a higher level of transcripts in MF compared to LL after HS treatment. Next, we randomly selected six bHLH genes from RNA-Seq data for qRT-PCR in detail. For the qRT-PCR assay, we detected bHLH expression after exposure to 12, 24, and 36 h HS and under normal conditions (CK). The results showed that these 6 bHLHs were significantly induced after 24 h HS treatment (Figure 6). *HG10006168*, *HG10004527*, and *HG10015989* showed increased expression in MF compared with LL after 12 and 24 h of heat treatment. Among them, the expression of *HG10004527* was significantly higher in MF than LL under normal conditions. *HG10008149*, *HG10020341*, and *HG10006039* were highly expressed in LL in comparison with MF after 24 h heat stress treatment. *HG10006039* also exerted higher expression in MF than LL under normal conditions (Figure 6).

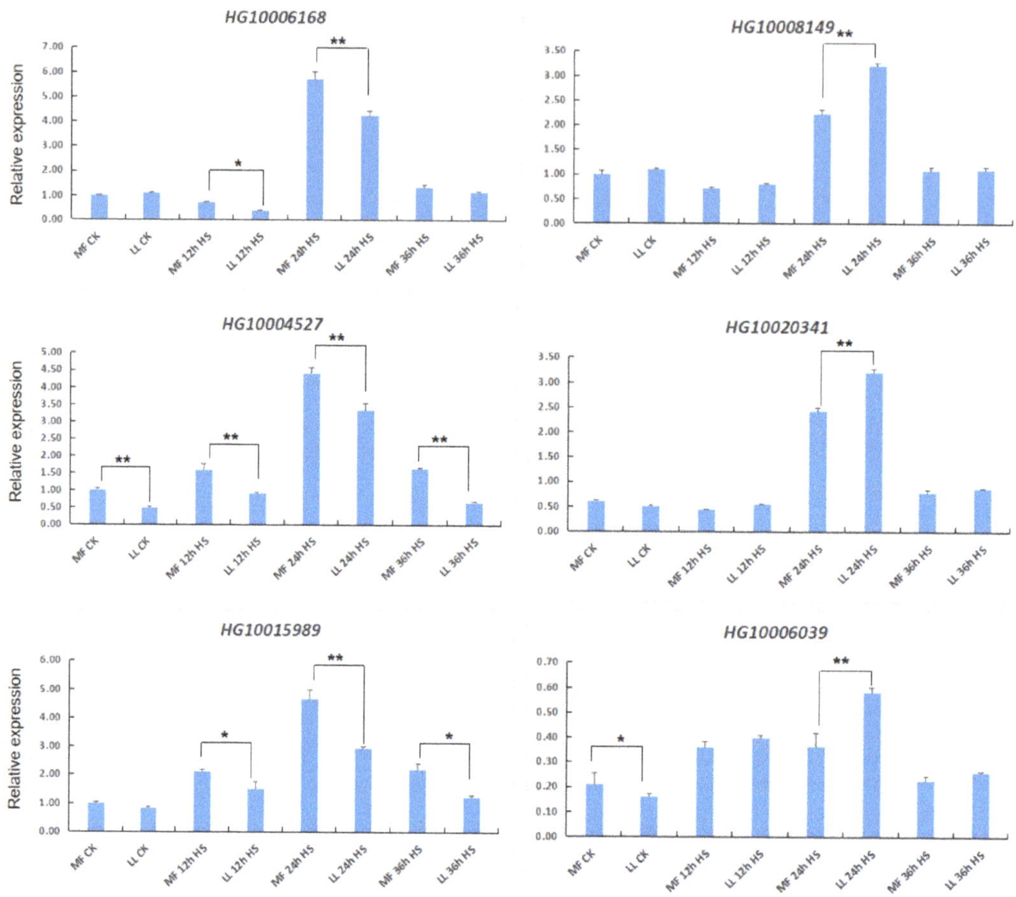

Figure 6. Relative expression of bHLH transcription factors. Total RNA was extracted from the true leaves of MF and LL growing under normal conditions (CK) and heat stress (H). *TuB-α* was used as an internal control. Data are the means ± SD of three independent replicates. Asterisks indicate a significant difference between treatments (* $0.01 \leq p \leq 0.05$, ** $p \leq 0.01$, Student's *t* test).

3.5. Expression of MAPK-Related Genes under Heat Stress

Based on KEGG analysis, we found that the MAPK (mitogen-activated protein kinase) signaling pathway was enriched during heat stress. Consistent with this observation, the expression of MAPK-related genes was significantly changed in the MF and LL seedlings under HS treatment (Table S7). Next, in order to understand the MAPK gene expression level in detail, 6 genes (Table S7) involved in the MAPK signal pathway were selected for qRT-PCR assay. The results showed that except *HG1000206*, the other five genes were significantly up-regulated in LL compared to MF after 24 h of heat stress treatment (Figure 7). Among them, gene expression of *HG10000206*, *HG10006437*, *HG10008395*, and *HG10005867* was prominently higher in MF compared with LL after 36 h heat stress. All of these genes showed no differences in expression between MF and LL under normal conditions (Figure 7).

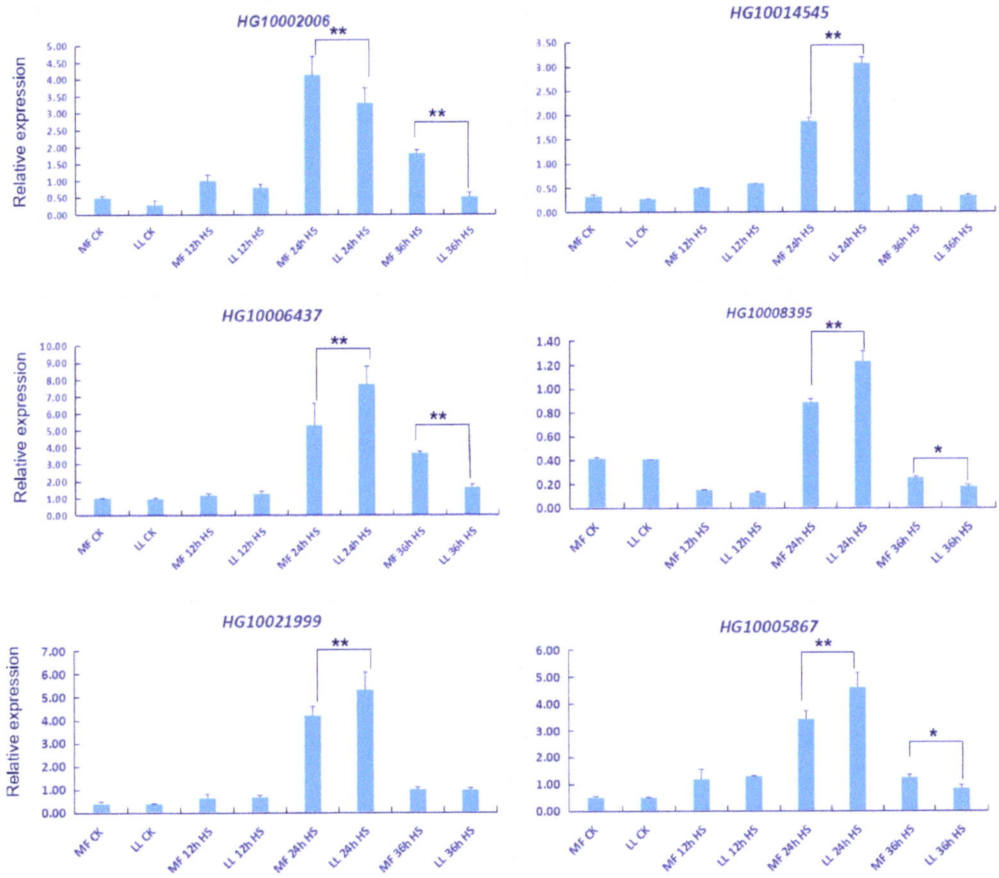

Figure 7. Relative expression of MAPK genes. Total RNA was extracted from the true leaves of MF and LL growing under normal conditions (CK) and heat stress (H). *TuB-α* was used as an internal control. Data are means ± SD of three independent replicates. Asterisks indicate a significant difference between treatments (* $0.01 \leq p \leq 0.05$, ** $p \leq 0.01$, Student's *t*-test).

4. Discussion

Heat stress severely limits crop production [26] by directly causing damage to enzymes and tissues [27], impairing flowering [28], and triggering oxidative stress at the plant reproductive stage [29]. It has been reported that many crop species are sensitive to heat stress, which ultimately impacts the economies of countries by reducing crop production [30]. Bottle gourd, as one of the most popular vegetables in South China, is pretty sensitive to heat stress, which leads to a decrease in production and quality. In our study, we used heat-resistant (MF) and heat-sensitive (LL) varieties to perform an RNA-sequencing assay. We found that the number of DEGs increased significantly 24 h after heat stress treatment. Among them, several DEGs related to the bHLH and MAPK signaling pathways showed prominent expression changes between MF and LL.

4.1. Analysis of bHLHs under Heat Stress

Transcription factors (TFs), such as WRKY [31], MYB [32], basic leucine zipper (bZIP) [33], and basic helix-loop-helix (bHLH) families, exert crucial roles in resistance to biotic and abiotic stress [34]. Among them, bHLHs are the second largest families and positively or negatively regulate plant resistance to various stressors [24,25]. For instance, bHLH2 positively regulates cold stress in Tartary buckwheat [35]. Overexpression of rice *OsbHLH001* confers freezing and salt tolerance in transgenic *Arabidopsis* [36]. Overexpressing grape *VvbHLH1* significantly improves salt and drought tolerance in transgenic *Arabidopsis* [2]. *OsbHLH148* increases drought resistance in rice [37]. Our study detected that the expression of several bHLHs changed significantly between MF and LL after heat stress treatment. Among them, three genes, *HG10006168*, *HG10004527*, and *HG10015989*, exerted high expression in MF compared with LL after 24 h of heat stress treatment. Similar findings have been observed in other plants. For example, the expression of Suaeda salsa *bHLH35* (the homolog of *HG10006168*) was significantly up-regulated after HS treatment [38]. In eggplant, the transcript levels of bHLH35 were also found to be up-regulated at 38 °C for 3 h and 6 h [39]. Additionally, heterologous expression of *Anthurium andraeanum bHLH35* has been shown to improve tolerance to cold and drought stress in *Arabidopsis* [40]. Collectively, these results suggest that *HG10006168* may play a positive role in HS resistance in bottle gourds. Like *bHLH35*, the expression of *bHLH93* (the homolog of *HG10004527*) was also induced by different stressors [41,42]. However, functional analysis in *Arabidopsis* showed that *bHLH93* is a key factor required for the promotion of flowering under non-inductive short-day conditions through the GA signaling pathway [43]. Consistent with these data, the expression of *HG10004527* was found to be higher in the ovary compared to roots and leaves, suggesting the potential role of this gene during the development of bottle gourd. Nevertheless, further research is needed to explore the functions of these genes in detail.

4.2. Analysis of the MAPK Signal Pathway under Stress

When plants are exposed to abiotic stress, many regulatory processes are controlled by many signaling pathways. Among them, MAPK (mitogen-activated protein kinase) exerts central roles during the stress resistance process [44,45]. MAPKs, as the evolutionarily conserved proteins in all eukaryotes, can be activated by substantially diverse stressors, and the activation of multiple MAPK pathways orchestrates fundamental cellular processes [46,47]. Previous studies have shown that the expression of MAPK-related genes, such as *MAPK2*, *MAPK3*, and *MAPK16*, is induced by abiotic stress treatments. Overexpression of *MAPK16* increases drought tolerance in *Arabidopsis* [48]. MAPK3 positively enhances resistance under cold, drought, and salt stress in *Arabidopsis* [49]. In tobacco, MAPK2 positively regulates salt and drought tolerance in tobacco [50]. In our study, we detected 6 MAPK-related genes, which were differentially expressed between MF and LL. Among them, five genes (*HG10014545*, *HG10006437*, *HG10008395*, *HG10021999*, and *HG10005867*) were significantly increased much more in LL than MF after 24 h heat stress treatment, indicating that these five genes might be involved in heat resistance of bottle gourd seedlings. Indeed, *Arabidopsis MAPK9*, which is the homolog of *HG10014545* and *HG10008395*, functions redundantly with another member of the MAPK family, *MAPK12*, as a positive regulator in ABA- and JA-induced stomatal closure [51]. This suggests that *MAPK9* and *MAPK12* regulate guard cell function in ABA signaling downstream of ROS. Moreover, further studies imply that *MAPK9* and *MAPK12* are positive regulators of SA signaling in *Arabidopsis* guard cells [52]. This finding is consistent with the role of *MAPK9* and *MAPK12* in regulating stomatal apertures, which contribute to the plant's defense against pathogens [53]. *Arabidopsis MAP KINASE KINASE KINASE 1* (*MAPKKK1*), which is the homolog of *HG10006437*, is a key transduction element that coordinates damage- and pathogen-associated molecular pattern-triggered immunity and orchestrate reactive oxygen species accumulation and signaling [54]. A subsequent study indicated a change in the localization pattern of *MAPKKK1* from the cytosol to plastids and nuclei after elicitor perception [55]. The function of *MAPK9*, *MAPK12*, and *MAPKKK1* in biotic stress response suggested that *HG10014545*, *HG10008395*, and *HG10006437* may also have a similar function in bottle gourd. Further study is required to investigate the specific role of these genes in bottle gourd and their potential involvement in the plant's defense mechanisms against pathogens.

5. Conclusions

In this study, we first identified a heat-resistant variety of bottle gourd that exhibited a higher survival rate, lower relative electrolytic leakage, and decreased stomatal aperture under HS. Next, RNA-Seq data showed that approximately 2000 genes were significantly altered under HS treatment compared to normal conditions in both heat-resistant and heat-sensitive varieties. Among these genes, several were found to be involved in the MAPK (mitogen-activated protein kinase) signaling pathway, as well as members of bHLH transcription factors, which showed significant up- or down-regulation after HS, which was further confirmed by real-time PCR (qRT-PCR) analysis. Our work may provide crucial genes for understanding the molecular regulation of heat resistance, which would be helpful in genetically improving heat tolerance in bottle gourd breeding.

Supplementary Materials: The following supporting information can be downloaded at: https://www.mdpi.com/article/10.3390/agronomy14020299/s1, Figure S1: GO analysis of DEGs under normal condition and 24 h after heat stress; Figure S2: Transcription factor analysis under normal conditions and 24 h after heat stress; Table S1: Primers used in this study; Table S2: The summary of sequence analysis; Table S3: The DEGs under normal conditions between MF and LL; Table S4: The DEGs under HS between MF and LL; Table S5: GO analysis of DEGs under normal conditions between MF and LL; Table S6: GO analysis of DEGs under HS between MF and LL; Table S7: DEGs of MAPK-related genes and members of bHLH transcription factors between MF and LL before and after treatment.

Author Contributions: Conceptualization, S.Y.; methodology, Q.P., B.J. and Y.L.; formal analysis, S.S., Y.W. and L.C.; investigation, T.Z. and X.C.; writing—original draft preparation, M.W.; writing—review and editing, M.W. and W.L. All authors have read and agreed to the published version of the manuscript.

Funding: This work was supported by the Agricultural Competitive Industry Discipline Team Building Project of Guangdong Academy of Agricultural Sciences (202114TD, 202103TD) and the Key-Area Research and Development Program of Guangdong Province (2022B0202110003). The funding agencies played no role in the design of the study, data collection, analysis, interpretation, or writing of the manuscript.

Data Availability Statement: The datasets generated for this study are accessible upon request from the authors.

Acknowledgments: We are grateful to Fachao Shi (Institute of Fruit Tree Research, Guangdong Academy of Agricultural Sciences) for the assistance in data analysis and phenotype identification.

Conflicts of Interest: The authors declare no conflicts of interest.

References

1. Wahid, A.; Gelani, S.; Ashraf, M.; Foolad, M.R. Heat tolerance in plants: An overview. *Environ. Exp. Bot.* **2007**, *61*, 199–223. [CrossRef]
2. Wang, D.; Qin, B.X.; Li, X.; Tang, D.; Zhang, Y.; Cheng, Z.K.; Xue, Y.B. Nucleolar DEAD-Box RNA helicase TOGR1 regulates thermotolerant growth as a Pre-rRNA chaperone in rice. *PLoS Genet.* **2016**, *12*, e1005844. [CrossRef]
3. McClung, C.R.; Davis, S.J. Ambient thermometers in plants: From physiological outputs towards mechanisms of thermal sensing. *Curr. Biol.* **2010**, *20*, 1086–1092. [CrossRef]
4. Mittler, R.; Finka, A.; Goloubinoff, P. How do plants feel the heat? *Trends Biochem. Sci.* **2012**, *37*, 118–125. [CrossRef]
5. Bita, C.E.; Gerats, T. Plant tolerance to high temperature in a changing environment: Scientific fundamentals and production of heat stress-tolerant crops. *Front. Plant Sci.* **2013**, *4*, 273. [CrossRef]
6. Zhu, J.K. Abiotic stress signaling and responses in plants. *Cell* **2016**, *167*, 313–324. [CrossRef]
7. Zandalinas, S.I.; Damián, B.; Vicent, A.; Aurelio, G.C.; Inupakutika, M.A.; Ron, M. ABA is required for the accumulation of APX1 and MBF1c during a combination of water deficit and heat stress. *J. Exp. Bot.* **2016**, *67*, 5381–5390. [CrossRef]
8. Larkindale, J.; Hall, J.D.; Knight, M.R.; Vierling, E. Heat Stress Phenotypes of Arabidopsis Mutants Implicate Multiple Signaling Pathways in the Acquisition of Thermotolerance. *Plant Physiol.* **2005**, *138*, 882–897. [CrossRef]
9. Shinozaki, K.; Yamaguchi-Shinozaki, K. Gene networks involved in drought stress response and tolerance. *J. Exp. Bot.* **2007**, *58*, 221–227. [CrossRef]
10. Liu, J.P.; Zhang, C.C.; Wei, C.C.; Liu, X.; Wang, M.G.; Yu, F.F.; Xie, Q.; Tu, J.M. The RING finger ubiquitin E3 ligase OsHTAS enhances heat tolerance by promoting H2O2-induced stomatal closure in rice. *Plant Physiol.* **2015**, *170*, 429–443. [CrossRef]

11. Takabatake, R.; Karita, E.; Seo, S.; Mitsuhara, I.; Kuchitsu, K.; Ohashi, Y. Pathogen-induced calmodulin isoforms in basal resistance against bacterial and fungal pathogens in tobacco. *Plant Cell Physiol.* **2007**, *48*, 414–423. [CrossRef]
12. Yang, X.H.; Liang, Z.; Lu, C.M. Genetic engineering of the biosynthesis of glycinebetaine enhances photosynthesis against high temperature stress in transgenic tobacco plants. *Plant Physiol.* **2005**, *138*, 2299–2309. [CrossRef]
13. Zhu, W.; Lu, M.H.; Gong, Z.H.; Chen, R.G. Cloning and expression of a small heat shock protein gene CaHSP24 from pepper under abiotic stress. *Afr. J. Biotechnol.* **2011**, *10*, 4968–4976.
14. Li, H.; Liu, S.S.; Yi, C.Y.; Wang, F.; Zhou, J.; Xia, X.J.; Shi, K.; Zhou, Y.H.; Yu, J.Q. Hydrogen peroxide mediates abscisic acid-induced HSP70 accumulation and heat tolerance in grafted cucumber plants. *Plant Cell Environ.* **2015**, *37*, 2768–2780. [CrossRef]
15. Yu, B.W.; Yan, S.S.; Zhou, H.Y.; Dong, R.; Lei, J.J.; Chen, C.M.; Cao, B.H. Overexpression of CsCaM3 Improves High Temperature Tolerance in Cucumber. *Front. Plant Sci.* **2018**, *9*, 797. [CrossRef]
16. Erickson, D.L.; Smith, B.D.; Clarke, A.C.; Sandweiss, D.H.; Tuross, N. An Asian origin for a 10,000-year-old domesticated plant in the Americas. *Proc. Natl. Acad. Sci. USA* **2005**, *102*, 18315–18320. [CrossRef]
17. Mashilo, J.; Odindo, A.O.; Shimelis, H.A.; Musenge, P.; Tesfay, S.Z.; Magwaza, L.S. Photosynthetic response of bottle gourd [Lagenaria siceraria (Molina) Standl.] to drought stress: Relationship between cucurbitacins accumulation and drought tolerance. *Sci. Hortic.* **2018**, *231*, 133–143. [CrossRef]
18. Yang, Y.; Yu, L.; Wang, L.; Guo, S. Bottle gourd rootstock-grafting promotes photosynthesis by regulating the stomata and non-stomata performances in leaves of watermelon seedlings under NaCl stress. *J. Plant Physiol.* **2015**, *186*, 50–58. [CrossRef]
19. Singh, A.P.; Singh, T.; Rickard, C.L. Visualising impregnated chitosan in Pinus radiata early wood cells using light and scanning electron microscopy. *Micron* **2010**, *41*, 263–267. [CrossRef]
20. Ye, J.; Fang, L.; Zheng, H.K.; Zhang, Y.; Chen, J.; Zhang, Z.J.; Wang, J.; Li, S.T.; Li, R.Q.; Bolund, L.; et al. WEGO, A web tool for plotting GO annotations. *Nocleic Acids Res.* **2006**, *34*, 293–297. [CrossRef] [PubMed]
21. Kanehisa, M.; Goto, S.; Kawashima, S.; Okuno, Y.; Hattori, M. The KEGG resource for deciphering the genome. *Nucleic Acids Res.* **2004**, *32*, 277–280. [CrossRef]
22. Li, Y.W.; Wang, Y.; Wu, X.Y.; Wang, J.; Wu, X.H.; Wang, B.G.; Lu, Z.F.; Li, G.J. Novel genomic regions of fusarium wilt resistance in bottle gourd [*Lagenaria siceraria* (Mol.) Standl.] Discovered in Genome-Wide Association Study. *Front. Plant Sci.* **2021**, *12*, 650157. [CrossRef]
23. Xu, P.; Wang, Y.; Sun, F.S.; Wu, R.L.; Du, H.L.; Wang, Y.H.; Jiang, L.B.; Wu, X.H.; Wu, X.Y.; Yang, L.M.; et al. Long-read genome assembly and genetic architecture of fruit shape in the bottle gourd. *Plant J.* **2021**, *107*, 956–968. [CrossRef]
24. Duek, P.D.; Fankhauser, C. bHLH class transcription factors take centre stage in phytochrome signalling. *Trends Plant Sci.* **2005**, *10*, 51–54. [CrossRef]
25. Feller, A.; Machemer, K.; Braun, E.L.; Grotewold, E. Evolutionary and comparative analysis of MYB and bHLH plant transcription factors. *Plant J.* **2011**, *66*, 94–116. [CrossRef]
26. Cao, Z.B.; Tang, H.W.; Cai, Y.H.; Zeng, B.H.; Zhao, J.L.; Tang, X.Y.; Lu, M.; Wang, H.M.; Zhu, X.J.; Wu, X.F.; et al. Natural variation of HTH5 from wild rice, Oryza rufipogon Griff., is involved in conferring high-temperature tolerance at the heading stage. *Plant Biotechnol.* **2022**, *20*, 1591–1605. [CrossRef] [PubMed]
27. Moore, C.E.; Meacham-Hensold, K.; Lemonnier, P.; Slattery, R.A.; Benjamin, C.; Bernacchi, C.J.; Lawson, T.; Cavanagh, A.P. The effect of increasing temperature on crop photosynthesis: From enzymes to ecosystems. *J. Exp. Bot.* **2021**, *72*, 2822–2844. [CrossRef]
28. Tian, Y.L.; Chen, J.; Chen, C.Q.; Deng, A.X.; Song, Z.W.; Zheng, C.Y.; Willem, H.M.; Zhang, W.J. Warming impacts on winter wheat phenophase and grain yield under field conditions in Yangtze delta plain, China. *Field Crops Res.* **2012**, *134*, 193–199. [CrossRef]
29. Khan, A.H.; Ma, Y.Z.; Wu, Y.L.; Akbar, A.D.; Shaban, M.H.; Ullah, A.B.; Deng, J.W.; Khan, A.S.; Chi, H.B.; Zhu, L.F.; et al. High-temperature stress suppresses allene oxide cyclase 2 and causes male sterility in cotton by disrupting jasmonic acid signaling. *Crop J.* **2023**, *11*, 33–45. [CrossRef]
30. Abbas, S. Climate change and major crop production: Evidence from Pakistan. *Environ. Sci. Pollut. Res.* **2022**, *29*, 5406–5414. [CrossRef]
31. Yokotani, N.; Sato, Y.; Tanabe, S.; Chujo, T.; Shimizu, T.; Okada, K.; Yamane, H.; Shimono, M.; Sugano, S.; Takatsuji, H.; et al. WRKY76 is a rice transcriptional repressor playing opposite roles in blast disease resistance and cold stress tolerance. *J. Exp. Bot.* **2013**, *64*, 5085–5097. [CrossRef]
32. Chen, N.; Chi, X.Y.; Pan, L.J.; Yu, S.L. Advances in MYB transcription factors during salt-stress regulation in plants. *Plant Physiol. J.* **2015**, *51*, 1395–1399.
33. Ma, H.Z.; Liu, C.; Li, Z.X.; Ran, Q.J.; Xie, G.N.; Wang, B.M.; Fang, S.; Chu, J.F.; Zhang, J.R. ZmbZIP4 contributes to stress resistance in maize by regulating ABA synthesis and root development. *Plant Physiol.* **2018**, *178*, 753–770. [CrossRef]
34. Toledo-Ortiz, G.; Huq, E.; Quail, P.H. The Arabidopsis basic/helix-loop-helix transcription factor family. *Plant Cell* **2003**, *15*, 1749–1770. [CrossRef]
35. Yao, P.F.; Sun, Z.X.; Li, C.L.; Zhao, X.R.; Li, M.F.; Deng, R.Y.; Huang, Y.J.; Zhao, H.X.; Chen, H.; Wu, Q. Overexpression of Fagopyrum tataricum FtbHLH2 enhances tolerance to cold stress in transgenic Arabidopsis. *Plant Physiol. Biochem.* **2018**, *125*, 85–94. [CrossRef]
36. Li, F.; Guo, S.Y.; Zhao, Y.; Chen, D.Z.; Chong, K.; Xu, Y.Y. Overexpression of a homopeptide repeat-containing bHLH protein gene (OrbHLH001) from Dongxiang Wild Rice confers freezing and salt tolerance in transgenic Arabidopsis. *Plant Cell Rep.* **2010**, *29*, 977–986. [CrossRef]

37. Seo, J.S.; Joo, J.; Kim, M.J.; Kim, Y.K.; Nahm, B.H.; Song, S.I.; Cheong, J.J.; Lee, J.S.; Kim, J.K.; Choi, Y.D. OsbHLH148, a basic helix-loop-helix protein, interacts with OsJAZ proteins in a jasmonate signaling pathway leading to drought tolerance in rice. *Plant J.* **2011**, *65*, 907–921. [CrossRef]
38. Li, M.; He, P.; Zhao, Z.; Liu, J.; Liu, H.; Ma, S.; Shen, Y.; Li, B. Effect of temperature on betacyanins synthesis and the transcriptome of Suaeda salsa. *Front. Plant Sci.* **2023**, *14*, 1203089. [CrossRef]
39. Zhang, S.; Zhang, A.; Wu, X.; Zhu, Z.; Yang, Z.; Zhu, Y.; Zha, D. Transcriptome analysis revealed expression of genes related to anthocyanin biosynthesis in eggplant (*Solanum melongena* L.) under high-temperature stress. *BMC Plant Biol.* **2019**, *19*, 387. [CrossRef]
40. Jiang, L.; Tian, X.; Li, S.; Fu, Y.; Xu, J.; Wang, G. The AabHLH35 transcription factor identified from Anthurium and raeanum is involved in cold and drought tolerance. *Plants* **2019**, *8*, 216. [CrossRef]
41. Wang, Y.; Wang, S.; Tian, Y.; Wang, Q.; Chen, S.; Li, H.; Ma, C.; Li, H. Functional characterization of a sugar beet BvbHLH93 transcription factor in salt stress tolerance. *Int. J. Mol. Sci.* **2021**, *22*, 3669. [CrossRef]
42. Samarina, L.S.; Bobrovskikh, A.V.; Doroshkov, A.V.; Malyukova, L.S.; Matskiv, A.O.; Rakhmangulov, R.S.; Koninskaya, N.G.; Malyarovskaya, V.I.; Tong, W.; Xia, E. Comparative expression analysis of stress-inducible candidate genes in response to cold and drought in tea plant [*Camellia sinensis* (L.) Kuntze]. *Front. Genet.* **2020**, *11*, 611283. [CrossRef]
43. Sharma, N.; Xin, R.; Kim, D.-H.; Sung, S.; Lange, T.; Huq, E. No flowering in short day (NFL) is a bHLH transcription factor that promotes flowering specifically under short-day conditions in Arabidopsis. *Development* **2016**, *143*, 682–690.
44. Nadal, E.D.; Alepuz, P.M.; Posas, F. Dealing with osmostress through MAP kinase activation. *Emro Rep.* **2002**, *3*, 735–740. [CrossRef]
45. Lin, L.; Wu, J.; Jiang, M.; Wang, Y. Plant mitogen-activated protein kinase cascades in environmental stresses. *Int. J. Mol. Sci.* **2021**, *22*, 1543. [CrossRef]
46. Cargnello, M.; Roux, P.P. Activation and function of the MAPKs and their substrates, the MAPK-activated protein kinases. *Microbiol. Mol. Biol.* **2011**, *75*, 50–83. [CrossRef]
47. Kyriakis, J.M.; Avruch, J. Mammalian MAPK signal transduction pathways activated by stress and inflammation: A 10-year update. *Physiol. Rev.* **2012**, *92*, 689–737. [CrossRef]
48. Shi, J.; Zhang, L.; An, H.L.; Wu, C.G.; Guo, X.Q. GhMPK16, a novel stress-responsive group D MAPK gene from cotton, is involved in disease resistance and drought sensitivity. *BMC Mol. Biol.* **2011**, *12*, 22. [CrossRef]
49. Sadau, S.B.; Ahmad, A.; Tajo, S.M.; Ibrahim, S.; Kazeem, B.B.; Wei, H.; Yu, S. Overexpression of GhMPK3 from cotton enhances cold, drought, and salt stress in Arabidopsis. *Agronomy* **2021**, *11*, 1049. [CrossRef]
50. Zhang, L.; Xi, D.M.; Li, S.W.; Gao, Z.; Zhao, S.L.; Shi, J.; Wu, C.G.; Guo, X.Q. A cotton group C MAP kinase gene, GhMPK2, positively regulates salt and drought tolerance in tobacco. *Plant Mol. Biol.* **2011**, *77*, 17–31. [CrossRef]
51. Jammes, F.; Song, C.; Shin, D.; Munemasa, S.; Takeda, K.; Gu, D.; Cho, D.; Lee, S.; Giordo, R.; Sritubtim, S. MAP kinases MPK9 and MPK12 are preferentially expressed in guard cells and positively regulate ROS-mediated ABA signaling. *Proc. Natl. Acad. Sci. USA* **2009**, *106*, 20520–20525. [CrossRef]
52. Khokon, M.A.R.; Salam, M.A.; Jammes, F.; Ye, W.; Hossain, M.A.; Okuma, E.; Nakamura, Y.; Mori, I.C.; Kwak, J.M.; Murata, Y. MPK9 and MPK12 function in SA-induced stomatal closure in Arabidopsis thaliana. *Biosci. Biotechnol. Biochem.* **2017**, *81*, 1394–1400. [CrossRef]
53. Jammes, F.; Yang, X.; Xiao, S.; Kwak, J.M. Two Arabidopsis guard cell-preferential MAPK genes, MPK9 and MPK12, function in biotic stress response. *Plant Signal. Behav.* **2011**, *6*, 1875–1877. [CrossRef]
54. Savatin, D.; Gigli, B.; Marti, L.; Fabbri, C.; Cervone, F.; De Lorenzo, G. The Arabidopsis NPK1-related protein kinases ANPs are required for elicitor-induced oxidative burst and immunity. *Plant Physiol.* **2014**, *165*, 1188–1202. [CrossRef]
55. Marti, L.; Savatin, D.V.; Gigli-Bisceglia, N.; De Turris, V.; Cervone, F.; De Lorenzo, G. The intracellular ROS accumulation in elicitor-induced immunity requires the multiple organelle-targeted Arabidopsis NPK1-related protein kinases. *Plant Cell Environ.* **2021**, *44*, 931–947. [CrossRef]

agronomy

MDPI

Article

Regional NDVI Attribution Analysis and Trend Prediction Based on the Informer Model: A Case Study of the Maowusu Sandland

Hongfei Hou [1], Ruiping Li [1,*], Hexiang Zheng [2], Changfu Tong [2,*], Jun Wang [2], Haiyuan Lu [2], Guoshuai Wang [2], Ziyuan Qin [2] and Wanning Wang [2]

[1] College of Water Conservancy and Civil Engineering, Inner Mongolia Agricultural University, Huhhot 010018, China; 2869216428@emails.imau.edu.cn

[2] Water Conservancy Research Institute of Pastoral Area, China Research Institute of Water Resources and Hydropower, Huhhot 010020, China; zhenghx@iwhr.com (H.Z.); wangjun@iwhr.com (J.W.); luhy@iwhr.com (H.L.); imau_wgs@163.com (G.W.); qinzy@iwhr.com (Z.Q.); wangwn@iwhr.com (W.W.)

* Correspondence: nmglrp@163.com (R.L.); tcf110110@163.com (C.T.)

Abstract: Terrestrial ecosystems depend heavily on their vegetation; it is possible to forecast future growth trends of regional vegetation by keeping an eye on changes in vegetation dynamics. To circumvent the potential reduction in prediction accuracy caused by the non-stationarity of meteorological changes, we analyzed the characteristics of NDVI (Normalized Difference Vegetation Index) spatial and temporal changes and the influencing factors over the past 20 years in the Maowusu Sandland of China via attribution analysis. We also constructed a comprehensive analysis system for vegetation pre-restoration. Moreover, we combined meteorological data from 2000 to 2018 and presented a deep-learning NDVI-Informer prediction model with a self-attentive mechanism. We also used distillation operation and fusion convolutional neural network for NDVI prediction. Incorporating a probsparse self-attention method successfully overcomes Transformer weaknesses by lowering the memory use and complexity of large time series. It significantly accelerates the inference speed of long time series prediction and works well with non-smooth data. The primary findings were: (1) the Maowusu Sandland's 20-year average showed a consistent increasing trend in the NDVI at 0.0034 a^{-1}, which was mostly caused by climate change, with a relative contribution rate of 55.47%; (2) The Informer-based model accurately forecasted the NDVI in the research region based on meteorological elements and conducted a thorough analysis of the MAPE (mean absolute percentage error) (2.24%). This suggests that it can effectively lower the data's volatility and increase prediction accuracy. The anticipated outcomes indicate that the trend will stabilize during the following ten years. To attain more sustainable and efficient agricultural production, the results of this study may be used to accurately estimate future crop yields and NDVI using previous data.

Keywords: NDVI; attribution analysis; self-attention mechanism; convolutional neural network; Informer model

1. Introduction

One of the key elements of terrestrial ecosystems is vegetation, which influences the earth's ecological balance, water cycle, and temperature [1,2] and contributes significantly to information transit, energy movement, and material cycle, among other things [3]. Jiazheng Zhang performed visualization calculation analysis and processing using the Google Earth Engine (GEE), which can monitor the dynamic changes of large-scale and long-time sequence vegetation and subsequently serve as a useful resource for local environmental management [4]. One of China's four main sandy zones, Maowusu Sandland, is surrounded by agricultural and livestock industries with a low forest cover and vulnerable ecosystems [5]. Maowusu Sandland had more than 90% desertified and sandy land in the

1950s, with sandy area making up 40% of the total area and only 2.6% of the area covered by forests [6–8]. The vegetation in the Maowusu Sandland has deteriorated considerably due to man-made and natural forces [9]. For the study of NDVI extraction in the Maowusu Sandland, it is crucial to understand the spatial and temporal dynamics of vegetation in the region as well as trend prediction. Vegetation degradation is accompanied by a drop in the NDVI [10]. This study examines the spatiotemporal aspects of NDVI changes over the past 20 years in the study region and investigates trend changes using remote sensing technologies.

The prediction of NDVI based on Moderate Resolution Imaging Spectroradiometer (MODIS) and Landsat data has gained considerable attention in recent years. It is now used to interpolate the Landsat NDVI data throughout an entire year to complete the evaluation of the vegetation state [11]. There are, however, fewer studies on the primary factors influencing vegetation change in dry and semi-arid regions and precise long-term projections, with earlier research mainly concentrating on how vegetation responds to the environment and human activity [12]. Precipitation and air temperature are currently the primary parameters used by many researchers, which are relatively varied concerning vegetation in space due to the effect of climate change on vegetation change [13]. The two areas where human activity has the greatest impact on vegetation change are: (1) implementing ecological initiatives, such as converting farms to forests and shutting down mountains to grow grass [14]; (2) human production and living activities, together with the urbanization process, have the potential to upset the ecosystem's delicate balance, preventing plants from growing normally [15]. We can gain a deeper understanding by thoroughly analyzing the effects of climate change and human activities on vegetation change. Currently, global change research focuses on the quantitative analysis of the impact of vegetation change drivers and the prediction of future trends [16]. Additionally, the multivariate linear residual analysis [17], which carefully examines the relative influence of human actions [18], dominates the quantitative analysis. The major models used for trend prediction include the stochastic Forest, Support Vector Machine (SVM), and Support Vector Regression (SVR) models [19], as well as machine learning techniques for forecasting future trends. For instance, Guo et al. [20] used the residual method, which is the most popular model for separating climate change and human activities, and is appropriate for long time series analyses, to quantitatively analyze the effects of climate change and human activities on changes in vegetation cover. The Support Vector Regression (SVR) approach was successfully used by Yang Wei et al. [21] and others to increase the NDVI inversion accuracy, however, the model suffers from a sluggish convergence speed and a very complicated optimization procedure [22]. To simulate and quantify the effects of climate change and human activity on vegetation cover, Shengzhi Huang et al. [23] introduced a Support Vector Machine (SVM) model. However, the limitations of machine learning, including its difficulty in feature learning, difficulty in determining hyper-parameters, and slow computation speed, decrease prediction accuracy.

Deep learning offers superior learning and fitting capabilities compared to statistical and machine learning approaches, and it can fully exploit large-scale time series data for improved short-term NDVI forecasting [24]. Convolutional neural networks (CNNs), recurrent neural networks (RNNs), deep belief networks (DBNs), and deep neural networks (DNNs) are examples of deep learning techniques [25]. Peiqiang Gao et al. [26] use CNNs to create an NDVI prediction model to thoroughly examine how the NDVI responds to environmental conditions, which significantly increases the accuracy of NDVI prediction. However, when dealing with very volatile and non-stationary data, CNN prediction accuracy suffers. The cyclic structure introduced by RNN is better able to handle time series data and extract NDVI data for fitting, however, it also has issues with fading gradients and the inability to record long-term dependencies [27]. As a specific type of RNN, the long short-term memory (LSTM) network outperforms current neural networks (RNNs) in short-term NDVI prediction tasks. However, LSTM also has issues with many parameters and sluggish model convergence [28]. The prediction accuracy of the aforementioned deep learning models is significantly better than that of conventional statistical models and machine learning techniques; however, it is not optimal when dealing with nonlinear time series data. In the field of time series prediction, the Transformer model [29] has recently been proposed and applied. However, despite producing predictions with better accuracy, the Transformer model has a high memory occupancy rate and time complexity.

This study focuses on fixing the issues with the earlier Transformer model, while the Informer model [30] is suggested to address the Transformer's weaknesses. This study aims to address three issues of the earlier Transformer: (1) The self-attention mechanism's complexity and memory requirements; (2) The memory bottleneck with lengthy input stacked layers; and (3) The performance deterioration of forecasting long output time series. In addition to exhibiting excellent potential in long-time series input and output dependencies, the improved Informer model introduces the probabilistic sparse self-attention mechanism, which can efficiently reduce the spatial complexity of the long sequence time-series forecasting (LSTF) problem.

Determining the long-term spatial and temporal NDVI changes in the Maowusu Sandland and the pattern of such changes is crucial for ensuring ecological quality [31]. There are currently few persistent analyses for trend prediction, and the geographical and temporal variations of NDVI in the Maowusu Sandland are in the monitoring stage. The Maowusu Sandland's NDVI trend in the future is crucial for addressing drought.

This study is based on the Google Earth Engine (GEE) cloud platform. It integrates multiple change analysis and attribution algorithms over 20 years to construct a set of comprehensive analysis systems for vegetation pre-restoration. These systems are used to predict future changes in vegetation trends, to thoroughly reveal through in-depth analysis the relationships and roles of vegetation, climate, and human activities, and to provide trustworthy scientific data (Figure 1) [32–34]. Therefore, the study's specific goals are to (1) thoroughly explore the relationship between the dynamic changes in vegetation and the spatial and temporal changes in the climate under artificial restoration and natural restoration, taking into account the context of 20 years of climate warming and aridification of severe ecological restoration projects as an entry point; (2) conduct a regional ecological assessment, evaluate the spatial and temporal changes, and forecast future NDVI trends to examine the distinct effects of artificial restoration and natural restoration on vegetation changes in the Maowusu Sandland area as an example; (3) compile and train 19 years' worth of NDVI data in the research region, as well as meteorological parameters, and propose a technique based on Informer's accurate long-term NDVI prediction, which will serve as a scientific foundation for ecological environmental conservation and sustainable development.

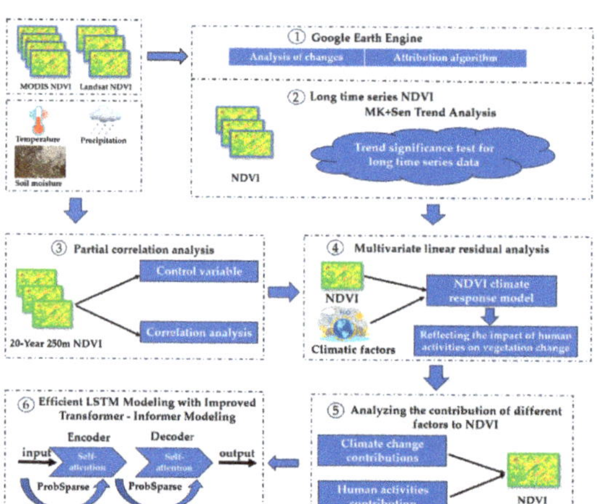

Figure 1. Methodology flowchart.

2. Materials and Methods

2.1. Study Area

Maowusu Sandland, with an average elevation of 1347 m above sea level, is situated in the semi-arid desert region of northwest China, between the Yulin area of Shaanxi Province and Ordos City of Inner Mongolia Autonomous Region (Figure 2). It has an administrative area that includes the Wushen Banner of Ordos City, Ejin Horo Banner, Ertok Banner, and Ertokqian Banner. The region's geography is high in the north and south and low in the east and west. It has a land area of roughly 2.98 × 10^4 km^2 [35]. With an average annual temperature of 6.0–8.5 °C, average annual precipitation of 220–400 mm, average annual evaporation of 2100–2600 mm, and average annual wind speed of 4.5 m/s, the Maowusu Sandland is classified as having a moderate continental monsoon climate. The region's climatic features include a dry environment, significant evaporation, uneven distribution of precipitation, powerful insolation, windy weather, and sandy terrain [36].

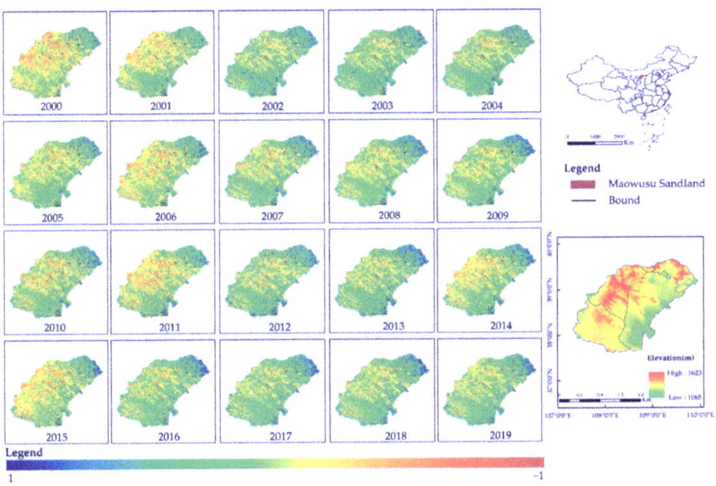

Figure 2. NDVI long time series data and elevation of the Maowusu Sandland.

2.2. Data and Processing

The method effectively removed the temporal and spatial constraints in the geomorphological study. The Google Earth Engine (GEE) cloud platform was used to generate long time series and multi-year mean analyses based on the Landsat series, which was used to construct the NDVI dataset of 250 m spatial resolution long time series for remotely sensed data in the study area.

The GEE platform (https://earthengine.google.com/) (accessed on 30 June 2023)was used to obtain NDVI data from the Moderate Resolution Imaging Spectroradiometer (MODIS) data. The Maowusu Sandland was subjected to a quantitative investigation of plant cover changes utilizing the monthly temporal and geographical resolution of 250 m MOD13Q1 NDVI. NDVI data were created using the average synthesis approach with data from January through December. Additionally, Landsat remote image NDVI obtained each year of the research period underwent pixel-by-pixel yearly 95th quantile synthesis, which can partially mitigate the uncertainty caused by the year's limited number of picture acquisitions.

Using ASTERGDEM 30M resolution digital elevation data, digital elevation model (DEM) data were chosen from the geospatial data cloud (https://www.gscloud.cn/) (accessed on 5 July 2023). ArcGIS was used to analyze the digital elevation model (DEM) data to acquire elevation information.

The two primary sources of meteorological data were the National Tibetan Plateau Science Data Center and the fifth-generation European Centre for Medium-Range Weather Forecasts (ECMWF) atmospheric reanalysis (ERA5) dataset from the GEE cloud platform. The mean_2m_air_temperature band temperature data at 2 m above ground level per hour and total_precipitation band calculations were the foundation of the European Centre for Medium-Range Weather Forecasts Fifth Generation Atmospheric Reanalysis (ERA5) dataset (resolution 27,830 m), which gathers daily mean temperature and daily precipitation data from 2000 to 2019. Meanwhile, researchers from Peng Shouzhang obtained data from the National Tibetan Plateau Science Data Center, including differences in mean annual temperature and rainfall through time and space. These data were created using the worldwide high-resolution climate dataset from World Clim and the global 0.5° climate dataset from Climatic Research Unit Time Series (CRU), which were both downscaled in the Chinese region using the Delta spatial downscaling tool. Data from 496 different meteorological observation stations were utilized for validation to confirm the accuracy of the data, and the validated findings demonstrated its reliability.

Three periods of land-use data were gathered for 2000, 2010, and 2020. The land-use data were acquired from the Globeland30 dataset (30 m resolution), a worldwide surface cover dataset created in China. Bare land, grassland, farmland, shrubland, wetland, water bodies, woodland, and man-made land surfaces were among the several types of land.

The surface soil moisture covariate from the ERA5 reanalysis data was used to interpolate the 0.25° resolution European Space Agency Climate Change Initiative (ESA-CCI) surface soil moisture product to obtain the spatiotemporally continuous day-by-day 0.25° resolution soil moisture. Machine learning algorithms were then used to downscale the daily 0.25°-resolution surface soil moisture to a daily 1-km resolution [37] using data from the International Soil Moisture Observation Network's global soil moisture observatory, along with other high-resolution optical remote sensing data.

The accuracy of the aforementioned data was strictly controlled and was collected from reputable sources, including the National Aeronautics and Space Administration (NASA), the Computer Network Information Center of the Chinese Academy of Sciences, the European Centre for Medium-Range Weather Forecasts (ECMWF), and the Global Surface Coverage Remote Sensing Mapping and Key Technology Research Project of the National High-Technology Research and Development Program of China.

2.3. Research Methodology

2.3.1. MK + Sen Trend Analysis

The temperature–rainfall time series was examined for noteworthy patterns using nonparametric approaches. The Mann–Kendall (MK) and Sen slope estimators, which are computationally effective, were employed to determine the trend and slope, respectively, as temperature–rainfall data were not normally distributed [38].

The Sen trend is a more reliable non-parametric statistic for trend calculations. It is insensitive to measurement error and outlier data. In contrast, the MK test does not require the samples to follow a particular distribution and is less affected by outliers, making it suitable for significant tests of trends in long time series data [39]. The Sen trend and MK tests can be combined to analyze long vegetation time series [32]. The formula for calculating the significant trend is as follows:

$$\beta = Median\left(\frac{x_j - x_i}{j - i}\right) \forall j > i \tag{1}$$

where X_j and X_i are time series data; β is an indicator that represents the trend of a time series; $\beta > 0$ indicates an upward trend in the time series; $\beta < 0$ indicates a downward trend in the time series.

2.3.2. Partial Correlation Analysis

In addition to applying to multivariate data, partial correlation analysis permits hypothesis testing to ascertain whether there is a statistically significant relationship between two variables by controlling for the impact of one or more confounding variables on the relationship between two variables [40]. The *t*-test, which statistically analyzes each pixel at a significance threshold of 0.05, was used for the correlation study [41]. This study examined the association between three variables—temperature, precipitation, and soil moisture—and the change in NDVI from 2000 to 2019 using MATLAB2016-based partial correlation analysis.

2.3.3. Multivariate Linear Residual Analysis

To isolate the portion of the climatic effect that impacts vegetation development, residual analysis is a method for modeling regression between climate parameters and the NDVI (Normalized Vegetation Index) [42,43]. Regression modeling is used in this approach to determine the expected values of the NDVI, which in turn produces the residuals between the real and predicted values. These values are then used to assess the influence of human activity on vegetation growth [44]. This paper establishes a multiple regression model between the NDVI and climate factors before the implementation of the ecological project (2000–2010) (i.e., a metric scale to calculate the NDVI climate response model) [45] to better reflect the impact of human activities on vegetation changes.

The NDVI predictions—driven solely by climate change—were calculated image-by-element from 2000 to 2020 using the established regression equations. The NDVI observations were then subtracted from the NDVI predictions—driven solely by climate change—to obtain the NDVI residual time series from 2000 to 2020 [34]. The following formula was used to calculate the residual series:

$$\varepsilon = NDVI_{abs} - NDVI_{pre} \tag{2}$$

where ε is the NDVI residual; $NDVI_{abs}$ is the observed NDVI value; $NDVI_{pre}$ is the predicted NDVI value.

2.3.4. Relative Contribution Rate

Relative contribution analysis is a technique for determining the precise percentage that each explanatory variable contributes to the variance of the explained variables. The challenge and concern regarding the relevant research domestically and internationally has been identifying and quantifying the determinants of vegetation change on a wide regional scale [46]. In this study, we adopted a technique to calculate the relative contribution rate of climate change and human activity to vegetation change using multivariate linear residual analysis [45]. We also identified the drivers of vegetation change on an image metric scale and quantified their relative contributions to vegetation change [46]. The associated formula is:

$$K = \frac{Slope(NDVI_{pre})}{Slope(NDVI_{obs})} \tag{3}$$

$$K = \frac{Slope(NDVI_{res})}{Slope(NDVI_{obs})} \tag{4}$$

where K is the relative contribution rate; $Slope(NDVI_{obs})$ is the trend value for the time series of actual NDVI values; $Slope(NDVI_{pre})$ is the trend value for the time series of predicted NDVI values based on multivariate regression analysis; $Slope(NDVI_{res})$ is the trend value for the NDVI residual time series based on residual analysis.

2.3.5. Accurate Long-Term NDVI Prediction Model Based on Informer

Although Transformer can predict long time sequences and demonstrates strong prediction ability, the model's multilayered encoder–decoder structure means that the longer the sequence, the slower the speed, and the worse the effect of the model in predicting long time sequences [47]. Herein, we address this issue and the inherent issues with the encoder–decoder architecture by introducing an effective LSTM model based on an improved Transformer, the Informer model. This model uses probabilistic sparsity inside the encoder—a self-attention mechanism—trained and learned according to sequences with high probability in the data and filtering the Q-value with high weight [48]. As a result, the complexity may be significantly reduced. Figure 3 presents the structure of this model.

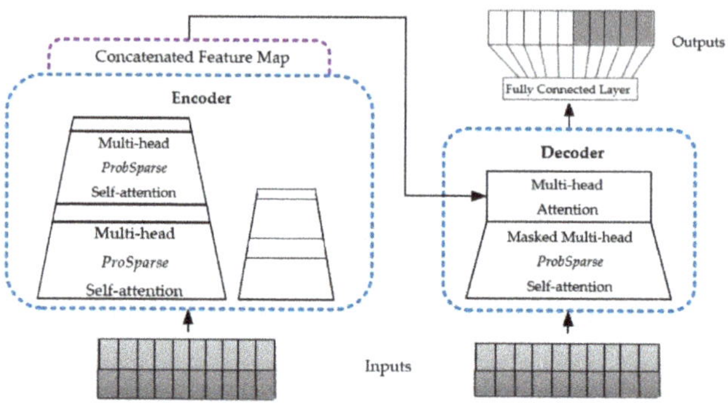

Figure 3. Schematic diagram of the Informer model.

Based on the encoder–decoder structure of the Informer prediction model system, we developed precise long time series NDVI prediction in this article. For the temporal dimension of the inputs, a distillation operation prioritizes high-level features with dominant features, generates a centralized self-attentive feature mapping in the following layer, and adds a max-pooling layer of two after stacking one layer, which drastically reduces the memory in the downsampling [30].

3. Results

3.1. Spatiotemporal Characteristics of Land Use Types

As illustrated in Figure 4, the land use categories for the Maowusu Sandland from 2000 to 2020 were plow, woodland, grassland, shrubland, wetland, wave, bare land, and man-made surface, in sequence. In the research area, the categories of land use include, in decreasing order, grassland, bare land, farmland, shrubland, water body, wetland, man-made land surface, and forest. Grassland, bare land, and plow all exhibited an upward tendency between 2000 and 2020, with grassland showing the largest increase, wave exhibited the slightest increase, and wetland and forest remaining relatively unchanged.

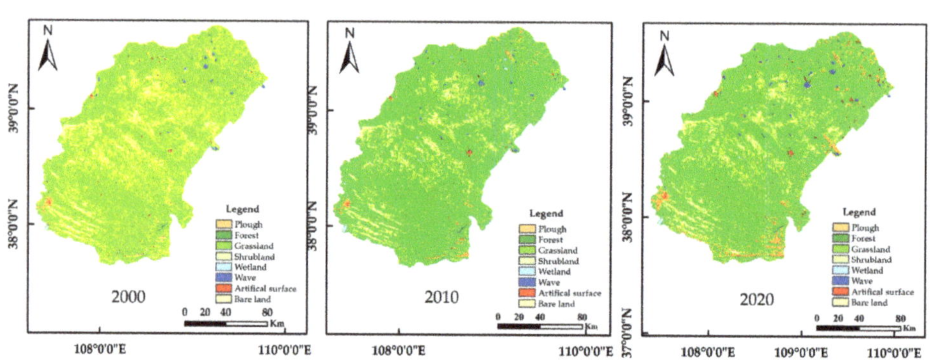

Figure 4. Land use type maps for 2000, 2010, and 2020.

In general, the Maowusu Sandland exhibits a gradual decline in cultivated land and an increase in woodland and grassland from the northwest to the southeast (Figure 5). The man-made land surface is sporadically distributed throughout the entire region but is primarily concentrated in the eastern part of the Maowusu Sandland and develops with a fast growth trend. After a field investigation, it was discovered that the predominant vegetation in the sandy land was sandy cypress. The Northwest region exhibited an increasing trend in grassland from 2000 to 2020, and the man-made land surface and arable land also showed a slight growth trend, moving in the direction of a trend in restoring ecological well-being.

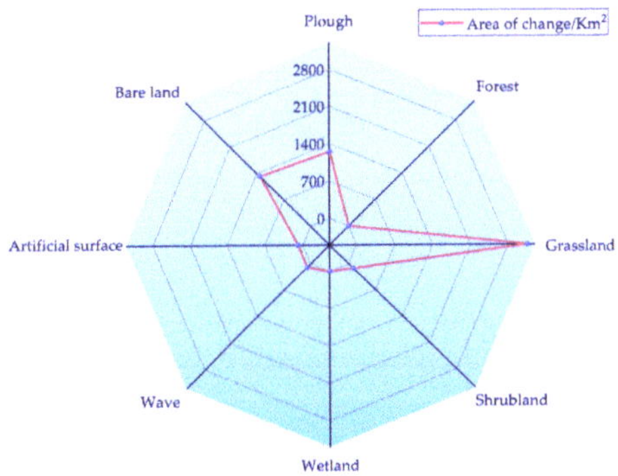

Figure 5. Land use type change map for 2000, 2010, and 2020.

3.2. *Analysis of Groundwater Impact on NDVI*

The 2019 groundwater depth spatial distribution map for the Maowusu Sandland was produced using the Spatial Analyst spatial analysis module of the ArcGIS 10.8 software. This map was used to analyze the inverse distance weights of the 2019 groundwater level depth based on the 2019 groundwater level depth data in Maowusu Sandland.

The Ordos Hydrological Bureau, located in China's Inner Mongolia Autonomous Region, supplied all information on burial depths and groundwater levels. Within the spectrum of places with data recordings, the 2019 groundwater level depth in the Maowusu Sandland exhibited more pronounced variances, as illustrated in Figure 6. In the research area, the depth of the groundwater varied widely, with shallow depths in the middle and northern areas and depths below 10 m in the northern and southern regions. According to the 2019 NDVI spatial distribution map of the Maowusu Sandland, this area had a wide range of variable groundwater depths. We discovered that vegetation was well-developed in areas with shallow groundwater table burial depths, which suggests that groundwater had a beneficial promoting effect on vegetation. However, most places had groundwater buried deeper than 7 m, and the relationship between groundwater and vegetation was weak.

Our daily monitoring of the Maowusu Sandland groundwater level changes in 2019 revealed that only points 4 and 7 had more obvious variations in groundwater level (Figure 7). This was largely because these points are situated in an area of cultivated land, where the groundwater is primarily used for irrigation water for agricultural development and where the groundwater level is buried deeper.

3.3. *Analysis of Vegetation Change Trends*

The Maowusu Sandland's average Sen slope value from 2000 to 2019 was 0.18, and its slope was 0.0034 a^{-1}, showing that the NDVI increased steadily while growing at a relatively modest rate (Figure 8). Most of the quicksand regions were located in the hinterland of the sandy land, which contributed to the study area's low vegetation cover in 2000. Most of these areas had <30% vegetation cover. The Maowusu Sandland has undergone construction and ecological protection projects in recent years from national and local governments, which have improved the ecological environment and increased the plant cover, although the geographical disparity is still evident [49].

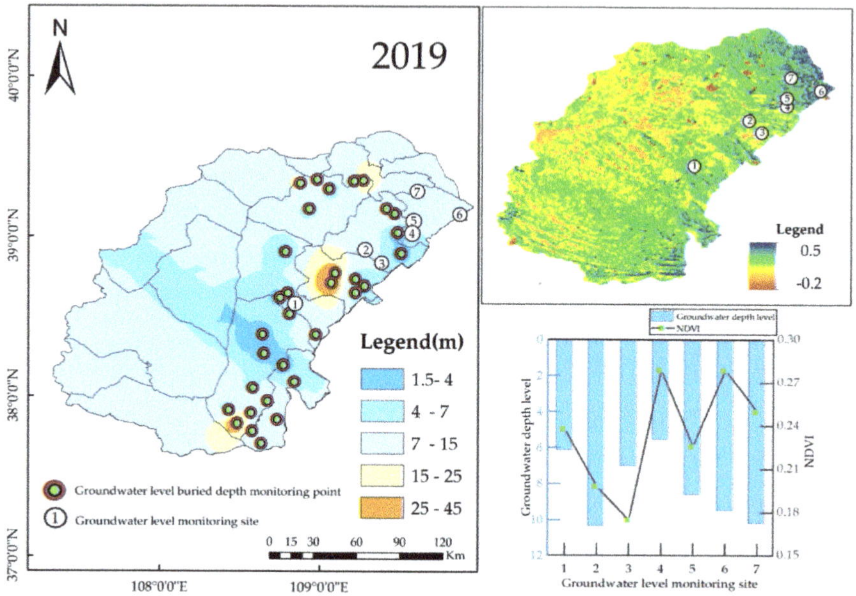

Figure 6. Spatial distribution of groundwater level changes in 2019.

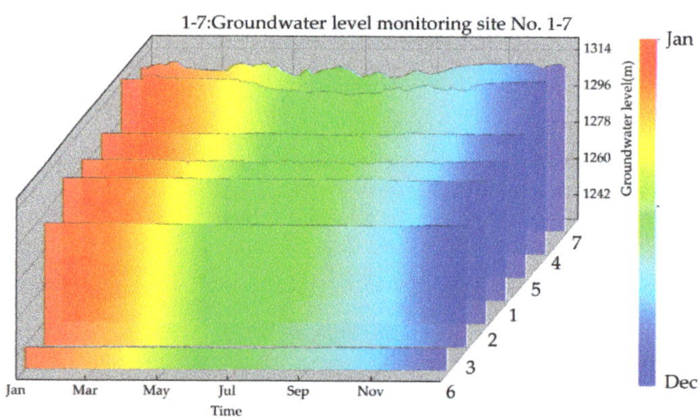

Figure 7. Monitoring maps of daily changes in groundwater levels.

The Maowusu Sandland area's vegetation cover exhibited a decreasing trend from northeast to southwest; however, because the ecological protection construction project's vegetation growth requires time, the vegetation cover in the eastern region is increasing significantly (Figure 9). This indicates that the project's ecological protection management is having an impact. Among these, the maximum NDVI in August 2020 is significantly greater than the highest NDVI in August 2000, which is closely related to China's local governments' participation in protection and restoration.

The change in vegetation cover was divided into nine classes based on the Theil-Sen Median trend value analysis and the Mann–Kendall nonparametric test at a significance level of 0.05: dramatic increase, significant increase, slightly significant increase, insignificant increase, unchanged, insignificant reduction, slightly significant reduction, significant reduction, and very significant reduction.

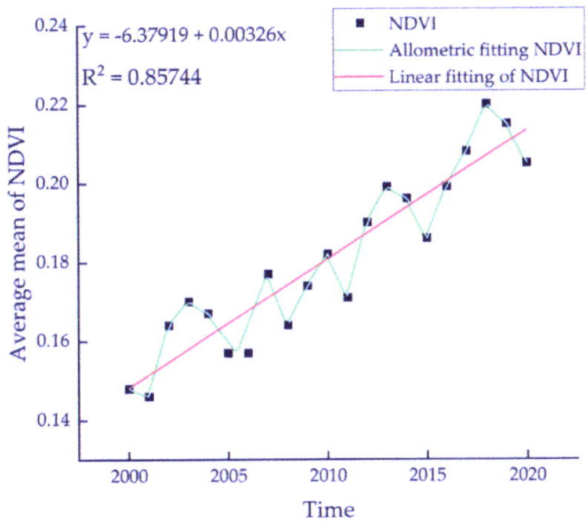

Figure 8. NDVI trend chart from 2000 to 2019.

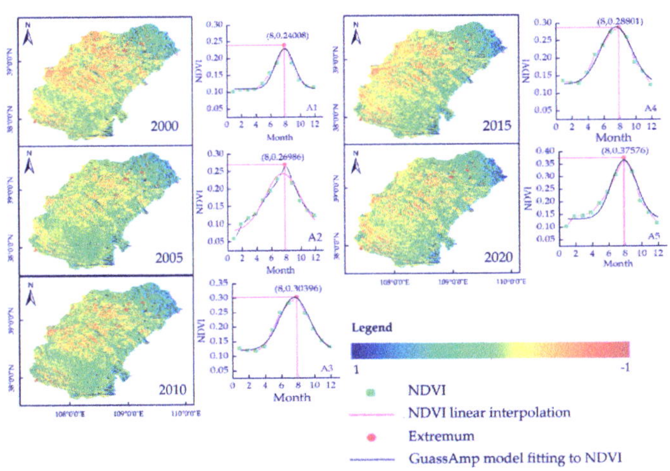

Figure 9. Spatial distribution of average NDVI in the Maowusu Sandland in 2000, 2005, 2010, 2015, and 2020, and linear interpolation and fitting of NDVI values. Annotation A1 corresponds to 2000, A2 corresponds to 2005, A3 corresponds to 2010, A4 corresponds to 2015, and A5 corresponds to 2020.

The Maowusu Sandland region significantly expanded, accounting for more than 90% of the region, as shown by the NDVI trend change map (Figure 10). Between 2000 and 2020, there was a clear difference in the amount of land, with an increasing trend and an obvious decreasing trend. Of these, 95.28% of the area had significantly improved vegetation cover, while only a small portion had degraded vegetation. This difference is likely due to the implementation of national ecological conservation and construction projects. Throughout the 20 years, the NDVI had a notable increasing trend, and the number of regions with a large increase in plant cover grew significantly. However, these areas were primarily concentrated in areas with high populations.

The 20-year NDVI trends were compared, revealing that plant cover rose gradually from 2000 to 2020 with no discernible variation in growth rate. Given that the population is more concentrated in the eastern half of the Maowusu Sandland, the growth rate of vegetation cover is significantly larger than it is in the western region. This finding suggests that the ecological restoration project has a substantial impact on the increase in vegetation cover.

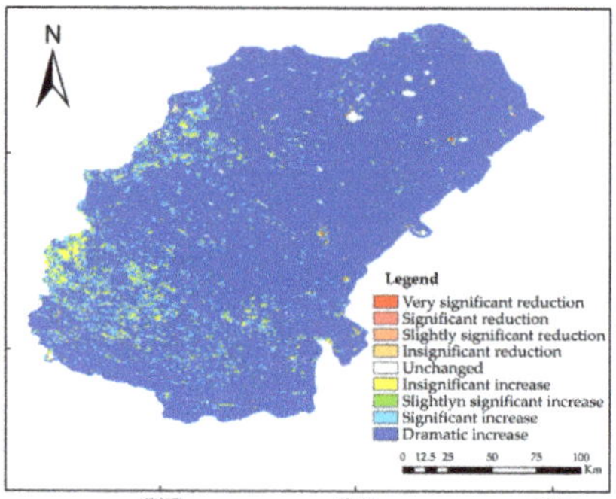

Figure 10. NDVI change trend chart from 2000 to 2019.

3.4. Characterization of Inter-Annual Variability of Climate Factors

In the study area, the temperature and precipitation change rates showed substantial declining trends between 2000 and 2010 (Figure 11). This suggests that the environment was drier, which was detrimental to the growth and recovery of vegetation. Meanwhile, the rates of temperature and precipitation changes showed substantial upward trends from 2013 to 2019 (Figure 11). During the 2011 to 2013 period, the temperature decreased, and the quantity of precipitation increased. The humid weather created ideal circumstances for the NDVI to thrive.

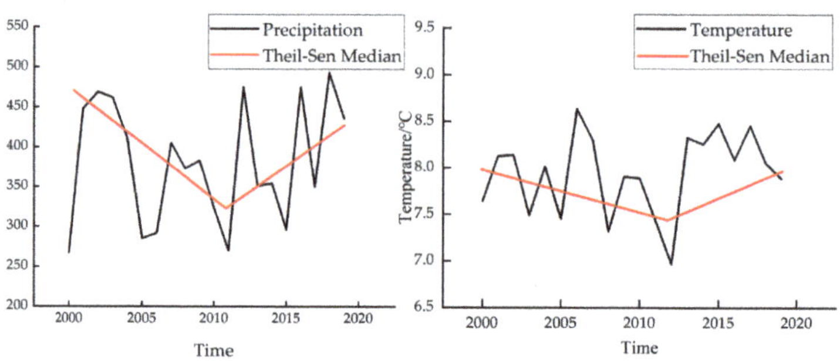

Figure 11. Characterization of inter-annual variability in precipitation and temperature.

From 2003 to 2018, it was discovered that July through September saw the most rainfall, while June through July saw high temperatures. This helps to keep high temperatures from accelerating drought and causing severe soil moisture loss due to crop transpiration and soil evaporation. There is a greater chance of very heavy precipitation because of the atmosphere's enhanced capacity to store water vapor before saturation, which is why spring and autumn rainfall has significantly risen. Increased precipitation, sufficient soil moisture, and appropriate temperature contributed to gains in plant cover and growth, creating the ideal environment for vegetation development and the NDVI (Figure 9).

3.5. Relationship between Climate Factors and NDVI

The elevation map indicates that the southeast region has more mountain ranges, high vegetation cover, low surface specific heat capacity, a tendency for stable temperatures, and vegetation that can produce water vapor for precipitation through transpiration and encourage infiltration to replenish the groundwater. This is favorable for the significance of the NDVI in the eastern region (Figure 9). Meanwhile, the western region has a flat terrain predominately made up of deserts, a dry climate, little vegetation cover, high evaporation, low air humidity, and a high water infiltration rate. As a result, transpiration and groundwater absorption by plants are significantly impacted, leading to a prolonged period of low vegetation cover. The Maowusu Sandland's typical flowing dunes were monitored for soil water seepage at various soil layer depths. The results show that deep water seepage is more affected by heavy rainfall than by changes in soil water storage; moreover, when rainfall is low, the relationship between deep water seepage and soil water storage is stronger.

In most locations, air temperature impeded plant development, as shown by the partial correlation coefficients between vegetation and climatic parameters (Figure 12). The rainfall in most regions encouraged vegetation growth, and the soil moisture in all sites encouraged plant growth. Overall, soil moisture had the greatest impact on vegetation changes, as indicated by the mean values of the partial correlation coefficients for temperature, precipitation, and soil moisture with vegetation NDVI, which were −0.394, 0.108, and 0.726, respectively. We discovered that temperature was the primary driver of the Maowusu Sandland, whereas rainfall was the primary driver in the eastern and southern parts. Among these, precipitation, air temperature, plant type, and topographic variables were the key determinants of soil moisture and accounted for the main drivers of the research region. Precipitation is particularly important for herbaceous plants growing in chestnut calcium soil. It is the main driving factor for the Maowusu Sandland's large expanse of the steppe zone, which contains high calcium ions and carbonates, loose texture, good permeability, and poor water-holding capacity. In particular, the primary cause of the Maowusu Sandland region's healthy vegetation development is climate change.

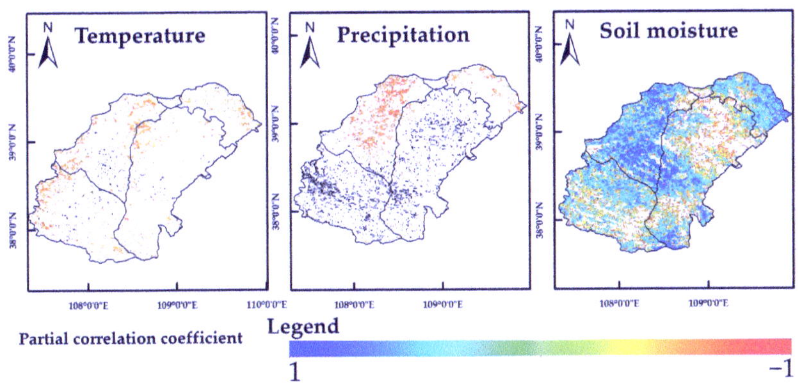

Figure 12. Partial correlation coefficients for temperature, precipitation, and soil moisture.

3.6. Changes in Plant Cover as a Result of Climatic Conditions and Human Activity

The primary drivers of the growth trends in vegetation NDVI from 2000 to 2019 were human activity and climate change, with the former predominating over the latter (Figure 13). The study area's vegetation NDVI improvement accounted for 99.26% of the total area. Of these, most of the vegetation improvement areas—primarily located in the eastern and southern regions of the study area—were driven by meteorological changes, while the remaining portion was primarily controlled by human activity. The primary cause of the rising trend in vegetation cover is the effects of climate change. The primary causes of the decreasing trends in plant cover are human activity-influenced forces.

In terms of population density and gross domestic product (GDP) in the study area in 2020 (Figure 14), the population was primarily distributed in the region of vegetation cover NDVI growth, with the high human living standard also located in this region. This is primarily because ecological restoration work in urban suburbs supports development. In contrast, the desert area has a small population and a low human living standard, making the area difficult to develop.

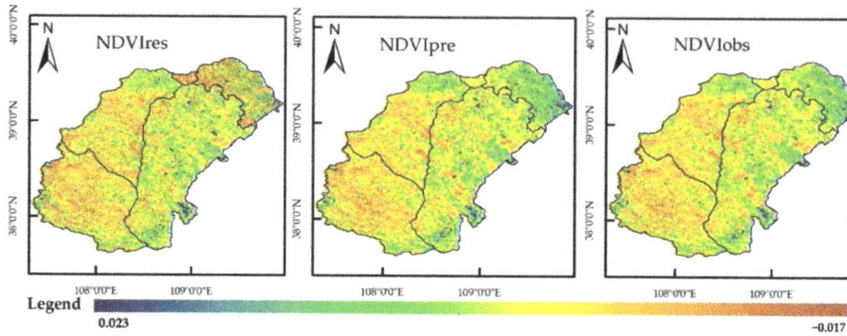

Figure 13. Trends in the residuals of the $NDVI_{res}$, $NDVI_{pre}$, and $NDVI_{obs}$.

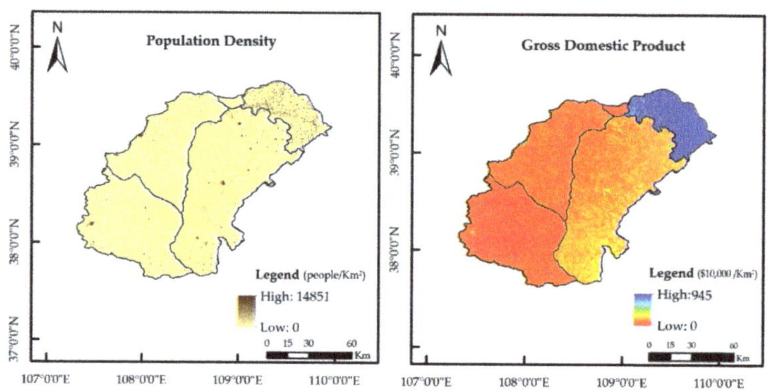

Figure 14. Population density and gross domestic product (GDP) in 2020.

In particular, the Maowusu Sandland region's vegetation improvement was primarily driven by climate change, with human-led ecological restoration projects accounting for a small portion of the region. Hence, the region's vegetation improvement was primarily caused by climate change, with human-led ecological restoration projects serving as a secondary driver. The combined impact of climate change is the primary cause of vegetation decline.

3.7. Relative Contribution of Climate and Human Activities to Vegetation Changes

In regions with increasing NDVIs, the average contribution of climate change to the enhancement of vegetation cover was 55.57%, whereas in regions with declining NDVI, the average contribution of human activities to the deterioration of vegetation cover was 88%. In terms of geographical distribution, the research region's most densely inhabited city center contributed less compared to vegetation cover. However, the highly populated region had a more favorable influence on ecological restoration efforts. We discovered that there are two main factors contributing to the NDVI decrease: first, the city center area has a greater negative contribution due to human activities compared with the desert area; second, the natural ecological restoration area system has more stable NDVI growth; the desert area has a narrow population distribution and climatic circumstances that make it more vulnerable to the harmful effects of climate change. Because the desert region has fewer people and a harsher environment that is not conducive to plant development, the negative effects of climate change are more pronounced.

Climate change and human activities have the greatest impact on vegetation changes, and climate change has the greatest impact on regions with growing trends in vegetation.

In the dry and semi-arid Mao Sandland, rainfall has the greatest impact on plant growth. Sufficient water is necessary to encourage the development and recovery of vegetation. The study area's shift in land use patterns is partially the result of frequent human activity; as urban sprawl has intensified, there has been a commensurate rise in the amount of man-made land and a corresponding decline in the plant cover in certain localized areas.

3.8. Accurate Long Sequence Prediction Based on Informer

To conduct this study, a total of 7305 days of daily meteorological data for 228 months, spanning from 1 January 2000 to 31 December 2019, were chosen for the Maowusu Sandland. These data included information on temperature and rainfall.

Python 3.8, the PyCharm 3.8.1 framework, Windows 11, and 16 GB comprised the experimental environment.

This study adjusts the hyperparameters according to the training results of its prediction model, and the final hyperparameters are set as follows: learning rate, 0.001; number of iterations, 20; batch size, 128; activation function, Gaussian error linear units (GELU). This study's hyperparameters were adjusted according to the meteorological data of the prediction model.

Four commonly used evaluation metrics were selected to assess the prediction performance of the model from different perspectives, i.e., mean absolute error (MAE), mean square error (MSE), root mean square error (RMSE), and mean absolute percentage error (MAPE). The smaller the value of these four evaluation indexes, the closer the prediction result is to the real value, and the more accurate the prediction accuracy of the model; the specific calculation formula is as follows:

$$e_{MAE} = \frac{1}{n}\sum_{i=1}^{n}|y_i - \hat{y}_i| \quad (5)$$

$$e_{MSE} = \frac{1}{n}\sum_{i=1}^{n}|y_i - \hat{y}_i|^2 \quad (6)$$

$$e_{RMSE} = \sqrt{\frac{1}{n\sum_{i=1}^{n}(|y_i - \hat{y}_i|)^2}} \quad (7)$$

$$e_{MAPE} = \frac{1}{n}\sum_{i=1}^{n}|y_i - \hat{y}_i|/y_i \quad (8)$$

where y_i is the true value of the original data; $\hat{y}_i$ is the predicted value of the original data; n is the test sample size; i is the test sample point sequence number.

The average daily temperature (calculated based on 24-h temperature observations), total daily precipitation (based on rainfall totals over 24 h), and NDVI data from 2000 to 2018 were used in the prediction validation experiment to assess the predictive model's accuracy. The prediction results are depicted in Figure 15.

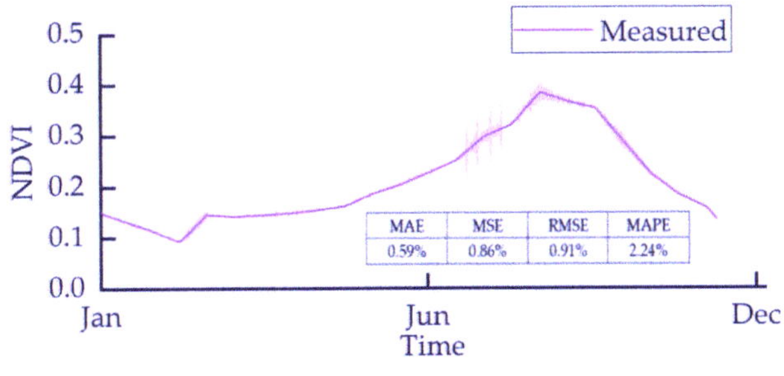

Figure 15. Informer model prediction error plot.

Figure 16 shows that the prediction error values of the Informer model's findings compared to the actual results are 0.59%, 0.86%, 0.91%, and 2.24% for mean absolute error

(MAE), mean square error (MSE), root mean square error (RMSE), and mean absolute percentage error (MAPE), respectively. This provides strong evidence that our enhanced Transformer-based Informer model closely approximates the real values' curve fit. Our prediction model will have a lower error value and use less computation time than the conventional model, thoroughly demonstrating the viability of the model. As a result, the model is more effective in forecasting NDVI's future trends.

In this study, we utilized the Informer model and input average daily temperature, total daily precipitation, and NDVI data from 2000 to 2020 for deep learning to forecast the NDVI trend for the following ten years (Figure 16). As further evidence of the reliability of the prediction findings, the peak NDVI for each year in the prediction results is approximately 0.3, which does not differ significantly from the peak input NDVI data in recent years. Figure 17 illustrates the forecast findings, which demonstrate that the NDVI trend will stabilize during the following ten years. The nonlinear fitted equation has a slope of 0.000294, indicating a minimal change over the following ten years (Figure 17).

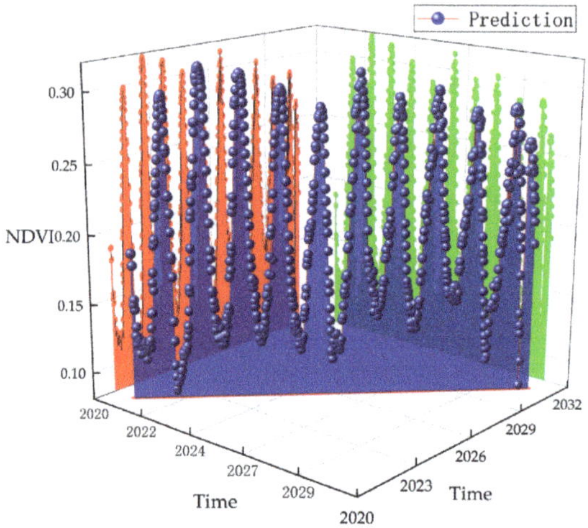

Figure 16. Map of NDVI projection results for the next ten years.

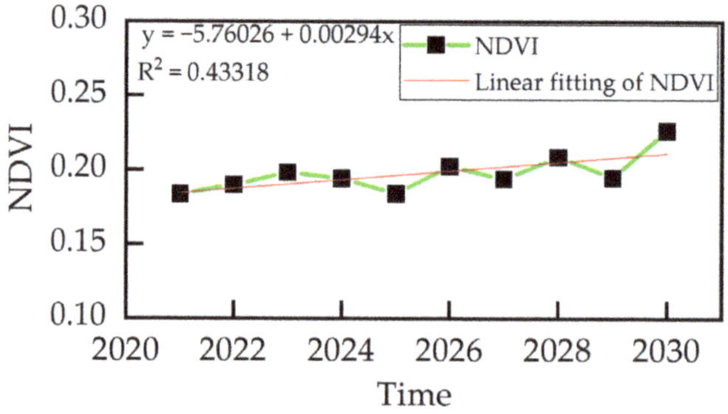

Figure 17. NDVI trend chart for the next ten years.

4. Discussion

4.1. Characteristics of Dynamic Changes in Land-Use Types and Changes in Groundwater Resources in the Maowusu Sandland Area

According to our research, the Maowusu Sandland area's primary land use categories from 2000 to 2019 were, in descending order, farmland, forest, grassland, shrubland, wetland, water body, bare ground, and man-made land surface (Figures 4 and 5). Man-made surfaces were sporadically distributed, mostly concentrated in the eastern region of the Maowusu Sandland, and developing rapidly, with a fast growth trend, the same as that reported in the past. The study area showed a trend of gradually increasing cropland and decreasing grassland from the southeast to the northwest [50]. Wetlands, cropland, bare land, and man-made land surfaces all exhibited an upward trend between 2000 and 2020, with cropland having the largest increase, water bodies showing a slight decrease, grassland exhibited the largest decline in 20 years, and forests and shrublands showing relatively no change. Grassland is a key land use type for the ecological repair and preservation of sandy land. According to the field research, the predominant vegetation was sandy cypress, and between 2000 and 2020, grassland had a decreasing tendency. However, man-made land surfaces and arable land also showed a minor growth trend, shifting toward reversing the ecologically unsound trend.

According to our 2020 daily monitoring data, from April through October, numerous monitoring sites located at agricultural sites within the study region exhibited decreasing groundwater levels. This shows that groundwater may be the area's primary source of agroecological water. In regions where the groundwater table is < 4 m deep, we discovered that the flora is well-developed. This supports earlier findings and implies that groundwater has a positive influence on promoting plants [51]. According to research by Zhang et al. [8], high-density vegetation in the Maowusu Sandland may have contributed to a drop in the groundwater table and retention of water in the soil. To maintain the sustainability of the ecosystem, we propose that water balance concerns be taken into account in upcoming ecological restoration initiatives in China and across the world.

4.2. Quantifying the Impacts of Human Activities and Climate Change on NDVI

Lin et al. [50] tracked the Maowusu Sandland region's sandy vegetation cover from 2000 to 2019 and linked changes in the local climate to NDVI-related land cover metrics. Using residual analysis, they assessed the various contributions and discovered that most of the NDVI exhibited an increasing trend, with only 0.08% in the highly damaged area. According to our study, which aligns with the other data, the vegetation cover in the studied region exhibited a decreasing trend from east to west. It also had a larger distribution of mostly low and medium vegetation cover [50]. Vegetation cover had a variable increasing trend of 0.0034 a^{-1}, which was primarily demonstrated by an ongoing but slower rate of increase in NDVI across the study period.

According to Tian et al. [52], in the Yellow River Basin, NDVI is positively associated with air temperature and precipitation, with precipitation having a greater impact on plant development than air temperature. According to Yang et al. [35], the Maowuosu Sandland region's response relationship between NDVI and climatic factors in one year was primarily in the spring, with most precipitation negatively correlated with NDVI and the areas of positive and negative correlation essentially flat in other months. In our study, the two climatic elements restricting plant growth—air temperature and precipitation—were both somewhat favorably and slightly negatively associated with the total NDVI growth rate in the study region.

Gao et al. [53] used the residual analysis of Maowusu Sandland from 1985 to 2020 to reveal that human activity contributed 56.44% of the NDVI, and climatic change contributed 43.56%. Lin et al. [50] measured the Maowusu Sandland's impact from 2000 to 2019, revealing that human activity was the primary driving force. In our research, we discovered that the Maowusu Sandland's plant cover was mostly favorably influenced by climate change, with a relative contribution rate of 55.47%, and that human activities and climate change had some heterogeneity. However, the improvement in vegetation was primarily attributed to the ecological restoration project driven by climate change, in which human activities only made up a small portion of the factors, and its implementation had a positive impact on governance. In contrast, the decline in vegetation was primarily attributed to the impact of climate change. Temperature changes and extreme weather have a greater impact on vegetation development. The ecosystem is more stable and resistant to the effects of the outside environment when the ecological environment is better.

4.3. Precision Long Time Series-Informer Model

To guarantee ecological quality and offer a scientific foundation for sustainable development, accurate NDVI predictions are crucial. Huang et al. [23] reported that the support vector machine (SVM) could accurately recreate the observed NDVI series. However, the normalized system criterion (NSC) value for the validation period was only guaranteed to be between 0.9 and 0.95. For the NPP spatial simulation distribution prediction, Wang et al. [54] utilized a CA-Markov model with a kappa coefficient of 0.8776. However, it can fully utilize large-scale time series data for improved short-term NDVI prediction compared to deep learning, which has higher learning and fitting capabilities. The NDVI and meteorological factor data from 2000 to 2018 were utilized for deep learning using the Informer model, an effective long short-term memory (LSTM) model enhanced based on the Transformer; the hyperparameters were ideally changed to provide a probabilistic sparse self-attention mechanism.

The model can recognize patterns in unsteady, smoothed meteorological factor data. Moreover, using a probabilistic sparse self-attention mechanism can remove the volatility of the meteorological factor to enhance the accuracy of NDVI predictions while retaining the original data's key elements. The mean absolute percentage error (MAPE) of the Informer model suggested in this study may be reduced to 2.24% after analyzing the NDVI prediction data, providing a strong scientific foundation for future NDVI predictions in China and throughout the world.

5. Conclusions

The Maowusu Sandland area has the following land use classifications, in descending order: grassland, bare land, farmland, shrubland, water body, wetland, man-made surface, and forest. The overall pattern is cropland progressively declining from east to west and grassland growing. The NDVI has a decreasing tendency from east to west and is mostly dispersed by low and medium vegetation cover. It swings upward by 0.0034 a^{-1}, primarily caused by the NDVI's constant increase throughout the period. In the Maowusu Sandland, which is primarily impacted by the climate change-led ecological restoration project, human activities account for a small portion of the factors. The ecological restoration project has a better management effect, with some heterogeneity in the contribution of vegetation cover due to both factors. The impact of climate change is primarily to blame for the decline in the vegetative cover.

The primary issues that the Informer model resolves with the Transformer are (1) not being directly affected by the long sequence time-series forecasting (LSTF) problem; (2) excessive memory occupancy; (3) encoder–decoder architectural restrictions; and (4) performance loss in long output time series prediction. We used this method for NDVI prediction because it can identify patterns in the unsteady smoothed weather factor data and, by removing the volatility of the weather factors through a probsparse self-attention mechanism, it can remove the volatility while maintaining the critical information of the original data, resulting in accurate long time series prediction (MAPE = 2.24%). Using historical data, the study's conclusions may accurately forecast future crop yields, crop pests and diseases, and NDVI. This opens up new agricultural production and management avenues and contributes to realizing more sustainable and productive agricultural practices. This discovery has broad ramifications for real-world applications and is significant for scientific research.

Author Contributions: The contributions of H.H., R.L., H.Z. and C.T. involved in designing the manuscript; H.H., H.L. and J.W. carried out this experiment; G.W., Z.Q. and W.W. analyzed the data and wrote the manuscript. All authors have read and agreed to the published version of the manuscript.

Funding: Key Special Project of "Science and Technology for the Development of Inner Mongolia": Project of "Creating Erdos National Sustainable Development Agenda Innovation Demonstration

Zone—Technology and Demonstration of Intensive and Efficient Utilization: 2021EEDSCXSFQZD010; Research on groundwater exploitation and utilization and ecological environment monitoring and early warning system development in Mu Us SandLand, China: MK20210222; Research on grassland drought monitoring and scale transformation based on multi-source and multi-temporal remote sensing: 52269004.

Data Availability Statement: The data presented in this study are available on request from the corresponding author. The data are not publicly available due to privacy.

Conflicts of Interest: The authors declare no conflict of interest.

References

1. Gong, Z.; Zhao, S.; Gu, J. Correlation analysis between vegetation coverage and climate drought conditions in North China during 2001–2013. *J. Geogr. Sci.* **2017**, *27*, 143–160. [CrossRef]
2. Zhao, J.; Du, Z.Q.; Wu, Z.T.; Zhang, H.; Guo, N. Seasonal variation of diurnal warming in temperate China and its impact on vegetation dynamics. *J. Geogr.* **2018**, *73*, 395–404.
3. Ye, W.; van Dijk, A.I.; Huete, A.; Yebra, M. Global trends in vegetation seasonality in the GIMMS NDVI3g and their robustness. *Int. J. Appl. Earth Obs. Geoinf.* **2021**, *94*, 102238. [CrossRef]
4. Zhang, J.; Li, C.; Wang, T. Dynamic changes of vegetation coverage on the Loess Plateau and its factors. *Res. Soil Water Conserv.* **2022**, *29*, 224–230, 241.
5. Li, X.; Yang, L.; Tian, W.; Xu, X.; He, C. A review of land use/cover change studies in northern China's agricultural and pastoral intertwined zone. *J. Appl. Ecol.* **2018**, *29*, 3487–3495.
6. Li, X. Analysis of the Desertification Reversal Process and Causes in the Mu Us Sandy Land. Ph.D. Thesis, Shaanxi Normal University, Xi'an, China, 2017.
7. Han, X.; Yang, G.; Qin, F.; Jia, G.; Ling, X.; Gao, G. Spatiotemporal Dynamic Evolution Pattern of Desertified Land in the Mu Us Sandy Land in the Past 30 Years. *Soil Water Conserv. Res.* **2019**, *26*, 144–157.
8. Zhang, M.; Wu, X. The rebound effects of recent vegetation restoration projects in Mu Us Sandy land of China. *Ecol. Indic.* **2020**, *113*, 106228. [CrossRef]
9. Dai, Z. Intensive agropastoralism; dryland degradation, the theGrain to Green Program and islands of sustainability in the US Sandy Land of China. *Agric. Ecosyst. Environ.* **2010**, *138*, 249–256. [CrossRef]
10. Nathan, S.D.; Wanger, J.; Zibrak, J.D.; Wencel, M.L.; Burg, C.; Stauffer, J.L. Using forced vital capacity (FVC) in the clinic to monitor patients with idiopathic pulmonary fibrosis (IPF): Pros and cons. *Expert Rev. Respir. Med.* **2021**, *15*, 175–181. [CrossRef]
11. Li, F.; Wang, G.; Du, H.; Li, M.; Liang, S.; Peng, H. Reconstruction of NDVI and Analysis of Annual Maximum Changes in a Typical Area of Qinghai Lake Basin by Integrating MODIS and Landsat Data. *Acta Prataculturae Sin.* **2023**, *32*, 28–39. [CrossRef]
12. Jiang, M.; He, Y.; Song, C.; Pan, Y.; Qiu, T.; Tian, S. Disaggregating climatic and anthropogenic influences on vegetation changes in Beijing-Tianjin-Hebei region of China. *Sci. Total Environ.* **2021**, *786*, 147574. [CrossRef]
13. Fang, J.; Yu, G.; Liu, L.; Hu, S.; Chapin, F.S., III. Climate change, human impacts, and carbon sequestration in China. *Proc. Natl. Acad. Sci. USA* **2018**, *115*, 4015–4020. [CrossRef] [PubMed]
14. Xu, Y.; Yang, D.; Tang, L.; Qiao, Z.; Ma, L.; Chen, M. Exploring the Impact of Grain-for-Green Program on Trade-Offs and Synergies among Ecosystem Services in West Liao River Basin, China. *Remote Sens.* **2023**, *15*, 2490. [CrossRef]
15. Guo, Z.; Xie, Y.; Guo, H.; Zhang, X.; Wang, H.; Bie, Q.; Ma, C. Do the ecosystems of Gansu Province in Western China's crucial ecological security barrier remain vulnerable? Evidence from remote sensing based on geospatial analysis. *J. Clean. Prod.* **2023**, *402*, 136740. [CrossRef]
16. Dong, X.; Zhou, Y.; Liang, J.; Zou, D.; Wu, J.; Wang, J. Assessment of Spatiotemporal Patterns and the Effect of the Relationship between Meteorological Drought and Vegetation Dynamics in the Yangtze River Basin Based on Remotely Sensed Data. *Remote Sens.* **2023**, *15*, 3641. [CrossRef]
17. Li, M.; Qin, Y.; Zhang, T.; Zhou, X.; Yi, G.; Bie, X.; Gao, Y. Climate Change and Anthropogenic Activity Co-Driven Vegetation Coverage Increase in the Three-North Shelter Forest Region of China. *Remote Sens.* **2023**, *15*, 1509. [CrossRef]
18. Wang, J.; Wang, S.; Li, Q.; Li, M. Determining the Contributions of Climate Change and Multiple Human Activities to Runoff and Sediment Reduction in the Eastern Loess Plateau, China. *Catena* **2023**, *232*, 107376. [CrossRef]
19. Goswami, S.; Mehta, A.; Verma, G. Trend and Time Series Analysis of Vegetation Dynamics Using Satellite Data: A Case Study of Uttarakhand, India. *Curr. J. Appl. Sci. Technol.* **2022**, *32*, 14–26.
20. Guo, J.; Wu, X.; Dong, G.; Li, Y.; Wu, R. Relative Contribution Analysis of Driving Factors for Vegetation Cover Change in the Tarim River Basin Based on MODIS/NDVI. *Arid Zone Res.* **2017**, *34*, 621–629. [CrossRef]
21. Yang, W.; Zhang, X.; Zhao, J. Remote Sensing Inversion of Leaf Area Index Based on Geospatial Model and SVR Algorithm. *Sci. Soil Water Conserv.* **2018**, *16*, 48–55. [CrossRef]
22. Luo, J.; Hong, T.; Gao, Z.; Fang, S.C. A robust support vector regression model for electric load forecasting. *Int. J. Forecast.* **2023**, *39*, 1005–1020. [CrossRef]
23. Huang, S.; Zheng, X.; Ma, L.; Wang, H.; Huang, Q.; Leng, G.; Guo, Y. The quantitative contribution of climate change and human activities to vegetation cover variations based on the GA-SVM model. *J. Hydrol.* **2020**, *584*, 124687. [CrossRef]

24. Blekanov, I.; Molin, A.; Zhang, D.; Mitrofanov, E.; Mitrofanova, O.; Li, Y. Monitoring of grain crops' nitrogen status from UAV multispectral images coupled with deep learning approaches. *Comput. Electron. Agric.* **2023**, *212*, 108047. [CrossRef]
25. Cheng, H.; Zhen, H. Short-term Electric Power Load Forecasting Model Based on Improved PSO Optimized RNN. *Electron. Meas. Technol.* **2019**, *42*, 94–98. [CrossRef]
26. Gao, P.; Du, W.; Lei, Q.; Li, J.; Zhang, S.; Li, N. NDVI Forecasting Model Based on the Combination of Time Series Decomposition and CNN–LSTM. *Water Resour. Manag.* **2023**, *37*, 1481–1497. [CrossRef]
27. Ahmad, R.; Yang, B.; Ettlin, G.; Berger, A.; Rodr guez-Bocca, P. A machine-learning-based ConvLSTM architecture for NDVI forecasting. *Int. Trans. Oper. Res.* **2023**, *30*, 2025–2048. [CrossRef]
28. Haq, M.A.; Ahmed, A.; Khan, I.; Gyani, J.; Mohamed, A.; Attia, E.A.; Pandi, D. Analysis of environmental factors using AI and ML methods. *Sci. Rep.* **2022**, *12*, 13267. [CrossRef]
29. Li, J.; Li, C.; Xu, W.; Feng, H.; Zhao, F.; Long, H.; Yang, G. Fusion of optical and SAR images based on deep learning to reconstruct vegetation NDVI time series in cloud-prone regions. *Int. J. Appl. Earth Obs. Geoinf.* **2022**, *112*, 102818. [CrossRef]
30. Zhou, H.; Zhang, S.; Peng, J.; Zhang, S.; Li, J.; Xiong, H.; Zhang, W. Informer: Beyond efficient transformer for long sequence time-series forecasting. In Proceedings of the AAAI Conference on Artificial Intelligence, Virtual, 2–9 February 2021; Volume 35, pp. 11106–11115.
31. Guo, Z.C.; Wang, T.; Liu, S.L.; Kang, W.P.; Chen, X.; Feng, K.; Zhi, Y. Biomass and vegetation coverage survey in the Mu Us sandy land based on unmanned aerial vehicle RGB images. *Int. J. Appl. Earth Obs. Geoinf.* **2021**, *94*, 102239. [CrossRef]
32. Yan, Y.; Yang, Y.; Wang, Y.; Yan, J. Trends and Attribution Analysis of Habitat Quality Changes in Beijing-Tianjin-Hebei Region over the Past 40 Years. *Remote Sens. Technol. Appl.* **2023**, *38*, 251–263.
33. He, Y.; Wang, L.; Niu, Z.; Nath, B. Vegetation recovery and recent degradation in different karst landforms of southwest China over the past two decades using GEE satellite archives. *Ecol. Inform.* **2022**, *68*, 101555. [CrossRef]
34. Dong, H.; Liu, Y.; Cui, J.; Zhu, M.; Ji, W. Based on GEE, spatial and temporal variations of vegetation cover and its influencing factors in Shandong Province. *Environ. Monit. Assess.* **2023**, *195*, 1023. [CrossRef]
35. Yang, M.; Jin, X.; Wang, T. Changes in Vegetation Phenology in the Mu Us Sandy Land and Its Response to Climate Change. *Bull. Soil Water Conserv.* **2022**, *42*, 242–249. [CrossRef]
36. Lian, H. Spatiotemporal Dynamics and Deep Leakage Characteristics of Soil Moisture in Four Land Use Types in the Mu Us Sandy Land. Ph.D. Thesis, Inner Mongolia Agricultural University, Hohhot, China, 2023. [CrossRef]
37. Cao, Y.; Fang, X.; Yang, L.; Jiang, X.; Liao, M.; Ren, L. Downscaling Study of CCI Soil Moisture in the Xiliao River Basin Based on Random Forest. *J. Geo-Inf. Sci.* **2023**, *25*, 1669–1681.
38. Frimpong, B.F.; Koranteng, A.; Molkenthin, F. Analysis of temperature variability utilizing Mann–Kendall and Sen's slope estimator tests in the Accra and Kumasi Metropolises in Ghana. *Environ. Syst. Res.* **2022**, *11*, 24. [CrossRef]
39. Wan, H.; Wang, Y.; Hou, P.; Liu, Y.; Li, G.; Zhao, S. Monitoring and analysis of vegetation and water dynamics in Chinese priority areas for biodiversity conservation during 2001–2016. *J. Ecol. Rural Environ.* **2019**, *35*, 273–282.
40. Lai, J.; Qi, S.; Liao, R.; Cui, R.; Li, P.; Tang, Y. Response of Vegetation Changes in the Southwest Mountainous Gorge Area from 2000 to 2019 to Climate Change and Human Activities. *J. Agric. Eng.* **2023**, *39*, 1–9.
41. Li, X.; Wang, X.; Liu, X.; Zhang, L.; Jin, X.; Chen, Y. Spatiotemporal Changes of NDVI in Shaanxi Province from 2000 to 2020 and Detection of Climate Factors. *Soil Water Conserv. Res.* **2023**, *39*, 1–11.
42. Xue, X.; Wang, Z.; Hou, S. NDVI-Based Vegetation Dynamics and Response to Climate Changes and Human Activities in Guizhou Province, China. *Forests* **2023**, *14*, 753. [CrossRef]
43. Evans, J.; Geerken, R. Discrimination between climate and human-induced dryland degradation. *J. Arid Environ.* **2004**, *57*, 535–554. [CrossRef]
44. Li, H.; Liu, G.; Fu, B. Response of vegetation to climate change and human activity based on NDVI in the Three-River Headwaters region. *Shengtai Xuebao/Acta Ecol. Sin.* **2011**, *31*, 5495–5504.
45. Jin, K.; Wang, F.; Han, J.; Shi, S.; Ding, W. Contribution of climatic change and human activities to vegetation NDVI change over China during 1982–2015. *Acta Geogr. Sin.* **2020**, *75*, 961–974.
46. Xu, H.; Cheng, J.; He, Y.; Wang, Y.; Zhang, M. The Impact of Climate Change and Human Activities on the Net Primary Productivity of Vegetation in Yunnan Province. *Plateau Meteorol.* **2023**, *32*, 1–10.
47. Dong, Z.; Li, F.; Sun, H.; Zhu, M.; Fan, B. Competency Assessment of Workload Management in the Approach Phase Based on the Informer Model. *Flight Dyn.* **2023**, *41*, 81–87. [CrossRef]
48. Zhu, L.; Han, K.; Zhu, C. Short-term Electric Power Load Forecasting Method Combining Informer with Secondary Decomposition Strategy. *Foreign Electron. Meas. Technol.* **2023**, *42*, 23–32. [CrossRef]
49. Guo, Z.; Liu, S.; Kang, W.; Chen, X.; Zhang, X. Change Trend of Vegetation Coverage in the Mu Us Sandy Region from 2000 to 2015. *J. Desert Res.* **2018**, *38*, 1099–1107.
50. Lin, M.; Hou, L.; Qi, Z.; Wan, L. Impacts of climate change and human activities on vegetation NDVI in China's Mu Us Sandy Land during 2000–2019. *Ecol. Indic.* **2022**, *142*, 109164. [CrossRef]
51. Xi, H.; Feng, Q.; Si, J.; Chang, Z.; Su, Y. Study on the response of NDVI to groundwater level changes in the lower reaches of the Heihe River Oasis. *China Desert* **2013**, *33*, 574–582.
52. Tian, Z.; Ren, Z.; Wei, H. Driving mechanism of spatiotemporal evolution of vegetation in the Yellow River Basin from 2000 to 2020. *Environ. Sci.* **2022**, *43*, 743–751. [CrossRef]

53. Gao, W.; Zheng, C.; Liu, X.; Lu, Y.; Chen, Y.; Wei, Y.; Ma, Y. NDVI-based vegetation dynamics and their responses to climate change and human activities from 1982 to 2020: A case study in the Mu Us Sandy Land, China. *Ecol. Indic.* **2022**, *137*, 108745. [CrossRef]
54. Wang, L.; Zhang, H.; Liu, Z.; Zhang, S.; Kong, J.; Gao, L. A coupled model for simulating and predicting the pattern of net primary productivity. *Geomat. Sci. Wuhan Univ.* **2021**, *46*, 1756–1765. [CrossRef]

MDPI AG
Grosspeteranlage 5
4052 Basel
Switzerland
Tel.: +41 61 683 77 34

Agronomy Editorial Office
E-mail: agronomy@mdpi.com
www.mdpi.com/journal/agronomy

www.ingramcontent.com/pod-product-compliance
Lightning Source LLC
Chambersburg PA
CBHW061609210926
51850CB00009B/2267
* 9 7 8 3 7 2 5 8 5 2 5 3 6 *